METHODS IN MOLECULAR BIOLOGY

Series Editor
John M. Walker
School of Life and Medical Sciences
University of Hertfordshire
Hatfield, Hertfordshire, AL10 9AB, UK

For further volumes:
http://www.springer.com/series/7651

Protein Engineering

Methods and Protocols

Edited by

Uwe T. Bornscheuer

Department of Biotechnology and Enzyme Catalysis, Institute of Biochemistry, Greifswald University, Greifswald, Germany

Matthias Höhne

Protein Biochemistry, Institute of Biochemistry, Greifswald University, Greifswald, Germany

 Humana Press

Editors
Uwe T. Bornscheuer
Department of Biotechnology
  and Enzyme Catalysis
Institute of Biochemistry
Greifswald University
Greifswald, Germany

Matthias Höhne
Protein Biochemistry
Institute of Biochemistry
Greifswald University
Greifswald, Germany

ISSN 1064-3745   ISSN 1940-6029   (electronic)
Methods in Molecular Biology
ISBN 978-1-4939-8463-3   ISBN 978-1-4939-7366-8   (eBook)
DOI 10.1007/978-1-4939-7366-8

Preface

Since the discovery of proteins and their numerous roles in life, scientists are fascinated to study the molecular basis of how proteins function. It is amazing to see the plethora of protein structures and mechanisms that appeared during evolution, and the creativity, which is operating in nature's continuing process of tailoring and fine-tuning proteins and, thus, life itself.

Proteins, especially enzymes, are also the key players of biocatalysis and biotechnology and thus they are linked to the wealth of our modern society. Besides deepening basic understanding, scientists are attracted by the possibility of knowledge-guided tailoring of proteins to suit the needs of biotechnological applications (rational protein design) or to create novel protein functions. As an alternative to this rationally inspired approach, scientists mimic the process of evolution by introducing random mutations in the laboratory (directed evolution). Although we are far away from understanding and reliably predicting protein folding and function *de novo*, there are remarkable success stories in the field of protein engineering: Enzymes were created that catalyze reactions not observed in nature, they were highly stabilized for robustness in industrial processes, and proteins having superior pharmacological profiles have been successfully created. Hence, protein engineering has become an indispensable tool for pharmaceutical and industrial biotechnology.

Protein engineering is a complex and versatile process. With this book we aim to collect basic and advanced protocols for both stages of protein engineering: (i) the library design phase and (ii) the identification of improved variants by screening and selection. The focus of the book lies on enzyme engineering using rational and semirational approaches. Library creation protocols for random mutagenesis and recombining methods are a very diverse field, and a collection of protocols for this approach has been published recently in the excellent volume *Directed Evolution Library Creation* of this series. Hence, this area is not covered in this edition.

As an introduction, Chapter 1 presents a general introduction into protein engineering. The book is then structured into three parts: *Part I* describes computational protocols for rational protein engineering with the aid of case studies. A review (Chapter 2) summarizes different design approaches and methodologies. Protein tunnel inspection and basic steps of molecular modeling are exemplified using the user-friendly software packages CAVER (Chapter 3) and YASARA (Chapter 4). Chapter 5 demonstrates how to use the FRESCO algorithm to stabilize proteins. The presented guide allows to follow this more complex, but very powerful computational engineering protocol. To study structure–function relationships, one useful experimental approach is to study the so-called mutability landscape of a protein. By characterizing every possible single variant of each amino acid position of a protein, beneficial substitutions and nonmutable residues can be identified. Chapter 6 presents a laboratory protocol for an efficient way how to construct and analyze such a library.

Part II focuses on the high-throughput expression of libraries and summarizes common solutions for various problems (Chapters 7 and 8). As a more advanced technique, Chapter 9 presents the split-GFP complementation assay. This approach allows determining the amount of the desired protein via fluorescence measurements in the presence of the entire host proteins. Activity data can then be normalized to the amount of total proteins without the need of enzyme purification. Chapter 10 covers expression and functional studies of membrane proteins using *E. coli* and insect cell-free expression systems.

High-throughput screening and selection assays are covered in *Part III* of this book. This is a very broad research area. Consequently, only exemplary screening protocols can be given as an inspiration for the development of alternative screening assays. An introductory review (Chapter 11) provides an overview of currently existing approaches. The following chapters deal with microplate assays: Chapter 12 describes the design of photometric screening protocols with emphasis on hydrolytic enzymes. Exemplary protocols for screening transaminases, laccases, and β-glucosidase are presented in Chapters 13–15. As screening campaigns have to be well planned and need an efficient way to collect, process, and visualize the data, Chapter 16 describes an open-source software solution that aids experimental planning, but especially data processing and visualization.

The last protocols present solutions for screening and selection procedures. This part of the book covers techniques like solid phase agar plate assays (Chapter 17), droplet sorting (Chapter 18), selection by FACS (Chapter 19), and a growth assays for active and thermo-stable variants (Chapter 20).

We very much hope that this compilation of concepts, methods, and protocols will help readers to facilitate the planning and performance of their experiments, but most importantly, that they will easily create and discover the desired improved proteins or enzymes. We keep our fingers crossed for success!

Greifswald, Germany   *Matthias Höhne*
                                                                                    Uwe T. Bornscheuer

Contents

Contributors

CARLOS G. ACEVEDO-ROCHA • *Department of Biocatalysis, Max-Planck-Institut für Kohlenforschung, Mülheim an der Ruhr, Germany; Department of Chemistry, Philipps-Universität Marburg, Marburg, Germany; Biosyntia ApS, Copenhagen, Denmark*

VÉRONIQUE DE BÉRARDINIS • *CEA, DSV, IG, Genoscope, Evry, France; CNRS-UMR8030, Evry, France; Université d'Evry Val d'Essonne, Evry, France*

JAN BECK • *Institute for Organic Chemistry and Biochemistry, Technische Universität Darmstadt, Darmstadt, Germany; Protein Engineering and Antibody Technologies, Merck-Serono, Merck KGaA, Darmstadt, Germany*

UWE T. BORNSCHEUER • *Department of Biotechnology and Enzyme Catalysis, Institute of Biochemistry, Greifswald University, Greifswald, Germany*

JAN BREZOVSKY • *Loschmidt Laboratories, Department of Experimental Biology, Research Centre for Toxic Compounds in the Environment RECETOX, Faculty of Science, Masaryk University, Brno, Czech Republic*

SUSANA CAMARERO • *Centro de Investigaciones Biológicas, CSIC, Madrid, Spain*

FRANCK CHARMANTRAY • *Institut de Chimie de Clermont-Ferrand, Clermont Université, Université Blaise Pascal, Clermont-Ferrand, France; CNRS, UMR6296, ICCF, Aubière, France*

JÖRG CLAREN • *Clariant Produkte (Deutschland) GmbH, Planegg, Germany*

PIERRE-YVES COLIN • *Department of Biochemistry, University of Cambridge, Cambridge, UK; Department of Biochemical Engineering, University College London, London, UK*

MARK DÖRR • *Department of Biotechnology and Enzyme Catalysis, Institute of Biochemistry, Greifswald University, Greifswald, Germany*

JIRI DAMBORSKY • *Loschmidt Laboratories, Department of Experimental Biology, Research Centre for Toxic Compounds in the Environment RECETOX, Faculty of Science, Masaryk University, Brno, Czech Republic*

SRUJAN KUMAR DONDAPATI • *Branch Bioanalytics and Bioprocesses (IZI-BB), Fraunhofer Institute for Cell Therapy and Immunology (IZI), Potsdam, Germany*

MAXIMILIAN J. L. J. FÜRST • *Groningen Biomolecular Sciences and Biotechnology Institute, University of Groningen, Groningen, The Netherlands*

MATTEO FERLA • *Department of Biochemistry, Oxford University, Oxford, UK*

ALEXANDER FULTON • *Institute of Molecular Enzyme Technology, Heinrich-Heine – Universität Düsseldorf, Forschungszentrum Jülich, Jülich, Germany; Novozymes A/S, Bagsvaerd, Denmark*

THIERRY GEFFLAUT • *Institut de Chimie de Clermont-Ferrand, Clermont Université, Université Blaise Pascal, Clermont-Ferrand, France; CNRS, UMR6296, ICCF, Aubière, France*

FABRICE GIELEN • *Department of Biochemistry, University of Cambridge, Cambridge, UK; Living Systems Institute, University of Exeter, Exeter, UK*

JULIUS GRZESCHIK • *Institute for Organic Chemistry and Biochemistry, Technische Universität Darmstadt, Darmstadt, Germany*

MATTHIAS HÖHNE • *Protein Biochemistry, Institute of Biochemistry, Greifswald University, Greifswald, Germany*

MARC R. HAYES • *Institute of Bioorganic Chemistry, Heinrich-Heine - Universität Düsseldorf, Forschungszentrum Jülich, Jülich, Germany*

EGON HEUSON • *Institut de Chimie de Clermont-Ferrand, Clermont Université, Université Blaise Pascal, Clermont-Ferrand, France; CNRS, UMR6296, ICCF, Aubière, France*

AURELIO HIDALGO • *Department of Molecular Biology, Center for Molecular Biology "Severo Ochoa" (UAM-CSIC), Universidad Autónoma de Madrid, Madrid, Spain*

FLORIAN HOLLFELDER • *Department of Biochemistry, University of Cambridge, Cambridge, UK*

MARIA SVEDENDAHL HUMBLE • *School of Biotechnology, Industrial Biotechnology, KTH Royal Institute of Technology, AlbaNova University Center, Stockholm, Sweden; Pharem Biotech AB, Biovation Park, Södertälje, Sweden*

SAMANTHA M. IAMURRI • *Department of Chemistry, Emory University, Atlanta, GA, USA*

KARL-ERICH JAEGER • *Institute of Molecular Enzyme Technology, Heinrich-Heine - Universität Düsseldorf, , Forschungszentrum Jülich, Jülich, Germany; Institute of Bio- and Geosciences IBG-1: Biotechnology, Forschungszentrum Jülich GmbH, 52428 Jülich, Germany*

DICK B. JANSSEN • *Groningen Biomolecular Sciences and Biotechnology Institute, University of Groningen, Groningen, The Netherlands*

DOREEN KÖNNING • *Institute for Organic Chemistry and Biochemistry, Technische Universität Darmstadt, Darmstadt, Germany*

HARALD KOLMAR • *Institute for Organic Chemistry and Biochemistry, Technische Universität Darmstadt, Darmstadt, Germany*

IVAN V. KORENDOVYCH • *Department of Chemistry, Syracuse University, Syracuse, NY, USA*

ROBERT KOURIST • *Institute of Molecular Biotechnology, TU Graz, Graz, Austria*

BARBORA KOZLIKOVA • *Human Computer Interaction Laboratory, Faculty of Informatics, Masaryk University, Brno, Czech Republic*

SIMON KRAH • *Institute for Organic Chemistry and Biochemistry, Technische Universität Darmstadt, Darmstadt, Germany; Protein Engineering and Antibody Technologies, Merck-Serono, Merck KGaA, Darmstadt, Germany*

STEFAN KUBICK • *Branch Bioanalytics and Bioprocesses (IZI-BB), Fraunhofer Institute for Cell Therapy and Immunology (IZI), Potsdam, Germany*

HENRIK LAND • *School of Biotechnology, Industrial Biotechnology, KTH Royal Institute of Technology, AlbaNova University Center, Stockholm, Sweden; Ångström Laboratory, Department of Chemistry, Molecular Biomimetics, Uppsala University, Uppsala, Sweden*

MIN LIU • *State Key Laboratory of Bioreactor Engineering, Newworld Institute of Biotechnology, East China University of Science and Technology, Shanghai, People's Republic of China*

STEFAN LUTZ • *Department of Chemistry, Emory University, Atlanta, GA, USA*

CAROLIN MÜGGE • *Junior Research Group for Microbial Biotechnology, Ruhr-University Bochum, Bochum, Germany*

PHILIP MAIR • *Department of Biochemistry, University of Cambridge, Cambridge, UK*

MARIO MENCÍA • *Department of Molecular Biology, Center for Molecular Biology "Severo Ochoa" (UAM-CSIC), Universidad Autónoma de Madrid, Madrid, Spain*

ISABEL PARDO • *Centro de Investigaciones Biológicas, CSIC, Madrid, Spain*

JEAN-LOUIS PETIT • *CEA, DSV, IG, Genoscope, Evry, France; CNRS-UMR8030, Evry, France; Université d'Evry Val d'Essonne, Evry, France*

JÖRG PIETRUSZKA • *Institute of Bioorganic Chemistry, Heinrich-Heine - Universität Düsseldorf, Forschungszentrum Jülich, Jülich, Germany; Institute of Bio- and Geosciences IBG-1: Biotechnology, Forschungszentrum Jülich GmbH, Jülich, Germany*

MANFRED T. REETZ • *Department of Biocatalysis, Max-Planck-Institut für Kohlenforschung, Mülheim an der Ruhr, Germany; Department of Chemistry, Philipps-Universität Marburg, Marburg, Germany*

ANA LUÍSA RIBEIRO • *Department of Molecular Biology, Center for Molecular Biology "Severo Ochoa" (UAM-CSIC), Universidad Autónoma de Madrid, Madrid, Spain*

JAVIER SANTOS-ABERTURAS • *Department of Molecular Microbiology, John Innes Centre, Norwich, UK*

SANDY SCHMIDT • *Institute of Molecular Biotechnology, TU Graz, Graz, Austria*

CHRISTIAN SCHRÖTER • *Institute for Organic Chemistry and Biochemistry, Technische Universität Darmstadt, Darmstadt, Germany; Protein Engineering and Antibody Technologies, Merck-Serono, Merck KGaA, Darmstadt, Germany*

THOMAS SCHWAB • *Boehringer Ingelheim Pharma GmbH & Co. KG, Biberach, Germany*

ULRICH SCHWANEBERG • *Lehrstuhl für Biotechnologie, RWTH Aachen University, Aachen, Germany; DWI Leibniz-Institute for Interactive Materials, RWTH Aachen University, Aachen, Germany*

REINHARD STERNER • *Institute of Biophysics and Physical Biochemistry, University of Regensburg, Regensburg, Germany*

BERNHARD VALLDORF • *Institute for Organic Chemistry and Biochemistry, Technische Universität Darmstadt, Darmstadt, Germany*

DOREEN A. WÜSTENHAGEN • *Branch Bioanalytics and Bioprocesses (IZI-BB), Fraunhofer Institute for Cell Therapy and Immunology (IZI), Potsdam, Germany*

MARTIN S. WEIß • *Department of Biotechnology and Enzyme Catalysis, Institute of Biochemistry, Greifswald University, Greifswald, Germany*

HEIN J. WIJMA • *Groningen Biomolecular Sciences and Biotechnology Institute, University of Groningen, Groningen, The Netherlands*

LIDAN YE • *Department of Chemical and Biology Engineering, Institute of Bioengineering and State Key Laboratory of Chemical Engineering, Zhejiang University, Hangzhou, People's Republic of China*

HONGWEI YU • *Department of Chemical and Biology Engineering, Institute of Bioengineering and State Key Laboratory of Chemical Engineering, Zhejiang University, Hangzhou, People's Republic of China*

STEFAN ZIELONKA • *Institute for Organic Chemistry and Biochemistry, Technische Universität Darmstadt, Darmstadt, Germany*

Chapter 1

Protein Engineering: Past, Present, and Future

Stefan Lutz and Samantha M. Iamurri

Abstract

The last decade has seen a dramatic increase in the utilization of enzymes as green and sustainable (bio) catalysts in pharmaceutical and industrial applications. This trend has to a significant degree been fueled by advances in scientists' and engineers' ability to customize native enzymes by protein engineering. A review of the literature quickly reveals the tremendous success of this approach; protein engineering has generated enzyme variants with improved catalytic activity, broadened or altered substrate specificity, as well as raised or reversed stereoselectivity. Enzymes have been tailored to retain activity at elevated temperatures and to function in the presence of organic solvents, salts and pH values far from physiological conditions. However, readers unfamiliar with the field will soon encounter the confusingly large number of experimental techniques that have been employed to accomplish these engineering feats. Herein, we use history to guide a brief overview of the major strategies for protein engineering—past, present, and future.

Key words Protein engineering, Protein design, Rational design, Directed evolution, Biocatalysis

1 Introduction

Enzymes represent nature's solution to drive chemical processes at a timescale and under conditions relevant for cellular life. By exploiting elements of classic thermodynamics, macromolecular dynamics, and quantum mechanics, enzymes can accelerate chemical reactions by up to seventeen orders of magnitude. Moreover, these (bio)catalysts can reach such rate accelerations while maintaining exquisite chemoselectivity, stereoselectivity, and regioselectivity in aqueous environments, ambient temperature, and atmospheric pressure. Recognizing these highly desirable functional properties, scientists have been exploiting native enzymes as biocatalysts in the chemical laboratory for well over 100 years [1]. While some enzymes from natural sources remain highly relevant today, their broader application at the bench, in therapeutics and industrial processes is inherently limited as high specificity and selectivity often restrict an enzyme's use beyond its natural substrates. To adapt enzymes for unnatural substrates, reaction environments and novel chemistries, the ability to either remodel

Uwe T. Bornscheuer and Matthias Höhne (eds.), *Protein Engineering: Methods and Protocols*, Methods in Molecular Biology, vol. 1685, DOI 10.1007/978-1-4939-7366-8_1, © Springer Science+Business Media LLC 2018

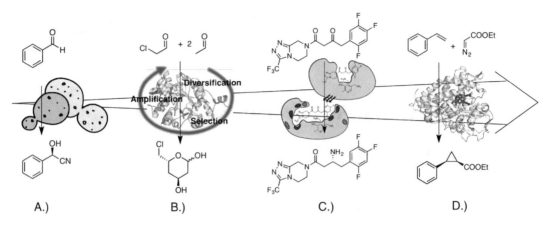

Fig. 1 Advances in protein engineering for tailoring biocatalysts. (**a**) A century ago, Rosenthaler used a crude enzyme preparation from almonds to convert benzaldehyde to mandelonitrile. (**b**) In the 1980s, advances in molecular biology and the introduction of directed evolution enabled generation of customized proteins as exemplified by an aldolase engineered for high selectivity and substrate tolerance in the synthesis of the atorvastatin side chain. (**c**) Semirational and computer-guided engineering offers new and effective strategies to tailor biocatalyts as demonstrated with transaminases for the asymmetric synthesis of sitagliptin. (**d**) Most recent protein engineering efforts focus on adopting biocatalysts for novel chemistry such as cyclopropanation reactions

existing natural enzymes or, more recently, to design entirely new biocatalysts has become both a challenge and an opportunity for protein engineers (*see* Fig. 1).

2   The First Step: Site-Specific Mutagenesis by Rational Design

The first successful attempts to remodel native enzymes in a controlled and reproducible fashion were reported in the early 1980s. These efforts were made possible by then recent advances in molecular biology, which introduced methods for oligonucleotide synthesis, DNA amplification by the polymerase chain reaction, site-specific cutting and pasting with restriction endonucleases and ligases, as well as techniques for extended DNA sequences analysis. Together, the new recombinant tools enabled scientists to deliberately and precisely substitute specific amino acid residues, replacing them with one of the other 19 natural amino acids [2, 3]. Suddenly, such rational protein modifications allowed for a hypothesis-driven approach toward answering fundamental questions related to individual amino acids' roles in protein structure and function. Beyond its application as an investigative tool for studying basic enzyme function, protein engineering soon found use for synthetic purposes as well. In a landmark paper published in 1985, Estell and coworkers modified subtilisin by rational protein engineering, replacing Met222, which had been identified as a site sensitive to oxidative damage [4]. While substitution of Met222 with Ser or Ala led to a reduction in catalytic activity to 30–50% of the native enzyme, both variants concurrently became resistant to 1 M hydrogen peroxide. As such, the authors not only experimentally verified the location of oxidative damage in the wild-type enzyme, but they also deliberately tailored the biocatalyst toward effectively operating in the desired reaction environment. This seminal subtilisin work, as well as subsequent site-specific mutagenesis studies of other enzymes, were largely guided by crystallographic information. While early successes by such rational design highlighted the potential of protein engineering, there was also plenty of anecdotal evidence for failed rational engineering attempts; a reflection of our limited understanding of enzymes' true structural and functional complexity.

3   Learning from Nature: Directed Evolution

In searching for more effective strategies for tailoring proteins in the laboratory, one can take cues from nature. Faced with the same challenge of functional complexity, nature's solution is Darwinian evolution: an iterative process consisting of (a) diversification through random variations in a parental gene sequence, followed by (b) selection for superior functional performance of the corresponding protein variant based on host cell fitness. Over time, this process represents an effective search algorithm to sample the vast array of possible protein sequences in order to repurpose existing proteins. Powerful demonstrations exemplifying the effectiveness of evolutionary mechanisms are the rapid emergence and resistance to antibiotics or of metabolic pathways for xenobiotics including herbicides, pesticides and synthetic polymers [5–8].

Protocols designed to harness Darwinian evolution for protein engineering at the bench first emerged in the late 80s. Early methods mostly relied on random mutagenesis for gene sequence diversification, generating multimillion member libraries via PCR with low fidelity DNA polymerases and suboptimal reaction conditions [9, 10]. Following the cloning of these libraries into a DNA vector and transformation into an expression host, the corresponding protein variants could be evaluated for "fitness" via selection in auxotrophic strains using agar plate or microtiter plate-based screening. Repeated over a dozen or more cycles, improved variants emerged as beneficial mutations accumulated with each round [11–14]. Nevertheless, the accumulation of beneficial mutations by random mutagenesis is complicated by the fact that each library member represents a distinct (clonal or asexual) evolutionary lineage. Beneficial amino acid substitutions can not be shared but must be found independently by each lineage. While the low probability for such an event could in theory be compensated for through more iterative cycles, in practice such a strategy is problematic due to simultaneous acquisition of neutral and, more importantly, deleterious amino acid changes. The introduction of DNA shuffling (also known as sexual DNA shuffling) by Stemmer and coworkers offered an elegant and effective approach to address the shortfalls of random mutagenesis [15, 16]. During each round of directed evolution by DNA shuffling, the genes of library members are fragmented into oligonucleotides and reassembled via homologous recombination, providing a mechanism to share (beneficial) and eliminate (deleterious) mutations laterally. In subsequent years, a variety of alternate experimental protocols have emerged, addressing potential technical problems [17, 18] and broadening the scope of parental sequences [19–22], without significant conceptual deviation from the original idea of in vitro recombination. Over two decades, DNA shuffling has remained a key method in protein engineering and likely will continue to play a central role in the field.

Despite many successful examples of directed enzyme evolution, random mutagenesis and DNA shuffling face a number of practical limitations which can greatly influence the outcome of a protein engineering experiment. For example, determining the optimal mutation frequency per gene sequence can prove tricky as too few nucleotide changes restricts the searchable sequence space (which represents all possible sequence variations for a given protein or gene sequence) [23]. On the other hand, function is thought to be sparse within sequence space, so too many mutations dramatically increase the chance for library members to lose all function.

The number of mutations per gene also determines library size. While directed evolution libraries with up to 10^{15} sequence variants have been reported, even they are insufficient to cover all possible variations if the average mutation frequency exceeds as few as two or three changes in a 1000-bp gene sequence [10, 24]. Worse still, gene libraries are routinely transformed into host organisms for expression of the corresponding protein variants. At maximum transformation efficiencies of 10^{10} colony-forming units for *E. coli*, the most common expression host, the creation of larger gene libraries becomes somewhat futile. Last but not least, each round of directed evolution must be concluded by functionally evaluating its members via either selection or screening methods. Again, library size is critical as the capacity to assess millions of variants is limited to methods such as in vivo complementation of auxotrophic host strains or fluorescence activated cell sorting (FACS) [25]. In light of the functional constraints of these methods, a majority of library analyses continues to be performed via screening in microtiter plates instead. Even with the help of high-end automated systems, microtiter plate screening is usually limited to no more than 10^{4} library members [26]. Given the sparsity of protein function in sequence space, sampling such a small percentage of library members greatly increases the risk of failure to capture variants with improved properties, thus compromising the success of the entire experiment.

Confronted with the experimental challenges of library size and analysis, more recent efforts on the technology side of protein engineering have shifted toward new library design strategies that allow for small, focused sequence pools with higher functional content. Additionally, more cost-effective and versatile high-throughput screening methods have concurrently emerged to aid with library analysis.

4 Does Size Really Matter? Small and Smart Focused Libraries

Departing from traditional directed evolution methods, an exciting new trend in protein engineering consists of semirational approaches [27]. Semirational approaches capitalize on information from ever-growing protein sequence and structure databases, as well as advanced computational and machine-learning algorithms, to guide the design of smaller, more focused libraries of protein variants. These smaller sequence pools not only offer potential time savings, but also reduce the dependency on high-throughput screening methods.

Briefly, the simplest semirational approaches utilize multiple sequence alignments to determine the degree of evolutionary variability of amino acids at each position in a protein sequence. Capitalizing on such information, Reetz and coworkers focused on amino acid residues in or near the active site (synonymous with their importance for enzyme function) to limit the number of residues targeted by randomization (CASTing). When applied iteratively, the multisite saturation mutagenesis approach dramatically reduced library size, yet proved highly effective for tailoring enzyme function [28–33]. Separately, information of entire enzyme superfamilies have been organized in searchable databases such as 3DM to effectively guide protein engineering [34, 35]. The integration of protein sequence, structure and functional data for native and engineered variants within an enzyme superfamily provides the basis for a comprehensive analysis of structure–function relationships, and has been shown to greatly facilitate the identification of beneficial positions to be targeted by protein engineering.

Rather than relying on experimental and sequencing data for guidance, the impact of amino acid substitutions on protein structure and function can also be presampled by computational methods. In silico tools such as the Rosetta Design software, YASARA and FoldX [36–38] utilize structural information and in conjunction with free energy state calculations and molecular dynamics simulations to predict the impact of amino acid substitutions, thereby dramatically reducing the number of variants that must be generated to identify improved enzymes [39–41].

Finally, another emerging strategy within semirational protein engineering constitutes the use of design-of-experiment methodologies, which employ smaller, functionally rich libraries to optimize amino acid sequences for arbitrary functions. These approaches algorithmically derive sequence-to-function relationships from pools of homologous protein sequences or experimental data and identify superior enzyme variants by systematic recombination of amino acid substitutions. With an explicit focus on efficiency and speed, methods such as ProSAR and ProteinGPS have successfully evolved enzyme variants by harboring up to 30 amino acid substitutions to meet a variety of design criteria, yet require preparation and functional evaluation of only a few hundred variants to achieve such functional gains [42–45].

5 "You Get What You Select For": Library Selection and Screening

Paralleling advances in smarter library design, and the development of highly efficient library analysis tools has also been an area of active research. Although established strategies including, library analysis by auxotrophic selection and microtiter plate-based screening, remain highly relevant, exciting new developments in robotics and microfluidics have introduced powerful new strategies for functional assessment of protein variants. Critical to both old and new methods, two fundamental aspects remain relevant for library analysis: (a) the need to link a library member's genotype with its phenotype and (b) the rule that "*you get what you select for*" [23, 46].

The necessity to maintain a tight connection between genotype and phenotype emerges from the fact that library diversity is typically introduced through modifications at the genetic level, i.e., mutations. Meanwhile, the functional consequences of these modifications must be evaluated at the protein level. By far, the simplest and most common way to establish such genotype–phenotype linkage is through transformation of the (DNA)-library into an expression host. In the presence of the appropriate selection markers, each host cell will maintain a single library variant and also facilitate the translation of the genetic information into protein [47]. While functional evaluation via auxotrophic complementation can be performed in bulk, it is far more common for individual host colonies to be grown and evaluated in isolation. Often assisted by robotic equipment and performed in microtiter plates, such approaches offer greater flexibility in the type of library analysis assay that can be performed. Yet as pointed out earlier, the throughput of such screening methods is limited and reaction volumes in the tens to hundreds of microliters per sample can drive up reagent costs. In addition, functional assays of the engineered protein with specific ligands or substrates can be problematic as the reagents must be effectively transported across the host cell membrane. Finally, endogenous host proteins can interfere with functional assays.

A clever strategy to circumvent membrane transport limitations and minimize host protein interference has been the development of surface-display systems [48–52]. Using host membrane proteins as fusion partners, engineered proteins have be effectively exported to the extracellular surface of viruses, bacteria, and yeast cells, yet remain covalently linked to the host cell containing their genetic information. While cell surface-display is extremely effective for the identification of high affinity binding proteins, which can be captured via column-immobilized ligands, the isolation of enzyme variants can be more challenging, as multiple catalytic turnover conditions often result in product diffusion and hence loss of a clear selection criterion.

A real paradigm shift in the methodology to analyze large protein libraries was the development of in vitro compartmentalization by Griffiths and Tawfik [53]. The creation of picoliter reaction vessels via a water-in-oil emulsion offered a simple, yet effective method to establish an artificial genotype–phenotype linkage. Subsequent studies demonstrated the tremendous versatility of in vitro compartmentalization to screen large protein engineering libraries and isolate variants with desired properties [54, 55]. More importantly, in vitro compartmentalization combined with microfluidics has now become one of the major technologies for high-throughput screening of protein engineering libraries [56–60].

Beyond the many established and emerging strategies to maintain a linkage of genotype and phenotype, a well-tested and proven aspect of library analysis is captured in the phrase "*you get what you select for.*" Too often, experimentalists rely on proxy substrates or do not pay close attention to all parameters that will factor into their functional assay, such as sample preparation conditions, buffer composition, changes in pH, and temperature. Inadvertently, such approximations and experimental oversight can result in the isolation of variants with undesirable properties or, in extreme cases, in engineered enzymes that do not display any of the targeted functional improvements. These factors are particularly important as screening throughput increases and reaction volumes decrease. Avoidance of proxy substrates, careful experimental design, and assay validation are critical aspects for planning the analysis of large combinatorial libraries.

6   Putting It All Together: New Tools, New Biocatalysts, New Challenges

Herein we have reviewed the progression of protein engineering techniques over the past two decades. When comparing these techniques retrospectively, a logical progression of engineering can be seen. Previously, this progression was classified as a series of waves by Bornscheuer et al., with each wave introducing a higher degree of engineering sophistication over the proceeding waves [61]. The first wave consisted of isolating enzymes from nature and utilizing them for their native activities. The second wave introduced two schools of thought: rational design versus directed evolution. The third wave incorporated structural data to create semirationally designed libraries. Recently, we have seen the emergence of a fourth wave.

Beyond improving existing properties, engineered variants are emerging that have novel activities not previously seen in nature [62]. The idea of creating novel catalysts for specific processes is not a new concept, but thus far has been limited to organic chemistry and small molecule catalysts. Protein engineering has advanced our understanding of basic protein function, elucidating new details regarding enzyme dynamics and affording new perspectives with respect to active site architecture. Together, these advances in technology and fundamental knowledge have set the stage for the next, fourth wave of biocatalysis, which utilizes the methods of directed evolution, rational and computational design to design novel enzymes possessing nonnative activities. Specifically, we have seen the application of these techniques toward the development of enzymes with nonnative activities for synthetic purposes.

Inspiration for these novel enzymes is derived from organic chemistry, specifically reactions that involve metal catalysts and are targeting processes that have not been previously discovered in nature. Enzymes are known for their efficiency, selectivity, and specificity while also being evolvable, a helpful trait when developing a new catalyst [62]. Arnold and Fasan have led the field by exploring the full capacity of heme proteins. Both groups have been able to completely expand the reaction scope of both cytochrome P450 BM3 and sperm whale myoglobin. Specifically, the Arnold group has worked with P450 BM3 extensively and has a directed evolution library of P450 BM3 variants [63]. From this library, roughly 100 variants were screened and the top variants were then subjected to further mutation in order to increase activity and alter stereoselectivity. Likewise, Fasan was able to design a myoglobin variant also capable of catalyzing the cyclopropanation of styrene and ethyl diazo acetate with high activity and enantioselectivity [64]. Conversely, his approach relied on the power of rational design: three active site residues were chosen based on their proximity to the distal face of the heme. While engineering heme proteins relies on the natural metallocofactors, others like Lewis and Ward have designed novel enzymes by creating artificial metalloenzymes (ArM) through cofactor insertion [65, 66]. By inserting a metal catalyst into a protein scaffold Lewis and Ward have been able to create ArMs also capable of performing C-H activation. Lewis was able to develop an ArM capable of olefin cyclopropanation from propyl oligopeptidase by covalently linking a dirhodium metal complex into the active site through the use of unnatural amino acids [66]. Rational design was instituted when deciding where the unnatural amino acid would be located and which residues needed to be mutated in order to expand the active site. With this method he was able to create an ArM with comparable activity and enantioselectivity to enzymes designed by Arnold and Fasan. Ward also relied on the power of rational design; more specifically, he used the computing power of Rosetta to identify residues of carbonic anhydrase II which would allow for tighter binding of the iridium metal complex [65]. Rosetta was able to identify a variant that bound the metal complex 64 times tighter than the wild-type enzyme and with increased activity and enantioselectivity.

Another good example of merging the latest technology for library design and analysis is the work by Baker, Hilvert and co-workers [67, 68]. They were able to explore the power of Rosetta-Match and Rosetta Design to find a protein scaffold that could support the desired active site shape. Once the scaffold was designed, Rosetta was used again to identify specific residues that would increase activity. The next round of optimization was achieved by multiple rounds of directed evolution. This combinatorial approach was repeated and produced a retro-aldolase with a total turnover number (TTN) 14-fold higher than the best commercially available aldolase antibody [68].

Though protein engineering has come a long way, there is still significant room for improvement, as evidenced by the fact that TTN values for laboratory designed enzymes still fall short of those presented by natural enzyme catalysts [62, 68]. Computational design can provide a good starting point for the laboratory based methods of rational design and directed evolution. As we ride the fourth wave of protein engineering, directed evolution, rational, and computational design will be utilized together to create better and focused smart libraries.

Acknowledgments

We thank the members of the Lutz lab for helpful comments and suggestions on the manuscript. Financial support in part by the US National Science Foundation (CBET-1159434 & CBET-1546790) is gratefully acknowledged.

References

1. Rosenthaler L (1908) Durch Enzyme bewirkte asymmetrische Synthese. Biochem Z 14:238–253

2. Winter G et al (1982) Redesigning enzyme structure by site-directed mutagenesis: tyrosyl tRNA synthetase and ATP binding. Nature 299:756–758

3. Wilkinson AJ et al (1983) Site-directed mutagenesis as a probe of enzyme structure and catalysis: tyrosyl-tRNA synthetase cysteine-35 to glycine-35 mutation. Biochemistry 22:3581–3586

4. Estell DA, Graycar TP, Wells JA (1985) Engineering an enzyme by site-directed mutagenesis to be resistant to chemical oxidation. J Biol Chem 260:6518–6521

5. Gorontzy T et al (1994) Microbial degradation of explosives and related compounds. Crit Rev Microbiol 20:265–284

6. Singh BK, Walker A (2006) Microbial degradation of organophosphorus compounds. FEMS Microbiol Rev 30:428–471

7. Davies J, Davies D (2010) Origins and evolution of antibiotic resistance. Microbiol Mol Biol Rev 74:417–433

8. Iredell J, Brown J, Tagg K (2016) Antibiotic resistance in Enterobacteriaceae: mechanisms and clinical implications. BMJ 352:h6420

9. Cadwell RC, Joyce GF (1992) Randomization of genes by PCR mutagenesis. Genome Res 2:28–33

10. Firth AE, Patrick WM (2005) Statistics of protein library construction. Bioinformatics 21:3314–3315

11. Arnold FH (1990) Engineering enzymes for non-aqueous solvents. Trends Biotechnol 8:244–249

12. Dube DK et al (1991) Artificial mutants generated by the insertion of random oligonucleotides into the putative nucleoside binding site of the HSV-1 thymidine kinase gene. Biochemistry 30:11760–11767

13. Chen KQ, Arnold FH (1993) Tuning the activity of an enzyme for unusual environments-sequential random mutagenesis of subtilisin-E for catalysis in dimethylformamide. Proc Nat Acad Sci USA 90:5618–5622

14. Moore JC, Arnold FH (1996) Directed evolution of a para-nitrobenzyl esterase for aqueous-organic solvents. Nat Biotechnol 14:458–467

15. Stemmer WP (1994) DNA shuffling by random fragmentation and reassembly: in vitro recombination for molecular evolution. Proc Natl Acad Sci U S A 91:10747–10751

16. Stemmer WP (1994) Rapid evolution of a protein in vitro by DNA shuffling. Nature 370:389–391

17. Zhao H, Giver L, Shao Z et al (1998) Molecular evolution by staggered extension process (StEP) in vitro recombination. Nat Biotechnol 16:258–261

18. Müller KM, Stebel SC, Knall S et al (2005) Nucleotide exchange and excision technology (NExT) DNA shuffling: a robust method for DNA fragmentation and directed evolution. Nucleic Acids Res 33:e117

19. Crameri A, Raillard SA, Bermudez E et al (1998) DNA shuffling of a family of genes from diverse species accelerates directed evolution. Nature 391:288–291

20. Ness JE, Welch M, Giver L et al (1999) DNA shuffling of subgenomic sequences of subtilisin. Nat Biotechnol 17:893–896

21. Kolkman JA, Stemmer WP (2001) Directed evolution of proteins by exon shuffling. Nat Biotechnol 19:423–428

22. Patnaik R, Louie S, Gavrilovic V et al (2002) Genome shuffling of Lactobacillus for improved acid tolerance. Nat Biotechnol 20:707–712

23. Romero PA, Arnold FA (2009) Exploring protein fitness landscapes by directed evolution. Nat Rev Mol Cell Biol 10:866–876

24. Patrick WM, Firth AE, Blackburn JM (2003) User-friendly algorithms for estimating completeness and diversity in randomized protein-encoding libraries. Protein Eng 16:451–457

25. Acevedo-Rocha CG, Agudo R, Reetz MT (2014) Directed evolution of stereoselective enzymes based on genetic selection as opposed to screening systems. J Biotechnol 191:3–10

26. Martis EA, Badve RR (2011) High-throughput screening: the hits and leads of drug discovery–an overview. J Appl Pharm Sci 1:2–10

27. Lutz S (2010) Beyond directed evolution--semi-rational protein engineering and design. Curr Opin Biotechnol 21:734–743

28. Clouthier CM, Kayser MM, Reetz MT (2006) Designing new Baeyer-Villiger monooxygenases using restricted CASTing. J Org Chem 71:8431–8437

29. Reetz MT, Carballeira JD, Peyralans J et al (2006) Expanding the substrate scope of enzymes: combining mutations obtained by CASTing. Chem Eur J 12:6031–6038

30. Reetz MT, Wang LW, Bocola M (2006) Directed evolution of enantioselective

enzymes: iterative cycles of CASTing for probing protein-sequence space. Angew Chem Int Ed 45:1236–1241

31. Agudo R, Roiban GD, Reetz MT (2012) Achieving regio- and enantioselectivity of P450-catalyzed oxidative CH activation of small functionalized molecules by structure-guided directed evolution. ChemBioChem 13:1465–1473

32. Gumulya Y, Sanchis J, Reetz MT (2012) Many pathways in laboratory evolution can lead to improved enzymes: how to escape from local minima. ChemBioChem 13:1060–1066

33. Parra LP, Agudo R, Reetz MT (2013) Directed evolution by using iterative saturation mutagenesis based on multiresidue sites. ChemBioChem 14:2301–2309

34. Kourist R, Jochens H, Bartsch S et al (2010) The α/β-hydrolase fold 3DM database (ABHDB) as a tool for protein engineering. ChemBioChem 11:1635–1643

35. Kuipers RK, Joosten HJ, van Berkel WJ et al (2010) 3DM: systematic analysis of heterogeneous superfamily data to discover protein functionalities. Proteins 78:2101–2113

36. Krieger E, Koraimann G, Vriend G (2002) Increasing the precision of comparative models with YASARA NOVA–a self-parameterizing force field. Proteins 47:393–402

37. Das R, Baker D (2008) Macromolecular modeling with Rosetta. Annu Rev Biochem 77:363–382

38. Richter F, Leaver-Fay A, Khare SD et al (2011) De novo enzyme design using Rosetta3. PLoS One 6:e19230

39. Bartsch S, Wybenga GG, Jansen M et al (2013) Redesign of a phenylalanine aminomutase into a phenylalanine ammonia lyase. ChemCatChem 5:1797–1802

40. Floor RJ, Wijma HJ, Colpa DI et al (2014) Computational library design for increasing haloalkane dehalogenase stability. ChemBioChem 15:1660–1672

41. Wijma HJ, Floor HJ, Jekel PA et al (2014) Computationally designed libraries for rapid enzyme stabilization. Protein Eng Des Sel 27:49–58

42. Liao J, Warmuth MK, Govindarajan S et al (2007) Engineering proteinase K using machine learning and synthetic genes. BMC Biotechnol 7:16

43. Ehren J, Govindarajan S, Morón B et al (2008) Protein engineering of improved prolyl endopeptidases for celiac sprue therapy. Protein Eng Des Sel 21:699–707

44. Midelfort KS, Kumar R, Han S et al (2013) Redesigning and characterizing the substrate specificity and activity of *Vibrio fluvialis* aminotransferase for the synthesis of imagabalin. Protein Eng Des Sel 26:25–33

45. Govindarajan S, Mannervik B, Silverman JA et al (2015) Mapping of amino acid substitutions conferring herbicide resistance in wheat glutathione transferase. ACS Synth Biol 4:221–227

46. Lutz S, Patrick WM (2004) Novel methods for directed evolution of enzymes: quality, not quantity. Curr Opin Biotechnol 15:291–297

47. Lodish H, Berk A, Zipursky SL et al (2000) Molecular cell biology, 4th edn. W. H. Freeman, New York, Section 7.1, DNA cloning with plasmid vectors. Available from: https://www.ncbi.nlm.nih.gov/books/NBK21498/

48. Boder ET, Wittrup KD (1997) Yeast surface display for screening combinatorial polypeptide libraries. Nat Biotechnol 15:553–557

49. Chao G, Lau WL, Hackel BJ et al (2006) Isolating and engineering human antibodies using yeast surface display. Nat Protocols 1:755–768

50. Bratkovič T (2009) Progress in phage display: evolution of the technique and its applications. Cell Mol Life Sci 67:749–767

51. Çelik E, Fischer AC, Guarino C et al (2010) A filamentous phage display system for *N*-linked glycoproteins. Protein Sci 19:2006–2013

52. Karlsson AJ, Lim HK, Xu H et al (2012) Engineering antibody fitness and function using membrane-anchored display of correctly folded proteins. J Mol Biol 416:94–107

53. Tawfik DS, Griffiths AD (1998) Man-made cell-like compartments for molecular evolution. Nat Biotechnol 16:652–656

54. Bernath K, Hai M, Mastrobattista E et al (2004) In vitro compartmentalization by double emulsions: sorting and gene enrichment by fluorescence activated cell sorting. Anal Biochem 325:151–157

55. Aharoni A, Griffiths AD, Tawfik DS (2005) High-throughput screens and selections of enzyme-encoding genes. Curr Opin Chem Biol 9:210–216

56. Agresti JJ, Antipov E, Abate AR et al (2010) Ultrahigh-throughput screening in drop-based microfluidics for directed evolution. Proc Natl Acad Sci U S A 107:4004–4009

57. Fischlechner M, Shaerli Y, Mohamed MF et al (2014) Evolution of enzyme catalysts caged in biomimetic gel-shell beads. Nat Chem 6:791–796

58. Ostafe R, Prodanovic R, Nazor J et al (2014) Ultra-high-throughput screening method for the directed evolution of glucose oxidase. Chem Biol 21:414–421

59. Zinchenko A, Devenish SRA, Kintses B et al (2014) One in a million: flow cytometric sorting of single cell-lysate assays in monodisperse picolitre double emulsion droplets for directed evolution. Anal Chem 86:2526–2533

60. Romero PA, Tran TM, Abate AR (2015) Dissecting enzyme function with microfluidic-based deep mutational scanning. Proc Natl Acad Sci U S A 112:7159–7164

61. Bornscheuer UT, Huisman GW, Kazlauskas RJ et al (2012) Engineering the third wave of biocatalysis. Nature 485:185–194

62. Prier CK, Arnold FH (2015) Chemomimetic biocatalysis: exploiting the synthetic potential of cofactor-dependent enzymes to create new catalysts. J Am Chem Soc 137:13992–14006

63. Coelho PS, Brustad EM, Kannan A et al (2013) Olefin cyclopropanation via carbene transfer catalyzed by engineered cytochrome P450 enzymes. Science 339:307–310

64. Bordeaux M, Tyagi V, Fasan R (2015) Highly diastereoselective and enantioselective olefin cyclopropanation using engineered myoglobin-based catalysts. Angew Chem Int Ed 54:1744–1748

65. Heinisch T, Pellizzoni M, Dürrenberger M et al (2015) Improving the catalytic performance of an artificial metalloenzyme by computational design. J Am Chem Soc 137:10414–10419

66. Srivastava P, Yang H, Ellis-Guardiola K et al (2015) Engineering a dirhodium artificial metalloenzyme for selective olefin cyclopropanation. Nat Commun 6:7789

67. Althoff EA, Wang L, Jiang L et al (2012) Robust design and optimization of retroaldol enzymes. Protein Sci 21:717–726

68. Giger L, Caner S, Obexer R et al (2013) Evolution of a designed retro-aldolase leads to complete active site remodeling. Nat Chem Biol 9:494–498

Part I

Computational Protocols

Chapter 2

Rational and Semirational Protein Design

Ivan V. Korendovych

Abstract

This mini review gives an overview over different design approaches and methodologies applied in rational and semirational enzyme engineering. The underlying principles for engineering novel activities, enantio-selectivity, substrate specificity, stability, and pH optimum are summarized.

Key words Rational protein design, Computational enzyme design, De novo enzyme design, Molecular dynamics, Molecular docking, Enantioselectivity, Substrate specificity, Thermostability, pH optimum

The ability to produce desired molecules in a direct, inexpensive and efficient fashion is the ultimate goal of applied chemistry. Despite the abundance of easy and inexpensive sources of energy (e.g., heat, electricity, and light) the complex task of taking available chemical building blocks to drive thermodynamically allowed processes in one particular direction is far from solved. Nature has found many ways to accomplish this task through enzymatic catalysis, promoted by proteins and nucleic acids. Thus, it is hardly surprising that ever since the discovery of the first enzyme chemists attempt to replicate their amazing efficiency by creating proteins capable of producing chemicals of industrial relevance. Many different approaches have been explored with various degrees of success (Table 1). Existing catalysts were repurposed to change the substrate scope and reactions specificity. Proteins that have no enzymatic function adopted new catalytic functions. Catalysts have been prepared from protein scaffolds not present in nature and proteins that have no observable enzymatic activity for the reaction of interest—this I refer to as de novo design. Finally, catalysts for reactions that were not observed in nature until now could be created in protein scaffolds by mutagenesis: novel activities were designed by a careful placement of chemical functionalities that are provided by nature's menu of amino acids to stabilize transition states, enable proton transfers, facilitate the interaction of

Uwe T. Bornscheuer and Matthias Höhne (eds.), *Protein Engineering: Methods and Protocols*, Methods in Molecular Biology, vol. 1685, DOI 10.1007/978-1-4939-7366-8_2, © Springer Science+Business Media LLC 2018

Table 1
Representative examples of proteins designed using various approaches

Design principles, methods[a]	Parameters introduced/optimized	Representative citations
Substitution of amino acids by rational design		
Visual inspection, Docking, ISM	Substrate specificity Stereoselectivity	[1]
CAVER, ISM	Activity, Stabiliy	[3, 4]
B-Fit, ISM	Thermostability	[5]
MD-simulations	Enantioselectivity	[6, 7]
Prediction of pK$_a$	pH Optimum	[8–11]
Computational design		
FRESCO	Thermostability	[12]
CASCO Rosetta Design	Enantioselectivity	[13]
Rosetta Design/Rosetta match	Introducing new chemical activities	[14–17]
Minimalist design	Introducing new chemical activities	[18–20]
De novo *design of protein folds*		
Semiempirical computation	Introducing catalysis	[21]
Introduction of noncanonical amino acids		
Rational, substrate docking	Introducing new chemical activities	[22]
Rosetta	Protein–peptide interface, metal cofactor binding	[23, 24]
Redesign of the existing or introduction of new cofactors		
Introducing metal cofactors into proteins		[25]
Substitution of metal ions in existing cofactors	Introducing new chemical activities	[26]
Transition metal complexes anchored by biotin conjugation		[27]

[a]Note the list is by no means exhaustive
ISM iterative saturation mutagenesis, *MD* molecular dynamics, *FRESCO* framework for rapid enzyme stabilization by computational libraries, *CASCO* catalytic selectivity by computational design

the substrate with the active site or with cofactors present in the protein, or modulate the chemical reactivity of natural cofactors. The spectrum of catalysis was further extended by introducing artificial cofactors or unnatural amino acids [28]. Table 1 gives examples for this large spectrum of design approaches.

Design tools have been very diverse: ranging from purely combinatorial [29] to highly rational [30]. Combinatorial methods (relying on random mutagenesis) have been successful in repurposing of existing proteins to adopt new functions and creation of new catalytic function from random sequences [29, 31, 32]. However, the enormity of sequence space to be explored in a design problem means that in practical terms some degree of rational input has to be made to limit the search space to a manageable size. Thus, a clear line between rational and combinatorial approaches is hard, if not impossible, to draw. One crucial requirement in rational design is the necessity to understand the molecular basis of the protein's property that is the subject of the design study (structure–function relationship). Table 1 lists specific rational design techniques and how they are used to modify a well-defined property of an enzyme.

Many application-oriented enzyme-engineering projects focus on creating or adapting the substrate scope of an enzyme to gain access to (a class of) compounds of interest. This often also involves tuning enantioselectivity or regioselectivity in the desired direction. Increasing the stability of the biocatalyst under process conditions is an equally important goal. For all these questions, a rational understanding has become available during the past few decades.

1 Semirational Tools for Engineering Substrate Specificity and Enantioselectivity

Certain features of catalysts can be modified relatively easily: Substrate specificity and enantioselectivity are often governed by steric factors of the active site [33]. Thus, the easiest approach to guide a semirational design is to use structural visualization to identify hot spot residues that are then targeted in a site-saturation mutagenesis experiment. The active site must be shape-complementary to the transition state of the reaction to accelerate formation of the desired product [34]. A well-defined geometry allows the preferred binding and positioning of one enantiomeric form of the substrate, or the preferred creation of one configuration of the chiral product. On the contrary, binding poses that lead to undesired regio or stereo isomers have to be blocked. Additionally, selectivity towards different substrates is affected during their passage of the entrance tunnel of the enzyme: modifications of tunnel residues influence the access of different compounds to the active site and thus induce selectivity [35].

Rational redesign of the active site is often easily possible, e.g., by blocking the productive binding of the undesired enantiomer by introducing a bulky residue. However, as enzymes are often more complex than it is apparent from the structural models, many effects cannot be predicted (due to protein dynamics or effects on protein folding). The more detailed the available information and knowledge of catalysis is, the better. While detailed structural information on the intermediates in the catalytic cycle *can* be obtained, most of X-ray and NMR structures present in the Protein Data Bank represent the structure *without* direct information about how substrate binds or is turned over. Additional studies that require crystallization of the enzyme with an appropriate inhibitor may require a long time without any definitive guarantee of success. Fortunately, several very successful algorithms have been developed to identify the location and possible poses of the substrate in the enzyme [36]. Cavity search and docking techniques give hints how and where the substrate might be bound. Even low-resolution information about how the substrate associates with the protein is often sufficient to make educated guesses in which positions mutagenesis needs to be done to achieve maximum desired effect. Especially when the active site or the substrate is large and can adopt multiple conformations, or when binding is based mainly on hydrophobic interactions, reliable predictions are not yet possible. Partial or complete randomization of identified hot spots is therefore an efficient approach, which often leads to success. Iterative site saturation mutagenesis has become a very popular engineering tool [37].

The CAVER software is an easy to handle tool for identification and analysis of tunnels and channels in protein structures [38]. CAVER is used as a plugin in Pymol, a program, which is employed frequently for protein visualizing [39]. It predicts the location of "hot spot" residues, which can be mutated to enhance enzyme activity, stability, specificity, and enantioselectivity. Another commonly used program YASARA [39] provides the user with a graphic, user-friendly interface to detect hotspots and to perform molecular mechanics based simulations for rational protein engineering. If no structure is available for the protein of interest, YASARA has a tool for the computer-aided construction of a homology model. Some structural information, although the accuracy of the model might be limited, can be obtained from related proteins with sequence identities as low as 30%. On the other hand, if a reliable structure is available, computational docking–which is also integrated in YASARA–has shown enormous predictive power in identifying residues to be modified in order to alter the selectivity and improve the reactivity of the existing proteins. However, much care has to be taken when interpreting results of docking experiments that rely on homology models.

It is universally accepted that enzymes are far from static and rely on concerted movement of amino acids to achieve function [40]. Semiempirical molecular dynamics, (MD) approaches have been extremely useful in deciphering the intricate details of protein-catalyzed chemical reactions [41]. Owing to the continuous improvement of computational hardware MD techniques are becoming more and more available to solve protein design problems [42]. MD simulations have been useful in improving the enzyme activity and enantioselectivity. This is the most difficult and time consuming aspect of rational design and much needs to be learned before our methods are efficient and accurate enough to reliably predict mutations that are likely to improve enzymatic efficiency.

MD simulations generate an ensemble of possible conformations and conformational transitions, as compared to a static picture provided by X-ray crystallography. Combined with knowledge of the reaction mechanism (e.g., from quantum mechanical modeling), MD simulations determine how frequently geometries that will promote catalysis according to the model are observed, as compared to "unproductive conformations" [43]. MD simulations are also used to identify dynamic, flexible regions of a protein. Changes in these regions can affect protein stability and activity, because catalysis requires certain flexibility of critical residues or parts of the protein. Loop flexibility can also determine reaction specificity, as was demonstrated by reengineering a phenylalanine mutase into a phenylalanine ammonia lyase by introducing a single mutation in a loop near the active site [44].

2 Advanced Computational Engineering for Optimizing Enantioselectivity and Thermostability

Computational engineering creates large virtual libraries of variants in silico. Designs are then evaluated and ranked automatically, e.g., by energy scoring functions or geometric restraints, and only a few hits (ten to some hundreds) are manually inspected and tested in the lab [30]. Different tools that often introduce several mutations at once are used for the creation of the libraries. Computational enzyme design allows for engineering highly enantioselective catalysts for a particular chemical transformation already catalyzed by the enzyme, complete redesign of active sites to fit substrate structures that are very different from the natural ones, and de novo design of enzymes, i.e., proteins catalyzing nonnatural reactions.

In the first step, optimal geometries of possible active site residues that stabilize the transition state of the reaction are predicted using QM simulations. The resulting arrangements of amino acid residues, called theozymes, are placed in suitable protein scaffolds identified by RosettaMatch. Finally, RosettaDesign optimizes the complete active site pocket to allow the precise positioning of the catalytic residues and the transition state. Designs are then evaluated and ranked in silico, and only a few (ten to some hundreds) are manually inspected and tested in the lab. This strategy was used for de novo engineering Kemp eliminases, retro-aldolase, Diels-Alderases, but also to generate highly enantioselective epoxide hydrolases [13–17]. For the latter study, in silico variants were screened using high-throughput multiple independent MD simulations [45], a technique that leads to a more complete sampling of protein conformational space in a shorter time (compared to long single-run MD simulations) and showed an improved correlation between predicted and observed enantioselectivity. This helped to reduce library size that had to be actually screened. Moreover, computational approaches can assist in improving enzymes using directed evolution: semirationally developed libraries produced up to 4–5-fold higher hit rate as compared to a full coverage libraries thus greatly limiting effort to identify productive mutations [46]. While many programs have been developed and successfully used for performing MD simulations, YASARA provides a user-friendly interface for a beginner.

A second very important engineering target is enzyme thermostability. It became clear early on that practical applicability (and evolvability!) of an enzymatic catalyst is related to its stability [47]. Enzymes from thermophilic organisms are commonly used in many different applications, but what if the catalyst to be repurposed/ improved has no obvious thermophilic analog? Homology modeling and rational evaluation of the structure has been very productive in identification of mutations to improve stability. This sometimes also leads to improvement in the yield of recombinant expression of soluble enzymes, although the evidence is somewhat anecdotal.

Several approaches have been successfully used to predict and improve thermostability [42, 48–50]. Most often protein stability is increased by rigidification of flexible sites. Analysis of B-factors in crystal structures (B-Fit Method) [5], high temperature unfolding MD simulations, and comparative MD simulations of homologous proteins from mesophilic and thermophilic organisms at different temperatures unravel flexible regions of the protein that are susceptible to unfolding to guide reengineering [42].

Alternatively, protein stability can be increased by improving hydrophobic packing of the protein core [51] and/or creating a favorable network of positive and negative charges at the protein surface [52]. Scoring of variants is then performed by evaluating differences in the free energy of folding using specifically parameterized energy functions [49, 53] such as one included in the FoldX suite. Finally, stabilizing disulfide bridges can be engineered into the protein using the FRESCO algorithm in YASARA.

In summary, the path to developing the ability to create *functional* proteins for a particular purpose has been long, windy, and full of obstacles. Decades of research in biochemistry, enzymology, and biotechnology produced a number of exciting discoveries that advance our understanding of enzymatic catalysis, nonetheless we still fall short from being able to create a single unique tool that will allow us to create efficient protein catalysts from scratch [54].

Despite the disappointment with the overall progress of the field, fueled in part by the overzealous promises that could not be fulfilled thrown around so easily, many amazing stories of success that apply rational principles to (re)design of proteins have emerged through the years [55–59]. Advances in computation led to an explosive growth of structural information and the development of robust tools for building protein structures of predefined fold. Creating a crucial link between a (re)designed well-defined structure and catalytic function is the next major milestone for the field.

References

1. Ghislieri D, Green AP, Pontini M et al (2013) Engineering an enantioselective amine oxidase for the synthesis of pharmaceutical building blocks and alkaloid natural products. J Am Chem Soc 135:10863–10869

2. Kille S, Zilly FE, Acevedo JP et al (2011) Regio- and stereoselectivity of P450-catalysed hydroxylation of steroids controlled by laboratory evolution. Nat Chem 3:738–743

3. Liskova V, Bednar D, Prudnikova T et al (2015) Balancing the stability-activity trade-off by fine-tuning dehalogenase access tunnels. ChemCatChem 7:648–659

4. Pavlova M, Klvana M, Prokop Z et al (2009) Redesigning dehalogenase access tunnels as a strategy for degrading an anthropogenic substrate. Nat Chem Biol 5:727–733

5. Reetz MT, Carballeira JD, Vogel A (2006) Iterative saturation mutagenesis on the basis of B factors as a strategy for increasing protein thermostability. Angew Chem Int Ed 45:7745–7751

6. Raza S, Fransson L, Hult K (2001) Enantioselectivity in *Candida antarctica* lipase B: a molecular dynamics study. Protein Sci 10:329–338

7. Rotticci D, Rotticci-Mulder JC, Denman S et al (2001) Improved enantioselectivity of a lipase by rational protein engineering. ChemBioChem 2:766–770

8. Tynan-Connolly BM, Nielsen JE (2007) Redesigning protein pK(a) values. Protein Sci 16:239–249

9. Pokhrel S, Joo JC, Yoo YJ (2013) Shifting the optimum pH of *Bacillus circulans* xylanase towards acidic side by introducing arginine. Biotechnol Bioprocess Eng 18:35–42

10. Pokhrel S, Joo JC, Kim YH et al (2012) Rational design of a *Bacillus circulans* xylanase by introducing charged residue to shift the pH optimum. Process Biochem 47:2487–2493

11. Xu H, Zhang F, Shang H et al (2013) Alkalophilic adaptation of XynB endoxylanase from *Aspergillus niger* via rational design of pKa of catalytic residues. J Biosci Bioeng 115:618–622

12. Wijma HJ, Floor RJ, Jekel PA et al (2014) Computationally designed libraries for rapid enzyme stabilization. Protein Eng Des Sel 27:49–58

13. Wijma HJ, Floor RJ, Bjelic S et al (2015) Enantioselective enzymes by computational design and *in silico* screening. Angew Chem Int Ed 54: 3726–3730

14. Rothlisberger D, Khersonsky O, Wollacott AM et al (2008) Kemp elimination catalysts by computational enzyme design. Nature 453:190–195

15. Siegel JB, Zanghellini A, Lovick HM et al (2010) Computational design of an enzyme catalyst for a stereoselective bimolecular Diels-Alder reaction. Science 329:309–313

16. Blomberg R, Kries H, Pinkas DM et al (2013) Precision is essential for efficient catalysis in an evolved Kemp eliminase. Nature 503:418–421

17. Jiang L, Althoff EA, Clemente FR et al (2008) De novo computational design of retro-aldol enzymes. Science 319:1387–1391

18. Korendovych IV, Kulp DW, Wu Y et al (2011) Design of a switchable eliminase. Proc Natl Acad Sci U S A 108:6823–6827

19. Moroz YS, Dunston TT, Makhlynets OV et al (2015) New tricks for old proteins: single mutations in a nonenzymatic protein give rise to various enzymatic activities. J Am Chem Soc 137:14905–14911

20. Raymond EA, Mack KL, Yoon JH et al (2015) Design of an allosterically regulated retroaldolase. Protein Sci 24:561–570

21. Burton AJ, Thomson AR, Dawson WM et al (2016) Installing hydrolytic activity into a completely de novo protein framework. Nat Chem 8:837–844

22. Pan T, Liu Y, Si C et al (2017) Construction of ATP-switched allosteric antioxidant selenoenzyme. ACS Catalysis 7(3):1875–1879

23. Renfrew PD, Choi EJ, Bonneau R et al (2012) Incorporation of noncanonical amino acids into Rosetta and use in computational protein-peptide interface design. PLoS One 7: e32637

24. Mills JH, Khare SD, Bolduc JM et al (2013) Computational design of an unnatural amino acid dependent metalloprotein with atomic level accuracy. J Am Chem Soc 135:13393–13399

25. Reetz MT, Jiao N (2006) Copper–phthalocyanine conjugates of serum albumins as enantioselective catalysts in Diels–Alder reactions. Angew Chem Int Ed 45:2416–2419

26. Key HM, Dydio P, Clark DS et al (2016) Abiological catalysis by artificial haem proteins containing noble metals in place of iron. Nature 534:534–537

27. Lo C, Ringenberg MR, Gnandt D et al (2011) Artificial metalloenzymes for olefin metathesis based on the biotin-(strept)avidin technology. Chem Commun 47:12065–12067

28. Bornscheuer UT, Huisman GW, Kazlauskas RJ et al (2012) Engineering the third wave of biocatalysis. Nature 485:185–194

29. Seelig B, Szostak JW (2007) Selection and evolution of enzymes from a partially randomized non-catalytic scaffold. Nature 448:828–831

30. Kiss G, Celebi-Olcum N, Moretti R et al (2013) Computational enzyme design. Angew Chem Int Ed 52:5700–5725

31. Reetz MT (2013) Biocatalysis in organic chemistry and biotechnology: past, present and future. J Am Chem Soc 135:12480–12496

32. Reetz MT (2011) Laboratory evolution of stereoselective enzymes: a prolific source of catalysts for asymmetric reactions. Angew Chem Int Ed 50:138–174

33. Reetz MT (2004) Controlling the enantioselectivity of enzymes by directed evolution: practical and theoretical ramifications. Proc Natl Acad Sci U S A 101:5716–5722

34. Schramm VL (2005) Enzymatic transition states: thermodynamics, dynamics and analogue design. Arch Biochem Biophys 433:13–26

35. Butler CF, Peet C, Mason AE et al (2013) Key mutations alter the cytochrome P450 BM3 conformational landscape and remove inherent substrate bias. J Biol Chem 288:25387–25399

36. Sousa SF, Fernandes PA, Ramos MJ (2006) Protein–ligand docking: current status and future challenges. Proteins Struct Funct Bioinf 65:15–26

37. Gumulya Y, Sanchis J, Reetz MT (2012) Many pathways in laboratory evolution can lead to improved enzymes: how to escape from local minima. ChemBioChem 13:1060–1066

38. Chovancova E, Pavelka A, Benes P et al (2012) CAVER 3.0: a tool for the analysis of transport pathways in dynamic protein structures. PLoS Comput Biol 8:e1002708

39. Available from: http://www.pymol.org

40. Eisenmesser EZ, Bosco DA, Akke M et al (2002) Enzyme dynamics during catalysis. Science 295:1520–1523

41. Adcock SA, McCammon JA (2006) Molecular dynamics: survey of methods for simulating the activity of proteins. Chem Rev 106:1589–1615

42. Childers MC, Daggett V (2017) Insights from molecular dynamics simulations for computational protein design. Mol Sys Des Eng 2:9–33

43. Wijma HJ, Floor RJ, Janssen DB (2013) Structure- and sequence-analysis inspired engineering of proteins for enhanced thermostability. Curr Opin Struct Biol 23:588–594

44. Bartsch S, Wybenga GG, Jansen M et al (2013) Redesign of a phenylalanine aminomutase into a phenylalanine ammonia lyase. ChemCatChem 5:1797–1802

45. Wijma HJ, Marrink SJ, Janssen DB (2014) Computationally efficient and accurate enantioselectivity modeling by clusters of molecular dynamics simulations. J Chem Inf Model 54:2079–2092

46. Chen MMY, Snow CD, Vizcarra CL et al (2012) Comparison of random mutagenesis and semi-rational designed libraries for improved cytochrome P450 BM3-catalyzed hydroxylation of small alkanes. Protein Eng Des Sel 25:171–178

47. Bloom JD, Labthavikul ST, Otey CR et al (2006) Protein stability promotes evolvability. Proc Nat Acad Sci U S A 103:5869–6874

48. Reetz MT, Soni P, Fernández L (2009) Knowledge-guided laboratory evolution of protein thermolability. Biotechnol Bioeng 102:1712–1717

49. Seeliger D, de Groot BL (2010) Protein thermostability calculations using alchemical free energy simulations. Biophys J 98:2309–2316

50. Zeiske T, Stafford KA, Palmer AG (2016) Thermostability of enzymes from molecular dynamics simulations. J Chem Theory Comput 12:2489–2492

51. Borgo B, Havranek JJ (2012) Automated selection of stabilizing mutations in designed and natural proteins. Proc Natl Acad Sci U S A 109:1494–1499

52. Gribenko AV, Patel MM, Liu J et al (2009) Rational stabilization of enzymes by computational redesign of surface charge–charge interactions. Proc Natl Acad Sci U S A 106:2601–2606

53. Van Durme J, Delgado J, Stricher F et al (2011) A graphical interface for the FoldX forcefield. Bioinformatics 27:1711–1712

54. Korendovych IV, DeGrado WF (2014) Catalytic effciency of designed catalytic proteins. Curr Opin Struct Biol 27:113–121

55. Wijma HJ, Janssen DB (2013) Computational design gains momentum in enzyme catalysis engineering. FEBS J 280:2948–2960

56. Yeung N, Lin YW, Gao YG et al (2009) Rational design of a structural and functional nitric oxide reductase. Nature 462:1079–1082

57. Kazlauskas RJ, Bornscheuer UT (2009) Finding better protein engineering strategies. Nat Chem Biol 5:526–529

58. Höhne M, Schätzle S, Jochens H et al (2010) Rational assignment of key motifs for function guides in silico enzyme identification. Nat Chem Biol 6:807–813

59. Yin H, Slusky JS, Berger BW et al (2007) Computational design of peptides that target transmembrane helices. Science 315:1817–1822

Chapter 3

Computational Analysis of Protein Tunnels and Channels

Jan Brezovsky, Barbora Kozlikova, and Jiri Damborsky

Abstract

Protein tunnels connecting the functional buried cavities with bulk solvent and protein channels, enabling the transport through biological membranes, represent the structural features that govern the exchange rates of ligands, ions, and water solvent. Tunnels and channels are present in a vast number of known proteins and provide control over their function. Modification of these structural features by protein engineering frequently provides proteins with improved properties. Here we present a detailed computational protocol employing the CAVER software that is applicable for: (1) the analysis of tunnels and channels in protein structures, and (2) the selection of hot-spot residues in tunnels or channels that can be mutagenized for improved activity, specificity, enantioselectivity, or stability.

Key words Binding, Protein, Tunnel, Channel, Gate, Rational design, Software, CAVER, Transport

---

1 Introduction

All biomolecules contain a complex system of voids—cavities, channels, tunnels, or grooves of various shapes and sizes (Fig. 1a). Cavities often host a site of functional importance, but in many cases, also tunnels and channels have functional roles. They secure the transport of ligands between different regions, e.g., connect buried cavities with the surface, different cavities, or even different cellular compartments, such as in membrane proteins. The presence of tunnels was already reported for numerous enzymes from all six Enzyme Commission classes as well as all main structural classes [1, 2]. The geometry, physicochemical properties, and dynamics of tunnels have been shown to determine the exchange rates of ligands between the active sites and a bulk solvent. Additional functions are often secured by the tunnels: (1) enabling the access of preferred substrates, while denying the access of nonpreferred ones and thus preventing the formation of nonproductive complexes, (2) avoiding the damage of the enzymes dependent on cofactors containing the transition metals by their poisoning, (3) preventing the damage to the cell by releasing toxic intermediates,

Uwe T. Bornscheuer and Matthias Höhne (eds.), *Protein Engineering: Methods and Protocols*, Methods in Molecular Biology, vol. 1685, DOI 10.1007/978-1-4939-7366-8_3, © Springer Science+Business Media LLC 2018

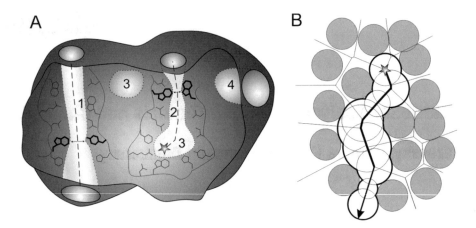

Fig. 1 (**a**) Schematic representation of complex voids in protein structures. Channel (1), tunnel (2), buried cavities (3), and surface groove (4) are in *light gray*. Channel and tunnel entrance is in *gray*. Channel and tunnel centerlines are shown as *dashed lines*, bottlenecks are indicated by *black arrows*, and bottleneck-lining residues are in *bold*. The *gray* star in the buried cavity represents the starting point for the tunnel calculation. Cavity-, channel-, and tunnel-lining residues are in *black* lines, while bottleneck residues are in *black* sticks. (**b**) Schematic representation of Voronoi diagram. Atoms are represented by *gray circles*, edges by *lines*, a starting point of tunnel by a *gray star*, a tunnel exit by an *arrow*, a tunnel surface by a *thick black contour*

(4) enabling reactions that require the absence of water, and (5) synchronizing reactions that require the contact of a large number of substrates, intermediates, or cofactors [1, 3]. Recognizing the importance of transport processes for enzymatic catalysis, the key–lock–keyhole model has recently been proposed [1] and experimentally validated by a number of protein engineering studies. These studies successfully exploited the tunnel modification for improving enzyme activity [4, 5], specificity [6], enantioselectivity [7], and stability [8, 9]. Identification of tunnels and channels in the complex voids present in protein structures is not a trivial task, and many software tools have recently been developed for this purpose, e.g., CAVER [10], MOLE [11], MolAxis [12], ChExVis [13], or BetaCavityWeb [14]. All these tools are based on computational geometry methods employing the Voronoi diagrams. These methods identify the pathways connecting a starting point in a buried cavity to a bulk solvent in the target protein structure (Fig. 1b). The starting point is defined by several atoms or residues surrounding an empty space of the occluded protein cavity. As the main output, the tools provide geometry of the tunnels, which have their minimal radius wider than the radius specified by the user. The tunnels' geometry is supplemented by the information about their characteristics, profiles, the tunnel-lining residues, and the residues forming the tunnel bottleneck, i.e., the narrowest part of the access tunnel [15]. The bottleneck residues represent particularly suitable hot-spots for the modification of tunnels' geometry since their

substitutions often have substantial impact on the function or stability of the protein [1, 3].

The software tools outlined in the previous paragraph enable fast identification of permanent tunnels in a single structure and provide information about their characteristics. The main limitation of such analysis is that protein dynamics is neglected and the transient tunnels are missed. In other words, only the tunnels that are open in the analyzed structure are identified, while the transient gated tunnels, which are temporarily closed in the investigated structure, can be overlooked [10, 16]. Additionally, it is often difficult to distinguish biologically relevant tunnels using the analysis of a single static structure [10, 16]. Both pitfalls can be overcome by analyzing the tunnels in an ensemble of protein conformations. These conformations can be obtained from: (1) NMR ensemble, (2) set of crystal structures, or (3) molecular dynamics simulations (*see* **Note 1**). From the aforementioned tools, only CAVER was designed for comprehensive analysis of tunnels in molecular ensembles and includes essential tunnel clustering [15]. However, setting up molecular dynamics simulations requires considerable expertise and knowledge of a studied protein. Hence, the description of molecular dynamics and the analysis of tunnels over an ensemble is beyond the scope of this chapter (*see* **Note 2**).

2 Overview of Implementations of CAVER 3.0

CAVER is a software tool widely used for the identification and analysis of transport tunnels in macromolecular structures. It is available in several implementations that provide various sets of features (Table 1), addressing needs of users with different level of experience and expectations: (1) a command-line application for analysis of both static and dynamic structures, (2) a PyMOL plugin for the analysis of static structures, and (3) a graphical user interface CAVER Analyst [17] for the analysis of both static and dynamic structures as well as the visualization of detected tunnels. Additionally, CAVER is also integrated into two web services: (4) CAVER Web (under preparation), (5) HotSpot Wizard [18] and (6) Caver-Dock (under preparation). These interactive web tools are accessible via the site: http://loschmidt.chemi.muni.cz/peg/software.

1. **CAVER Command-line application** is suited for advanced users who aim to analyze large ensembles of structures, typically counting thousands or tens of thousands of snapshots. The larger number of structures may require the use of supercomputers. Moreover, the command-line application is employed as tunnel analysis engine in all other implementations. The setting of the tunnel calculation is performed via a

Table 1
Main features of various implementations of CAVER 3.0

Feature	CAVER command-line	CAVER PyMOL plugin	CAVER Analyst	CAVER Web	HotSpot Wizard
Input format	PDB	Formats supported by PyMOL	PDB, mdcrd, xtc, and dcd	PDB	PDB
Analysis of ensembles	+	−	+	−	−
Assistance during preparation of input structure	−	−	+	+	+
Manipulation of protein structure	−	+	+	−	−
Additional analyses of protein structure	−	+	+	−	−
Advanced settings of calculation	+	+	+	−	−
Start from–user-defined point	+	+	+	+	−
Start from–buried cavity	−	−	+	+	+
Start from–catalytic residues	−	−	+	+	+
Direct visualization of tunnels	−	+	+	+	+
Interactive GUI	−	−	+	+	−
Interactive exploration of output statistics	−	−	+	+	−
Advanced visualization	−	−	+	−	−
Detailed information about tunnels properties	+	+	+	+	−
Physicochemical properties of tunnels	−	−	+	+	−
Evolutionary conservation	−	−	−	+	+

configuration file, and the visualization of the tunnels is enabled by scripts for PyMOL and VMD programs, or by using CAVER Analyst. Detailed information about the tunnels, the residues, and the atoms surrounding these tunnels, and the bottleneck residues are provided in the text files. The software is freely available at www.caver.cz.

2. **CAVER PyMOL plugin** is suited for unexperienced users who need a simple exploration of tunnels or channels in their

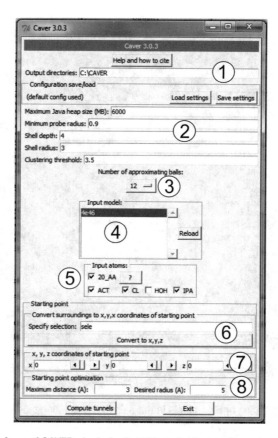

Fig. 2 Input form of CAVER plugin for PyMOL. Individual sections described in the protocol are numbered. 1–output directories, 2–tunnel calculation parameters, 3–number of approximating balls, 4–list of input model for analysis, 5–input atoms, 6–selection defining a starting point, 7–coordinates of the starting point, 8–parameters for the starting point optimization

biomolecular structures from crystallographic or NMR analyses. The plugin provides a graphical interface for setting up the calculation and interactive visualization of results (Figs. 2 and 3). The starting point position can be specified by using any PyMOL selection or alternatively by its coordinates. The calculation can be performed for any structure loaded into PyMOL. Similarly to the command-line application, additional information is available from the text files. The software is freely available at www.caver.cz.

3. **CAVER Analyst** provides the users with the interactive multi-platform environment for a comprehensive analysis on both static and dynamic structures (Fig. 4a). It allows interactive exploration of tunnels computed either by the command-line version or directly from the CAVER Analyst interface. It contains additional features for evaluation of the biological relevance of tunnels (calculation of physicochemical properties,

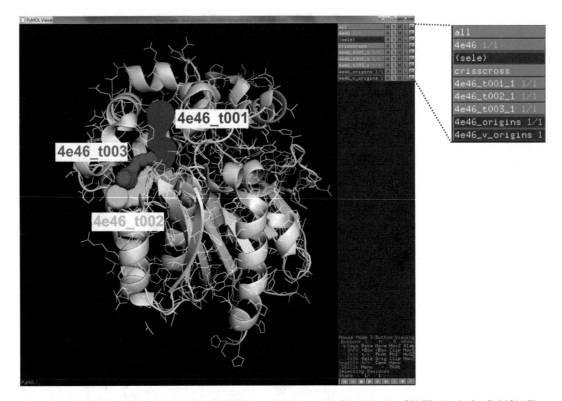

Fig. 3 Visualization of the tunnels in the haloalkane dehalogenase DhaA by the CAVER plugin for PyMOL. The protein is shown as *gray* cartoon and individual tunnels as colored spheres. The tunnels are available as separate PyMOL objects labeled as *4e46_tX*, where X represents the tunnel ID, and are ordered accordingly to their throughputs

filtering of tunnels by their parameters, etc.) and for comparative analysis of tunnels in homologous structures. The setup of the calculation is facilitated by cavity calculations and querying databases for functionally relevant residues. It accepts ensembles in Protein data bank (PDB) format, as well as formats of several molecular dynamics programs: *mdcrd* of AMBER [19], *xtc* of GROMACS [20], and *dcd* of CHARMM [21]. The software is freely available at www.caver.cz.

4. **CAVER Web** is a web server providing intuitive, straightforward, and fast analysis of tunnels for inexperienced users (Fig. 4b). The user is guided through the preparation of the input structure and setting of the tunnel calculations in a step-by-step manner, maximizing the biological relevance of obtained results with minimal effort. The visualization of tunnel geometry using JSMOL is provided along with several basic tables and graphs summarizing the most important properties of the identified tunnels. The full version is freely available at: http://loschmidt.chemi.muni.cz/caverweb.

A

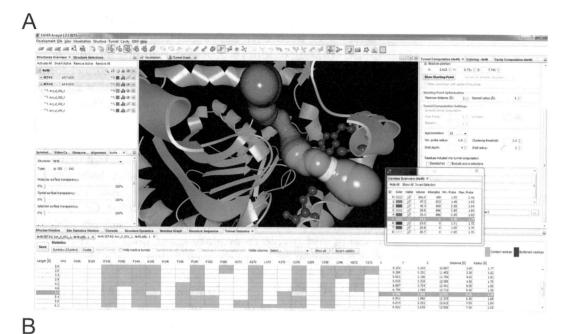

B

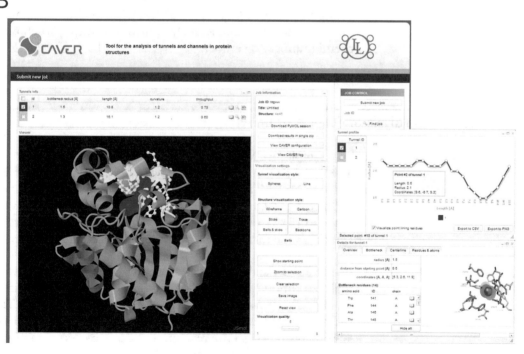

Fig. 4 Graphical user interfaces of CAVER for interactive analysis of tunnels and channels. (**a**) CAVER Analyst and (**b**) CAVER Web

5. **HotSpot Wizard** is a web server focused on automated pre-diction of hot-spot residues for mutagenesis that are likely to alter enzyme activity, selectivity, or stability. Within its work-flow, tunnels are calculated from the automatically identified

active site cavity. Tunnels are visualized with their basic parameters, e.g., bottleneck radius and length, and the identity of tunnel-forming residues is provided. Additionally, the information on the evolutionary conservation of the tunnel-forming residues is provided. The server is freely available at: http://loschmidt.chemi.muni.cz/hotspotwizard.

6. **CAVER Dock** is a software tool for rapid analysis of transport processes in proteins. It models the transportation of a ligand from outside environment into the protein active or binding site and vice versa. The input is a protein structure in PDB format and a ligand structure in the PDBQ format. The outputs are ligand's trajectory and energetic profile. CAVER Dock implements a novel algorithm which is based on molecular docking and is able to produce contiguous ligand trajectory and estimation of a binding energy along the pathway. It uses CAVER for pathway identification and heavily modified AUTODOCK VINA as a docking engine. The tool is much faster than molecular dynamic simulations (2-20 min per job). The software is easy to use as it requires in its minimalistic configuration the setup for AUTODOCK VINA and geometry of the tunnel. The tool is freely available at: http://loschmidt.chemi.muni.cz/caverdock.

3   Protocol for Analysis of Tunnels by CAVER

In this example, we will analyze tunnels in a crystal structure of haloalkane dehalogenase DhaA (PDB-ID 4e46) using the CAVER plugin for PyMOL. The aim of this analysis is to: (1) find all tunnels present in the structure, (2) select the most functionally relevant tunnels, and (3) identify residues forming these tunnels and their bottlenecks as the most promising targets for protein engineering.

3.1 Calculation Setup

1. In the PDB database, select a protein structure, ideally with high resolution, without missing atoms or residues and with a biologically relevant quaternary structure (*see* **Note 3**). The structure of the haloalkane dehalogenase DhaA with PDB-ID 4e46 determined to the resolution 1.26 Å by protein crystallography is a suitable choice.

2. In PyMOL, use *Plugin → PDB Loader Service → 4e46* to download the PDB file of the haloalkane dehalogenase DhaA from RCSB PDB. Alternatively, use *File → Open* to load the PDB file from the local computer (*see* **Note 4**).

3. Use *Plugin* → *Caver 3.0.x* to invoke the CAVER plugin for calculation of tunnels (Fig. 2). In case the plugin is not available, *see* **Note 5** for further instructions.

4. In the *Output directories* field specify the path to the directory where the results of the CAVER analysis should be stored.

5. Set the parameters for calculation of tunnels:

    (a) *Maximum Java heap size*—specifies the maximum memory allocated for computation (*see* **Note 6**). The default value of 6000 MB is more than enough to enable the analysis of the *4e46* structure, however this structure can be processed even with as little as 500 MB of the memory.

    (b) *Minimum probe radius*—defines the minimum radius of the identified tunnels. For permanently opened tunnels, this parameter should approximately correspond to the dimension of the transported ligands. However, smaller radii may be more appropriate for transient/gated tunnels to emulate missing protein dynamics. For discussion on drawbacks of using smaller probe, *see* **Note 7**. In our case, the default value of 0.9 Å will be sufficient to disclose all main ligand pathways.

    (c) *Shell radius* and *Shell depth* parameters—define the molecular surface of the protein and ultimately the endpoints of the identified tunnels. The protein surface is delineated by rolling a probe of the *Shell radius* over the protein structure. Smaller *Shell radius* provides less approximate description of the protein surface, but it could also disallow the identification of the appropriate tunnels (*see* **Note 8**). The *Shell depth* parameter disables the branching of tunnels within a given depth from the protein surface, defined by the *Shell radius*. Suitable values for these two parameters significantly differ between individual proteins and need to be optimized for each case (*see* **Note 8**). The default values (*Shell depth* 4 Å and *Shell radius* 3 Å) are suitable for our example.

    (d) *Clustering threshold*—defines the level of details for tunnel branches. The smaller the *Clustering threshold*, the more branches will be provided. A very small value can result in the identification of many nearly identical tunnel branches. The default value of 3.5 is suitable for our example.

    (e) *Number of approximating balls*—specifies the number of balls that are employed to represent individual atoms in the input structure of protein to enable the construction of an ordinary Voronoi diagram. Using a high number of balls increases the accuracy of results, but also the computational time and memory demands. For most of the cases, the usage of default 12 balls provides acceptable precision.

6. Select *4e46* as the *Input model* out of a list of input models available to perform the tunnel analysis of this structure.

7. In the *Input atoms* section, choose only the relevant atoms belonging to the macromolecule to be employed in the analysis (*see* **Note 9**). In this example, use 20 standard amino acid residues (20_AA) as the only atoms included in the calculation. Remaining ligands named ACT, CL, and IPA have to be deselected. Water molecules (HOH) are excluded automatically.

8. Specify the starting point for the tunnel calculation. The starting point is initially placed in the center of mass of the residues or atoms specified by PyMOL selection. Residues suitable for the creation of such a selection are: (1) residues forming the relevant cavity, (2) catalytic residues in the cavity, or (3) ligands in the cavity (*see* **Note 10**). It is important to stress that an appropriate starting point has to be located in an empty space of the cavity and cannot intersect with any protein atoms (*see* **Note 11**). In our example, select the ligand with residue name IPA, which is conveniently located at the center of the active site (*see* **Note 12**). Once a suitable selection is created in PyMOL, the user has to input the name of this selection into the *Specify selection* field and press the *Convert to x, y, z* button. This will show the initial location of the starting point as a white cross. At this stage, the user has to check if the point is located inside the empty space within the protein structure. If it is not, the user can either manually modify the position of the starting point by changing the absolute coordinates, or rely on the automatic optimization of the position that is controlled by the *Maximum distance* and *Desired radius* parameters. This optimization will relocate the starting point to the empty sphere of the *Desired radius* up to the *Maximum distance* from its initial position (*see* **Note 13**).

9. Press *Compute tunnels* button to start calculation of tunnels in the structure of the haloalkane dehalogenase DhaA.

3.2 Interpretation of Results

1. Explore the tunnels identified in the context of the haloalkane dehalogenase DhaA structure (Fig. 3). The identified tunnels become available as separate PyMOL objects labeled as *4e46_tX*, where X represents tunnel IDs, and ordered accordingly to the tunnel *throughput*, i.e., a metrics combining the length and width of the tunnel. The throughput reflects the predicted ability of a tunnel to transport small molecules. Individual tunnels can be shown/hidden and their visualization style modified by using the right control panel of PyMOL.

Table 2
Main geometric parameters of the tunnels identified in the haloalkane dehalogenase DhaA

ID	Avg_BR [Å]	Avg_L [Å]	Avg_C	Avg_throughput
1	1.57	9.18	1.13	0.76
2	1.28	16.11	1.49	0.62
3	1.06	10.69	1.25	0.59

Avg_BR average bottleneck radius, *Avg_L* average tunnel length, *Avg_C* average tunnel curvature, *Avg_throughput* average tunnel throughput. Please note that in the case of static protein structure, these values are averaged over a single tunnel, i.e., these parameters correspond directly to the bottleneck radius, tunnel length, and curvature of a particular tunnel

2. Extract the main geometric parameters of the identified tunnels from the *summary.txt* file, which is located in the *Output directories* (*see* **Note 14**). The most important parameters of each tunnel are the bottleneck radius (*Avg_BR*), the tunnel length (*Avg_L*), the tunnel curvature (*Avg_C*), and the tunnel throughput (*Avg_throughput*) (Table 2).

3. Combining the visualization of the tunnels with the information about their parameters, the relevance of each tunnel with respect to protein function can be estimated (*see* **Note 15**). In our case, the first tunnel (*4e46_t001*), with the bottleneck radius of nearly 1.6 Å, is the only tunnel wide enough to enable the transport of any molecule without a need for conformational changes in the protein structure. Therefore, this tunnel will be in focus of our further analyses. Note that two auxiliary tunnels could still represent the suitable targets for engineering—especially of protein stability [8] and activity [5].

4. Perform an analysis of the tunnel profile by plotting the tunnel radius against the tunnel length. This enables identification of tunnel bottlenecks. For this purpose, use the data from the *caver_output/calculation_id/analysis/tunnel_profiles.csv* file, located in the previously defined *Output directories* (*see* **Note 14**). Focusing on the tunnel cluster 1, plot *R* versus *length* parameters (Fig. 5a). One can easily identify the location of the tunnel bottleneck 5.5 Å from the starting point from this graph (*see* **Note 16**).

5. Identify the tunnel- and the bottleneck-forming residues representing the potential hot-spots for mutagenesis (*see* **Note 17**). Open the *caver_output/calculation_id/analysis/residues.txt* file from the *Output directories* (*see* **Note 14**). There are 22 residues lining the tunnel 1 (Fig. 5b). Out of these, 18 residues contribute to the formation of the tunnel by their side chains, while the remaining four residues participate only by backbone with their side chains oriented outward the tunnel. The former residues represent good hot-spots for modulation of the

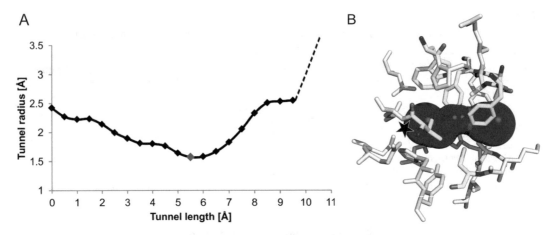

Fig. 5 Analysis of the main tunnel of the haloalkane dehalogenase DhaA. (**a**) Profile of the main tunnel. The position of the bottleneck is represented by the *red* diamond in the profile. The dashed-line illustrates the further widening of the pathway at the protein surface. (**b**) Tunnel-lining and bottleneck residues identified by CAVER. Residues are shown as *white* (tunnel-lining) and *yellow* (bottleneck) sticks

properties of the tunnel by substitutions. However, one has to be careful since some of these residues form also the wall of the buried cavity, and thus are often involved in other protein functions, such as ligand binding and catalysis. These indispensable residues must be excluded from further considerations to provide a viable design (*see* **Note 18**). A safer alternative is to focus on the bottleneck residues that are less frequently overlapping with the catalytic or ligand binding residues. At the same time, their modification has higher probability to alter the tunnel properties [10]. The list of residues forming the tunnel bottleneck is available in the file *caver_output/calculation_id/analysis/bottlenecks.csv*, located in the *Output directories* (*see* **Note 14**). In our case, there are eight bottleneck residues (Fig. 5b): Trp141, Phe144, Ala145, Thr148, Phe149, Ala172, Cys176, and Val245. Referring to the *residues.txt* file, the user can verify that all these residues contribute to the tunnel by their side chain, and thus could be considered as proper hot-spots.

4   Notes

1. Individual structures of the ensemble must be aligned to a selected reference structure, located in a single directory, and in the case of time-dependent ensemble also consecutively numbered in a computer-recognized order (e.g., the structure "10" could come before "2" thus it is necessary to number the

structures as "02" and "10" instead). For NMR structures available from the PDB database, the individual models present in the PDB file must be split into separate files. The detailed protocol for the analysis of ensemble data is described in the User guides of CAVER 3.0 and CAVER Analyst 1.0.

2. The molecular dynamics trajectories of some proteins could be procured from databases like MoDEL—http://mmb.pcb.ub.es/MoDEL [22].

3. The PDB file of the analyzed protein has to be of sound quality, hence special attention should be devoted to its resolution, information on missing atoms or residues, and its biological unit. Structures with resolution below 2 Å are preferred for the analysis. However, beware of the alternative conformation of residues that are frequently present in high-resolution structures, as CAVER retains only those conformations with the highest occupancy by default. Missing atoms or residues could lead to the identification of artificial tunnels. Using an asymmetric unit instead of a proper biological unit of the protein could also be a source of errors: (1) artificial tunnels are identified that lead through a space that is occupied by another protein chain of the biological unit, and (2) some relevant tunnels may not be identified because their openings at the surface can be blocked by other protein chain, that would not be present in the biological unit.

4. Aside from the PDB format supported by the CAVER plugin and the CAVER command-line application, additional input formats from molecular dynamics trajectories are supported in CAVER Analyst (AMBER—*mdcrd*, GROMACS—*xtc*, and CHARMM—*dcd*).

5. The caver_3.0_plugin.zip file containing CAVER plugin for PyMOL and the user guide with instructions for the installation procedure can be obtained from the download section at http://www.caver.cz.

6. *Maximum Java heap size* parameter should be increased in the case when the *out of memory* error is encountered. When only a limited memory is available, the requirements can be reduced by decreasing the *number of approximating balls* parameter.

7. Using too small probe radius results in the identification of many artificial tunnels since narrow tunnels can be identified nearly everywhere in the protein structure. Due to a large number of such tunnels, the time required for their analysis increases notably. It is therefore preferred to analyze transient tunnels from molecular dynamics simulation using a probe of more realistic radius.

8. *Shell depth* and *Shell radius* parameters need to be optimized to avoid artificial tunnels (Fig. 6a). While the smaller *Shell radius*

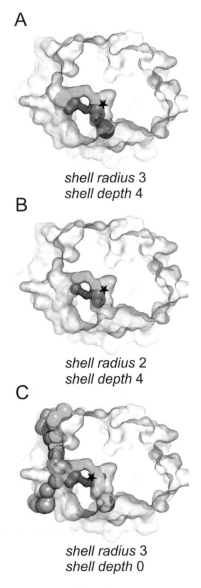

Fig. 6 Effect of different settings of the shell radius and shell depth parameters on tunnels of the haloalkane dehalogenase DhaA. (**a**) Optimal tunnels obtained by using the most suitable parameters. (**b**) Short tunnels poorly describing the real length due to too small *Shell radius*. (**c**) Artificial tunnels slithering on the protein surface due to too small *Shell depth*. The protein is shown as *grey surface*, individual tunnels as *colored spheres*, and the starting point is depicted as a *black star*

provides more realistic surface, it should also be larger than the maximum radius of the individual tunnels and must be larger than the bottleneck radius of each tunnel. Otherwise no tunnels or artificially shortened tunnels will be identified, since in such a case the protein interior or its part will be considered as bulk solvent (Fig. 6b). The user should increase the *Shell radius* whenever the obtained tunnels are ending prematurely before reaching the protein surface. A hallmark of too large *Shell radius* and too small *Shell depth* is the presence of many tunnels that seem to be slithering on the protein surface (Fig. 6c). In case the *Shell radius* cannot be further decreased and many slithering tunnels are still identified, the user should increase the *Shell depth* instead. Since the *Shell depth* prohibits branching of tunnels, it is necessary to check that all important tunnel branches are identified under such settings. Conversely, if some important tunnel or some important tunnel branch is missing, the *Shell depth* should be decreased.

9. In general, only atoms of the investigated biomolecules, i.e., amino acid residues or nucleotides, should be included in the analysis, as the ligand moieties present in the structures could occlude the tunnels and disallow their identification. However, special care has to be taken with covalently bound atoms of modified residues, cofactors, or metals that should be included in the analysis as integral components of the biomolecule.

10. Automatic identification of catalytic residues and occluded cavities can be performed by CAVER Analyst, CAVER Web, or Hotspot Wizard. Alternatively, the user can perform a manual analysis of cavities and important residues. There are many tools serving for the identification and analysis of pockets in protein structure [23, 24]. Some of these tools, e.g., CASTp [25], MetaPocket 2.0 [26], or Fpocket [27], are available as easy-to-use web servers, providing a list of residues forming each identified pocket. To recognize the relevant pockets, the user can employ the information on catalytic residues in CASTp, a consensus prediction of several tools in MetaPocket 2.0, or a druggability score provided by Fpocket. If no relevant pocket can be unambiguously selected, the information on functional or catalytic residues from UniProtKB [28] or Catalytic Site Atlas 2.0 [29] databases can be employed for specifying the functional pocket in which these residues are localized.

11. When using the protein residues to define the starting point, more than one residue from the protein is often required for the optimal placement of the point into an empty space. By clicking on three residues around a cavity, PyMOL will create the selection named "sele". Atoms included in this selection will be used by CAVER to calculate a starting point as their geometric center upon pressing the *Convert to x, y, z* button. The readers may want to try this approach, e.g., by selecting Asp106, Leu209 and Phe168 residues.

12. It is important to note that defining the starting point by the ligand IPA will work only in the case that the atoms of the

ligand were excluded from the analysis during the **step** 7, as exemplified by our protocol.

13. Increasing the *Desired radius* parameter will place the starting point in the center of the closest empty space large enough to harbor a sphere of at least *Desired radius*, within the *Maximum distance*. The potential pitfall of setting both *Maximum distance* and *Desired radius* parameters to relatively high values is that such settings could displace the calculation starting point outside the protein structure or to a different cavity, and thus no tunnels will be identified. After the completion of the tunnel calculation, the position of the optimized starting point will be available in the output folder (*see* **Note 14**) as the PyMOL object *4e46_v_origins*.

14. Individual output files with details about the CAVER analysis are located in the *Output directories* specified during the fourth step of the calculation setup, in the folder *caver_output/calculation_id*. *Calculation_id* subfolder with the highest number corresponds to the latest analysis.

15. The properties that may account for higher functional relevance of the tunnels are larger width, shorter length, and less curved trajectory [10]. The relevant tunnels are often composed of residues with physicochemical properties complementary to their cognate ligands [30]. The bottleneck residues are often flexible to enable a tunnel gating [16].

16. In some cases, the tunnel bottleneck may be detected at the beginning of the tunnel, which indicates that the starting point was incorrectly placed too close to the protein atoms. Such problem can be solved by moving the starting point into an empty space either manually or by adjusting the *Maximum distance* and the *Desired radius* parameters.

17. Both the tunnel-lining and the bottleneck-forming residues are identified by a predefined distance from the tunnel surface (3 Å by default). This approach may also provide false positive results, i.e., residues in the second shell of the tunnel or bottleneck. Therefore, without filtering over an ensemble of protein structures, the location and orientation of the identified residues should be visualized to verify their role in the tunnel or bottleneck formation.

18. The residues forming the walls of ligand-binding or catalytic pockets, as well as catalytic residues, can be identified by the tools discussed (*see* **Note 10**). The residues in a contact with bound ligands present in the structure could be identified by LigPlot+ [31], PoseView [32], or LPC [33]. Alternatively, users can employ an integrative analysis by HotSpot Wizard [18]. A detailed discussion on the construction of smart libraries can be found in the book chapter by Sebestova et al. in the recent Methods in Molecular Biology series [34].

Acknowledgments

The authors would like to express their thanks to Sergio Marques and David Bednar (Masaryk University, Brno) and to the editors Uwe Bornscheuer and Matthias Höhne (University Greifswald, Greifswald) for critical reading of the manuscript. MetaCentrum and CERIT-SC are acknowledged for providing access to super-computing facilities (LM2015042 and LM2015085). The Czech Ministry of Education is acknowledged for funding (LQ1605, LO1214, LM2015051, LM2015047 and LM2015055). Funding has been also received from the European Union Horizon 2020 research and innovation program under the grant agreement No. 676559.

References

1. Prokop Z, Gora A, Brezovsky J et al (2012) Engineering of protein tunnels: keyhole-lock-key model for catalysis by the enzymes with buried active sites. In: Lutz S, Bornscheuer UT (eds) Protein engineering handbook. Wiley-VCH, Weinheim, pp 421–464

2. Gora A, Brezovsky J, Damborsky J (2013) Gates of enzymes. Chem Rev 113:5871–5923

3. Kingsley LJ, Lill MA (2015) Substrate tunnels in enzymes: structure-function relationships and computational methodology. Proteins 83:599–611

4. Biedermannova L, Prokop Z, Gora A et al (2012) A single mutation in a tunnel to the active site changes the mechanism and kinetics of product release in haloalkane dehalogenase LinB. J Biol Chem 287:29062–29074

5. Pavlova M, Klvana M, Prokop Z et al (2009) Redesigning dehalogenase access tunnels as a strategy for degrading an anthropogenic substrate. Nat Chem Biol 5:727–733

6. Chaloupkova R, Sykorova J, Prokop Z et al (2003) Modification of activity and specificity of haloalkane dehalogenase from *Sphingomonas paucimobilis* UT26 by engineering of its entrance tunnel. J Biol Chem 278:52622–52628

7. Prokop Z, Sato Y, Brezovsky J et al (2010) Enantioselectivity of haloalkane dehalogenases and its modulation by surface loop engineering. Angew Chem Int Ed 49:6111–6115

8. Koudelakova T, Chaloupkova R, Brezovsky J et al (2013) Engineering enzyme stability and resistance to an organic cosolvent by modification of residues in the access tunnel. Angew Chem Int Ed 52:1959–1963

9. Liskova V, Bednar D, Prudnikova T et al (2015) Balancing the stability–activity trade-off by fine-tuning dehalogenase access tunnels. ChemCatChem 7:648–659

10. Chovancova E, Pavelka A, Benes P et al (2012) CAVER 3.0: a tool for the analysis of transport pathways in dynamic protein structures. PLoS Comput Biol 8:e1002708

11. Sehnal D, Svobodova Varekova R, Berka K et al (2013) MOLE 2.0: advanced approach for analysis of biomacromolecular channels. J Cheminform 5:39

12. Yaffe E, Fishelovitch D, Wolfson HJ et al (2008) MolAxis: efficient and accurate identification of channels in macromolecules. Proteins 73:72–86

13. Masood TB, Sandhya S, Chandra N et al (2015) CHEXVIS: a tool for molecular channel extraction and visualization. BMC Bioinformatics 16:119

14. Kim J-K, Cho Y, Lee M et al (2015) BetaCavityWeb: a webserver for molecular voids and channels. Nucleic Acids Res 43:W413–W418

15. Brezovsky J, Chovancova E, Gora A et al (2013) Software tools for identification, visualization and analysis of protein tunnels and channels. Biotechnol Adv 31:38–49

16. Kingsley LJ, Lill MA (2014) Ensemble generation and the influence of protein flexibility on geometric tunnel prediction in cytochrome P450 enzymes. PLoS One 9:e99408

17. Kozlikova B, Sebestova E, Sustr V et al (2014) CAVER analyst 1.0: graphic tool for interactive visualization and analysis of tunnels and channels in protein structures. Bioinformatics 30:2684–2685

18. Pavelka A, Chovancova E, Damborsky J (2009) HotSpot wizard: a web server for identification of hot spots in protein engineering. Nucleic Acids Res 37:W376–W383

19. Case DA, Cheatham TE, Darden T et al (2005) The Amber biomolecular simulation programs. J Comput Chem 26:1668–1688

20. Berendsen HJC, van der Spoel D, van Drunen R (1995) GROMACS: a message-passing parallel molecular dynamics implementation. Comput Phys Commun 91:43–56

21. Brooks BR, Bruccoleri RE, Olafson BD et al (1983) CHARMM: a program for macromolecular energy, minimization, and dynamics calculations. J Comput Chem 4:187–217

22. Meyer T, D'Abramo M, Hospital A et al (2010) MoDEL (molecular dynamics extended library): a database of atomistic molecular dynamics trajectories. Structure 18:1399–1409

23. Henrich S, Salo-Ahen OMH, Huang B et al (2010) Computational approaches to identifying and characterizing protein binding sites for ligand design. J Mol Recognit 23:209–219

24. Perot S, Sperandio O, Miteva MA et al (2010) Druggable pockets and binding site centric chemical space: a paradigm shift in drug discovery. Drug Discov Today 15:656–667

25. Dundas J, Ouyang Z, Tseng J et al (2006) CASTp: computed atlas of surface topography of proteins with structural and topographical mapping of functionally annotated residues. Nucleic Acids Res 34:W116–W118

26. Zhang Z, Li Y, Lin B et al (2011) Identification of cavities on protein surface using multiple computational approaches for drug binding site prediction. Bioinformatics 27:2083–2088

27. Schmidtke P, Le Guilloux V, Maupetit J et al (2010) Fpocket: online tools for protein ensemble pocket detection and tracking. Nucleic Acids Res 38:W582–W589

28. UniProt Consortium (2015) UniProt: a hub for protein information. Nucleic Acids Res 43:D204–D212

29. Furnham N, Holliday GL, de Beer TAP et al (2014) The catalytic site atlas 2.0: cataloging catalytic sites and residues identified in enzymes. Nucleic Acids Res 42:D485–D489

30. Pravda L, Berka K, Svobodova Varekova R et al (2014) Anatomy of enzyme channels. BMC Bioinformatics 15:379

31. Laskowski RA, Swindells MB (2011) LigPlot+: multiple ligand-protein interaction diagrams for drug discovery. J Chem Inf Model 51:2778–2786

32. Stierand K, Rarey M (2010) Drawing the PDB: protein-ligand complexes in two dimensions. ACS Med Chem Lett 1:540–545

33. Sobolev V, Sorokine A, Prilusky J et al (1999) Automated analysis of interatomic contacts in proteins. Bioinformatics 15:327–332

34. Sebestova E, Bendl J, Brezovsky J et al (2014) Computational tools for designing smart libraries. Methods Mol Biol 1179:291–314

Chapter 4

YASARA: A Tool to Obtain Structural Guidance in Biocatalytic Investigations

Henrik Land and Maria Svedendahl Humble

Abstract

In biocatalysis, structural knowledge regarding an enzyme and its substrate interactions complements and guides experimental investigations. Structural knowledge regarding an enzyme or a biocatalytic reaction system can be generated through computational techniques, such as homology- or molecular modeling. For this type of computational work, a computer program developed for molecular modeling of proteins is required. Here, we describe the use of the program YASARA Structure. Protocols for two specific biocatalytic applications, including both homology modeling and molecular modeling such as energy minimization, molecular docking simulations and molecular dynamics simulations, are shown. The applications are chosen to give realistic examples showing how structural knowledge through homology and molecular modeling is used to guide biocatalytic investigations and protein engineering studies.

Key words Biocatalysis, Enzyme, Energy minimization, Homology modeling, Molecular modeling, Molecular docking simulations, Molecular dynamics simulations, Protein engineering

1 Introduction

In Biocatalysis, enzymes are applied as catalysts to enhance the reaction rate as well as to control the reaction- and substrate specificity of a chemical transformation [1]. Enzymes are nature's own catalysts developed through evolution to perform chemical transformations fulfilling the native host requirement. The native host environmental conditions most often differ from those reaction conditions applied in biocatalytic syntheses. This limits the application of enzymes in their native form [2]. Generally, an effective biocatalytic process requires the enzyme to be stable, soluble and easy to produce [3]. Depending on the specific enzyme and the type of chemical reaction to be performed, substrate specificity as well as temperature and/or solvent stability of a native enzyme might not match the desired parameters. Hence, protein engineering is often applied to overcome these hurdles. In protein engineering, an enzyme is modified at the gene level.

Uwe T. Bornscheuer and Matthias Höhne (eds.), *Protein Engineering: Methods and Protocols*, Methods in Molecular Biology, vol. 1685, DOI 10.1007/978-1-4939-7366-8_4, © Springer Science+Business Media LLC 2018

Numerous papers have described protein engineering as a successful method to improve enzymes for specific biocatalytic applications. For protein engineering, three main experimental methods have been developed: directed evolution, semirational design, and rational design [4–11]. Directed evolution does not require structural knowledge and is based on iterative creation of gene libraries by a random mutagenesis method in combination with a suitable screening method to search through a large pool of created enzyme variants. Selected enzyme variants are chosen as templates for the next round of random mutagenesis followed by screening. Both semirational and rational design use structural information to guide the mutagenesis work. In semirational design, the amino acid residues to be altered by random or semi-random mutagenesis methods are generally determined using structural information. Knowledge regarding enzyme substrate interactions is often beneficial. A successful protein engineering study by (semi)rational design requires (1) knowledge about reaction mechanism, (2) a protein 3D structure, (3) a computer with a software to visualize the protein 3D structure, and (4) knowledge on how to interpret the structural information. Computational techniques can preferably be applied to complement practical laboratory experiments. Most often, theoretical experiments are less expensive and less time consuming than practical laboratory experiments.

Structural information requires a protein 3D structure, which may be found in the PDB protein data bank [12, 13]. If no structure of the specific enzyme is deposited in the protein data bank, a new structure can be experimentally determined by X-ray crystallography or nuclear magnetic resonance (NMR) techniques. Then, the precise arrangement of each atom in the protein structure may be determined. To harbor these two experimental techniques, the specific knowledge of a crystallographer or a computational biology scientist is demanded. X-ray crystallography is the most commonly applied technique, which is also shown by the number of solved structures: In August 2017, close to 111,000 X-ray protein crystal structures were deposited to the protein data bank [12], which is tenfold more than the number of protein structures solved by NMR-techniques. When no protein 3D structure is available and solving the specific 3D structure by the above-mentioned techniques is not an alternative, homology modeling can be applied. Homology modeling is a theoretical method to predict a protein's 3D structure using structural information from a related protein. This is theoretically possible, since the structural arrangement of atoms in a protein is known to depend on its amino acid sequence.

A computer program is required to visualize a protein 3D structure or to perform molecular modeling. There are various programs available for this purpose. Some programs only allow protein structure visualization, while others allow the user to do advanced molecular modeling.

Molecular modeling is the general technique to modify and simulate a protein structure in silico by computational techniques. Molecular modeling can give theoretical knowledge regarding an enzyme, specific enzyme substrate interactions to guide protein-engineering work or to support theoretical explanations to results received experimentally in the laboratory. The most commonly applied molecular modeling techniques within biocatalytic investigations are (1) homology modeling, (2) energy minimization, (3) molecular docking simulations, and (4) molecular dynamic simulations.

Homology modeling is a computational technique to predict and generate a protein 3D structure using protein sequence information and one or several related protein structures. The goal in homology modeling is to create a protein 3D structure that is of comparable quality to those high-resolution structures made by experimental techniques, such as X-ray or NMR. Still, to predict a protein's 3D structure using exclusively protein sequence information remains an unsolved challenge. Homology modeling is based on the observations that the folding of a protein structure is determined by its amino acid sequence [14, 15] and that evolution often has conserved the overall fold of amino acid sequences sharing a certain similarity [16, 17]. The latter observation has shown that this is depending on the number of aligned residues and the presence of identical residues [18]. In homology modeling, the enzyme to create a 3D-model structure of is called query or target. A related homologous protein with known 3D structure of high-resolution in pdb file format is called template. Templates are commonly found by simple alignments comparing sequence identity. Typically, higher sequence identity results in higher quality of the resulting model. As a rule of thumb, similarities of target and template protein should be >30% (better: >40%).

Homology modeling is a multistep process that commonly involves (1) finding a template to the target by initial sequence alignment, (2) alignment correction, (3) generation of the target protein backbone, (4) modeling of loops and side-chains, (5) optimization of the generated model, and (6) validation of the homology model to detect structural errors and iteration (if the structure is not satisfying enough) [19–21]. However, the use of a homology structure depends on the quality of the final structure.

Energy minimization is often applied in molecular modeling of protein systems to create or refine the hydrogen bond network, remove unfavorable contacts and lower the overall system energy. However, an energy minimization does not necessarily result in a structure that is closer to the native structure compared to the starting structure [20]. Systems can be trapped in a local minimum and consequently not reach the global energy minimum. To overcome this problem, molecular dynamics simulations can be applied. Molecular dynamics gives information regarding different motions in a system, such as intra- and intermolecular motions, over time. Each atom in a protein has a potential energy, which can be computed by Newton's second law of motion. The movement of particles in a system is then calculated in small time steps. For this, a force field needs to be applied. The use of force fields (molecular mechanics) is a simplification of quantum mechanics where electronic motions are ignored. The energy of the system is only described as a function of the nuclear positions [22].

Molecular docking simulations can be applied to explore and predict substrate binding modes or substrate coordination in an enzyme. This computational technique is often used within drug discovery for hit identification. In a docking simulation, the substrate is called ligand and the enzyme is called receptor. The ligand is commonly built manually or taken from a crystal structure. The docking of the ligand often starts from a random position outside the enzyme and may allow both ligand and receptor flexibility. Resulting binding modes of the ligand in the receptor are explored and the resulting binding energy can be calculated.

YASARA is a computer program with no specific system requirements for simple to advanced molecular modeling applications [23, 24]. The program is available in four different packages: YASARA View, YASARA Model, YASARA Dynamics, and YASARA Structure. The simplest version, YASARA View, is available for online download and is suitable for molecular graphics and analysis. For molecular modeling and molecular dynamics, more advanced versions are required: YASARA Model or YASARA Dynamics. YASARA Structure is required for homology modeling and molecular docking simulations. The more advanced program packages include all functions of the less advanced program packages including additional modeling tools.

This chapter is focusing on how to use YASARA Structure as a tool to gain important structural enzyme information for use in biocatalytic investigations. YASARA Structure contains a range of suitable features to aid the modeling work, such as (1) a graphical interface, (2) a user-friendly tutorial including learning movies and premade scripts. However, the user has to be alert and validate the obtained modeling results carefully. Minor mistakes during a method set up can result in major errors and misinterpretations. Therefore, critical usage and knowledge regarding the modeling limits are important.

2 Materials

A computer with internet connection and the YASARA Structure program.

YASARA runs on Linux, Microsoft Windows, and MacOS. Of course, a faster CPU decreases the calculation time. Further acceleration can be obtained if a suitable GPU card is available. For recommendations, please have a look at http://yasara.org/gpu.htm.

3   Methods

In this chapter, protocols for (1) homology modeling, (2) substrate building, (3) energy minimizations, (4) molecular docking simulations and (5) molecular dynamics simulations using the YASARA Structure program are given. These protocols will be described within the workflow of two specific research applications (Subheadings 3.1 and 3.2) to illustrate the use of YASARA Structure as a tool to gain structural information to complement biocatalytic investigations.

3.1   Create a 3D Model of an Amine Transaminase by Homology Modeling

The gene encoding a class III aminotransferase from *Bacillus* sp. Soil768D1 was found in the NCBI GenBank® database [25, 26].

>gi|950170211|ref|WP_057219702.1| aminotransferase class III [Bacillus sp. Soil768D1]

3.1.1   Find the Gene of Interest

MSLTVQKINWEQVKEWDRKYLMRTFSTQNEYQPVPIES
TEGDYLIMPDGTRLLDFFNQLYCVNLGQKNPKVNAAIKEA
LDRYGFVWDTYSTDYKAKAAKIIIEDILGDEDWPGKVRFVS
TGSEAVETALNIARLYTNRPLVVTREHDYHGWTGGAATVT
RLRSYRSGLVGENSESFSAQIPGSSYNSAVLMAPSPNMFQDS
NGNCLKDENGELLSVKYTRRMIENYGPEQVAAVITEVSQG
AGSAMPPYEYIPQIRKMTKELGVLWITDEVLTGFGRTGKW
FGYQHYGVQPDIITMGKGLSSSSLPAGAVLVSKEIAEFMDR
HRWESVSTYAGHPVAMAAVCANLEVMMEENFVEQAKNS
GEYIRSKLELLQEKHKSIGNFDGYGLLWIVDIVNAKTKTPYV
KLDRNFTHGMNPNQIPTQIIMKKALEKGVLIGGVMPNTM
RIGASLNVSREDIDKAMDALDYALDYLESGEWQQS

3.1.2   Homology Modeling

1. Open the YASARA Structure program. To set your working folder ("Your working directory"):

   Go to Options > Working directory

   From now on, all newly created files will be saved in this folder. Also, save your amino acid sequence (ex. ATSoil768D1. fasta), from which you want to build a homology model, in the "Your working directory" folder.

2. Prepare to start a homology modeling experiment "the easy way" (*see* **Note 1**).

   Go to Options > Macro & Movie > Set target

   The target is your fasta sequence file saved in the "Your working directory" folder.

   All premade scripts or macros are found in the YASARA folder > mcr or at www.yasara.org. For the homology modeling "the easy way", use the macro "hm_build.mcr" [27]. All default parameters are already set in the macro and modifications of those can be made before the experiment is started. The "hm_build.mcr" macro will create a tetrameric structure.

Here, a dimeric structure is preferred for the aminotransferase. Consequently, the script requires a modification to create a dimer instead of a tetramer (*see* **Note 2**).

3. Start the homology modeling experiment "the easy way":

   Go to Options > Macro & Movie > Play macro > hm_build.mcr

   The homology modeling experiment starts as soon as the macro is played.

4. Inspect the results: In the end of the homology modeling experiment, a report is created. It includes the following information: (1) the homology modeling target, (2) the homology modeling parameters, (3) the homology modeling templates, (4) the secondary structure prediction, (5) the target sequence profile, (6) the initial homology models, (7) the model ranking, and (8) the hybrid model.

YASARA is performing the homology modeling in several steps. Firstly, YASARA identifies the homology modeling target and modeling parameters set by the macro. Then, the program searches for modeling templates by sequence alignments using PSI-BLAST [28]. For the class III aminotransferase from *Bacillus* sp. Soil768D1, five homology modeling templates (as were set by the macro) were selected based on a sequence alignment total score (Table 1).

To refine the alignments and support loop modeling, a secondary structure prediction is made using PSI-BLAST [28] and the

Table 1
YASARA runs a sequence alignment to identify possible templates. Based on the highest total score[a] number, five templates (PDB ID) of 112 hits, were found by three PSI-BLAST iterations made to extract a position specific scoring matrix (PSSM) from UniRef90

No	PDB ID	Sequence identity (%)	Sequence cover (%)	BLAST E-value	Resolution (Å)	Total score[a]
1	4AH3-A	29	91	4e−130	1.57[b]	185.67
2	3HMU-B	27	92	3e−139	2.10[c]	172.68
3	3N5M-A	28	89	7e−131	2.05	169.08
4	3GJU-A	28	90	1e−125	1.55	167.35
5	4OKS-A	27	90	4e−122	1.80[d]	158.74

[a]The total score is the product of the BLAST alignment score, the WHAT_CHECK [29] quality score in the PDBFinder2 database and the sequence coverage
[b]YASARA downloaded this structure from PDB_REDO (www.cmbi.ru.nl/pdb_redo), since re-refinement improved the structure quality Z-score by 0.434
[c]YASARA downloaded this structure from PDB_REDO (www.cmbi.ru.nl/pdb_redo), since re-refinement improved the structure quality Z-score by 0.122
[d]YASARA downloaded this structure from PDB_REDO (www.cmbi.ru.nl/pdb_redo), since re-refinement improved the structure quality Z-score by 0.112

PSI-Pred secondary structure prediction algorithm [30]. PSI-BLAST will create a target sequence profile based on a multiple sequence alignment using UniRef90 sequences. The target sequence profile is made to improve alignment of the target to the template and the result will indicate the probability for the three secondary structure conformations: coil, helix, or strand.

Models are built for the target using the template structures. The number of models built on each template is depending on the quality of the alignment. For the class III aminotransferase from *Bacillus* sp. Soil768D1, five models were built per template giving a total number of 25 models. In addition, a number of loops needed to be modeled. As an example, in homology model 1/25 (based on template 4AH3, alignment variant 01) about 83% (394 of 476) of the target amino acid residues are aligned to template residues. The sequence identity is 29% and the sequence similarity is 50%. Twelve loops had to be modeled.

Each created model is refined in two steps, as all amino acid residues are built, optimized and fine-tuned. The newly modeled loops are refined by a combined steepest descent and simulated annealing minimization keeping the backbone atoms fixed. Thereafter, a full unrestrained simulated annealing minimization is performed on the half-refined model. After refinements, the quality of the model is evaluated by Check (*see* **Note 3**) and shown in Table 2.

As all models are generated and refined, a ranking of all models based on the overall quality Z-score is made. In the end, the best parts of all 25 models are combined to create one hybrid model. The hybrid model may score worse than the best model. However, since the hybrid model is based on all models it often has larger amino acid coverage, which may be preferred.

Table 2
The quality of the homology model 1/25 (based on template 4AH3, alignment variant 01) after refinement in two steps

Check type[c]	After half-refinement[a]		After full-refinement[b]	
	Z-score[d]	Comment[e]	Z-score	Comment
Dihedrals	−1.014	Satisfactory	−0.357	Good
Packing 1D	−2.385	Poor	−2.065	Poor
Packing 3D	−2.326	Poor	−2.118	Poor
Overall quality	−2.159	Poor	−1.842	Satisfactory

[a]Half-refinement by a combined steepest descent and simulated annealing minimization keeping the backbone atoms fixed
[b]Refinement by a full unrestrained simulated annealing minimization performed on the half-refined model[a]
[c]*See* **Note 3**.
[d]The quality Z-score is calculated as the weighted sum of: $0.145 \times$ "Dihedrals" $+ 0.390 \times$ "Packing1D" $+ 0.465 \times$ "Packing3D"
[e]The correlation between Z-score and comment is shown in **Note 4**

3.1.3 Validate the
Homology Modeling
Experiment

1. View the homology modeling result:

   Open YourModel.html saved in "Your working directory folder".

   Quality data for all models including the hybrid model is shown in the resulting report. However, the user should also validate the 3D models manually. This could be done by a simple comparison of the Check reports for one or several models to the corresponding data achieved using a crystal structure (Table 3).

2. Start the validation procedure by setting the force field:

   Simulation > Force field > YASARA2

   Then, go to:

   Analyze > Check > Dihedrals/Packing1D/Packing3D/ Model quality (*see* **Note 3**).

   A manual structural comparison by structural alignment to related enzymes (in terms of overall structure RMDS values, location of the important amino acid residues for inhibitor, substrate, and cofactor hydrogen bond coordination as well as the position of catalytically active amino acid residues) should be made to validate the 3D-model applicability.

3. For a structural alignment in YASARA, start by loading your files. Load yob files from "Your working directory" and select the pdb files from the internet:

   File > Load > yob file from "Your working directory"
   File > Load > Pdb file from internet

Table 3
Structural quality comparison between the crystal structure 4ah3.pdb (used as template 1, molecule A and C), the homology model 1/25 (based on template 4AH3, alignment variant 01) after refinement, and the hybrid model using Check Object (force field YASARA2)

Structures	4ah3_AC		Homology model 1/25[a]		Hybrid model	
Check type[b]	Z-score[c]	Comment[d]	Z-score	Comment	Z-score	Comment
Dihedrals	−0.570	Good	−0.357	Good	−0.157	Good
Packing 1D	0.289	Optimal	−2.065	Poor	−1.070	Satisfactory
Packing 3D	−0.989	Good	−2.118	Poor	−2.179	Poor
Overall	−0.575	Good	−1.842	Satisfactory	−1.453	Satisfactory

[a]Homology model 1/25 (based on template 4AH3, alignment variant 01 after refinement)
[b]*See* **Note 3**
[c]The quality Z-score is calculated as the weighted sum of: 0.145 × "Dihedrals" + 0.390 × "Packing1D" + 0.465 × " Packing3D"
[d]The correlation between Z-score and comment is shown in **Note 4**

To align the structures [31]:

Analyze > Align > Objects with MUSTANG.

The resulting data received from the structural alignments of the homology model 1/25, based on template 4AH3, alignment variant 01, and the five templates to the hybrid model is shown in Table 4. In Fig. 1, the structural alignment of the homology model 1/25 (based on template 4AH3, alignment variant 01) to 4AH3_AC to the hybrid model is shown. In a close-up view of the active site, the amino acid residues responsible for the cofactor, pyridoxal-5′-phosphate (PLP), coordination is correctly positioned. By showing the hydrogen bond pattern the important hydrogen bond coordination between Asp269 to the PLP pyridine nitrogen hydrogen and the phosphate anchor to "the phosphate group binding cup" [32] is verified:

View > Show interactions > Hydrogen bonds of > All > Extend selection to include hydrogen bond partners, showing all hydrogen bonds involving the selection.

A close up view of one active site of the hybrid model is shown in Fig. 2.

Table 4
Structural comparison of the hybrid model to the homology model 1/25 (based on template 4AH3, alignment variant 01) and the five templates. The RMSD[a], the number of aligned residues[b], and the sequence identity are shown

Structure	RMSD[a] (Å)	Number of aligned residues[b]	Sequence identity (%)
Homology model 1/25	1.376	851	100
4AH3_AC[c]	1.464	804	26.6
3HMU[d]	1.543	795	26.4
3N5M[e]	1.692	739	31.7
3GJU[f]	1.531	386[g]	29.8
3OKS[h]	1.702	717	27.8

[a]RMSD means the Root Mean Standard Deviation
[b]The target monomer contains 476 amino acid residues (the dimer contains 952)
[c]Crystal structure (455 residues) of a holo omega-transaminase from *Chromobacterium violaceum* [33]
[d]Crystal structure (460 residues) of a class III aminotransferase from *Silicibacter pomeroyi* (unpublished)
[e]Crystal structure (452 residues) of a *Bacillus anthracis* aminotransferase (unpublished)
[f]Crystal structure (458 residues) of a putative aminotransferase (Mll7127) from *Mesorhizobium loti* Maff303099 (unpublished)
[g]Monomer
[h]Crystal structure (444 residues) of a 4-aminobutyrate transaminase from *Mycobacterium smegmatis* [34]

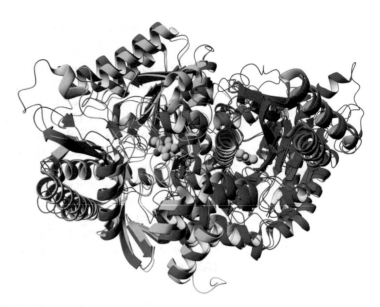

Fig. 1 Structural alignment of 4AH3_AC (*magenta*) to the hybrid model. The two subunits of the hybrid dimer are shown as *ribbon* in *green* and *blue* color, respectively, and PLP is shown in element colors

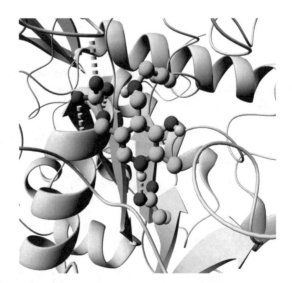

Fig. 2 A close-up view on one of two active sites of the hybrid model. The two subunits of the hybrid dimer are shown as *ribbon* in *grey* color. The cofactor, PLP, is coordinated to the catalytic lysine, K298, as an internal aldimine shown by *balls*-and-*sticks* in element colors. An aspartate, Asp269, is coordinating to the PLP pyridine nitrogen hydrogen. The phosphate group of PLP is coordinating into "the phosphate group binding cup" [32]. The hydrogen bond coordinations are shown by *yellow dashes*

3.2 Explore Enzyme Substrate Interactions by Investigating the Role of Arg417 in the Dual-Specificity of an Amine Transaminase

3.2.1 Prepare Protein Structure

Usually, protein X-ray structures obtained from the PDB databank [12] are not optimized in terms of energy and positioning of amino acid side chains. Cofactor-dependent enzymes do not always contain the specific cofactor. Other cofactor substitutes, in the form of inhibitors or different ions, can be present. Therefore, pdb-files need to be cleaned up and prepared before any modeling can be performed.

In this protocol, molecular docking simulations of pro-(S)-quinonoid complexes to a class III aminotransferase from *Silicibacter pomeroyi* will be performed. This common reaction intermediate of PLP-dependent enzymes arises after the condensation of an amino substrate to the cofactor PLP and abstraction of the hydrogen at the α-C atom by the catalytic base. However, the crystal structure (3HMU.pdb) of this aminotransferase contains two molecules of SO_4^{2-} instead of the cofactor PLP. The two ions need to be replaced by two PLP molecules. If the apo enzyme does not undergo significant structural changes during binding of the cofactor, and if a holo enzyme (with bound cofactor) from this superfamily is known, the coordinates for the cofactor can be obtained from this homologous structure. In this case, the structure 3FCR.pdb contains PLP and will be used for this purpose. After the addition of PLP, a series of steps to clean up and energy minimize the resulting structure will be performed before the docking simulations.

1. Download the two crystal structures, 3HMU.pdb and 3FCR.pdb, from the PDB databank and open them in YASARA.

   File > Load > PDB file > 3HMU.pdb

   File > Load > PDB file > 3FCR.pdb

2. The command "Clean" will add all missing hydrogen atoms and identify residues having several conformations.

   Edit > Clean > All

   Only one of the multiple conformations will be kept. It might be necessary to manually check, which of the conformations were deleted. If one does not agree with the choice of the program, it is also possible to delete a conformation manually by clicking successively on each of the atoms of the undesired side chain conformation followed by the 'del' key.

3. To save space, some pdb-files of biologically active multimeric proteins are saved as monomers. If a pdb-file contains a "REMARK 350", YASARA can use this data to build a multimeric protein structure. By using the command "Oligomerize", YASARA creates the second subunit of 3FCR. The second subunit (monomer) is created as a separate object. The two subunits need to be joined as one object.

   Edit > Oligomerize > Object > 3FCR

   Edit > Join > Object > 3FCR_2 > 3FCR

4. Adopt the coordinates of PLP from the dimeric FCR to 3HMU by a structural alignment.

Analyze > Align > Objects with MUSTANG > 3FCR > 3HMU

The 3FCR object contains several molecules. The object needs to be split in order to separate PLP from the rest of the object. This turns all molecules in the previous 3FCR object into their own objects. PLP can then be joined to 3HMU.

Edit > Split > Object > 3FCR

Locate PLP in both subunits (3FCRA Atom 23062-23085 and 3FCRB Atom 31050-31073) and join them to 3HMU:

Edit > Join > Object > Select Object 4 and 13 > Click OK > 3HMU

Delete all remaining parts of 3FCR that are not joined to 3HMU:

Edit > Delete > Object > Name > Select 3FCRA and 3FCRB > Click OK

5. The PLP molecules are now included in the 3HMU object. However, all molecules (SO_4^{2-} and H_2O) that occupy the same part of the active site as the new PLP molecules need to be deleted.
Delete the two SO_4^{2-} molecules bumping on PLP:

Edit > Delete > Residue > Name: SO_4

Delete the seven H_2O molecules bumping on PLP:

Edit > Delete > Residue > Name: Hoh > and this manual selection: 517 634 652 672 724 742 767 > OK

6. The PLP molecules need to be in the correct protonation state for catalysis (Fig. 3).
Add a hydrogen atom to the pyridine nitrogen of PLP:

Right click on the PLP pyridine nitrogen > Add > hydrogens to: Atom > 1 hydrogen > OK

7. In order to perform a simulation in YASARA, a simulation cell is required. The simulation cell is a boundary that limits the program to only perform calculations inside it.

Simulation > Define simulation cell > Set size automatically: Extend 5.0 A > around all atoms

Simulation > Cell boundaries > Periodic

Simulation > Fill cell with... > Water > OK

8. In YASARA, there are several force fields to choose from. The force field should be chosen carefully to suit the application. In this application, AMBER03 is chosen as force field.

Simulation > Force field > Select AMBER03 > OK, and if a force field is selected above, also set its default parameters

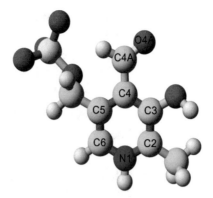

Fig. 3 One PLP molecule originated from 3FCR.pdb. The atoms are colored in element color: *cyan*—carbon, *red*—oxygen, *yellow*—phosphorus, *blue*—nitrogen, and *white*—hydrogen. The atoms are named and numbered according to YASARA. The chemical bonds, connecting the atoms in PLP, are assigned and shown in different colors: *grey* for single bond; *magenta* for resonance bond of order 1.33, *red* for resonance bond of order 1.5, and *yellow* for a double bond

9. Now, the structure is cleaned and a simulation cell is set up. An energy minimization will be performed to optimize the hydrogen bonding network and position amino acid residues to obtain as low energy in the system as possible.

   Options > Choose experiment > Energy minimization

   When the energy minimization is completed, YASARA can visualize hydrogen bonds to see if PLP is correctly positioned (e.g., that the phosphate group of PLP is hydrogen bonded to the phosphate binding cup (Ser124 and Thr322) and that the pyridine nitrogen is hydrogen bonded to Asp261). To show the hydrogen bonding network:

   Right click anywhere on PLP > Show > Hydrogen bonds > Residue > Extend selection to include hydrogen bonding partners, showing all hydrogen bonds involving the selection. > OK

10. As the protein structure is energy minimized the preparations for the docking can begin. All H_2O molecules occupying the active site may prevent productive docking and consequently needs to be removed.

    Delete H_2O molecules blocking the binding pocket in subunit A:

    Edit > Delete > Residue > Name: Hoh > Belongs to or has: Mol A > and this manual selection: 528 644 656 906

Delete H_2O molecules blocking the binding pocket in subunit B:

Edit > Delete > Residue > Name: Hoh > Belongs to or has: Mol B > and this manual selection: 539 591 644 671
Edit > Delete > Object > Sequence: 3 Water > OK

Now, the structure can be saved.

File > Save as > YASARA Scene > Filename: structure. sce > OK

11. When the active site of an enzyme is known, the simulation cell can be redefined to only include that part in the docking experiment. This will reduce the time of the experiment, as less computational power is used. The following steps (11 and 12) will be performed separately for both subunits (protein chains).

Define new simulation cell around the active site of subunit A:

Simulation > Define simulation cell > Set size automatically: Extend 10.0 A > around selected atoms > and this manual selection: 16025 (aldehyde carbon of PLP)

12. As the simulation cell is set, the PLP molecules can be removed. But, before both PLP molecules are deleted, one of them is saved as a YASARA Object for later use in ligand preparation.

Edit > Split > Object > 3HMU > OK

Save PLP as YASARA Object (this only has to be performed for one subunit):

File > Save as > YASARA Object > Object: 8 3HMUA (Or the object corresponding to PLP in subunit A) > Filename: PLP. yob > OK

Now, the protein without PLP is prepared and saved.

Edit > Delete > Object > Sequence: 8 3HMUA > OK
Edit > Join > Object > Sequence: Select Objects 3-9 > OK > Sequence: Select Object 1 > OK
File > Save as > YASARA Scene > Filename: subunit_a_receptor.sce > OK

Repeat the last two steps (11 and 12) but for subunit B.

3.2.2 Prepare Ligand   The two substrates L-alanine (Ala) and (S)-1-phenylethylamine (PEA) will be built, separately, based on the PLP molecule object (Fig. 3) to form two planar pro-(S)-quinonoid complexes (called PLP_ALA and PLP_PEA) (Figs. 4 and 5). The quinonoid structures should correspond to the first planar reaction intermediate in the proposed reaction mechanism [35]. The completed ligands will later be used for docking simulations. Building small molecules is

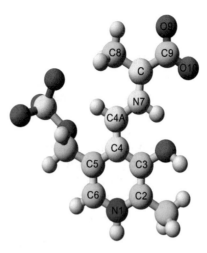

Fig. 4 The ligand PLP_Ala in planar pro-(*S*)-quinonoid structure conformation based on the PLP molecule (Fig. 3). The atoms are colored in element color: *cyan*—carbon, *red*—oxygen, *yellow*—phosphorus, *blue*—nitrogen, and *white*—hydrogen. The atoms are named and numbered according to YASARA. The chemical bonds, connecting the atoms in PLP, are assigned and shown in different colors: *grey* for single bond; *magenta* for resonance bond of order 1.33, *red* for resonance bond of order 1.5, and *yellow* for a double bond. The atoms C4, C4A, N7, and C are shown in *yellow* despite the different atom types to show that the geometry of this dihedral bond is fixed

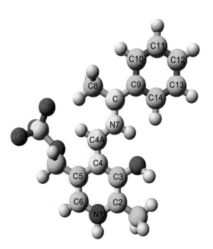

Fig. 5 The ligand PLP_PEA in planar pro-(*S*)-quinonoid structure based on the ligand PLP_Ala (Fig. 4). The atoms are colored in element color: *cyan*—carbon, *red*—oxygen, *yellow*—phosphorus, *blue*—nitrogen, and *white*—hydrogen. The atoms are named and numbered according to YASARA. The chemical bonds, connecting the atoms in PLP, are assigned and shown in different colors: *grey* for single bond; *magenta* for resonance bond of order 1.33, *red* for resonance bond of order 1.5, and *yellow* for a double bond. The atoms C4, C4A, N7, and C are shown in *yellow* despite the different atom types to show that the geometry of this dihedral bond is fixed

straightforward with YASARA. We recommend viewing the interactive tutorial video on building:

Help > Play help movie > 4.1 Building small molecules

With this knowledge and the structures given in Figs. 3, 4, and 5, you should be able to easily add the missing atoms to PLP, and to swap the bonds into the correct bond order (single/double bond). For displaying the bond order, press the F2 key—different bonds will be displayed in different colors. Usually, the final geometry will be optimized by an energy minimization step. This is especially important if cyclized molecules have been built or if many hetero atoms are included.

(a) **PLP_Ala**

1. Open the previously saved PLP Object:

   File > Load > YASARA Object > Browse: PLP.yob > OK (Fig. 3)

2. In the first steps, the bond orders of PLP will be adjusted properly. Select atoms N1 and C2 (multiple atoms can be selected by holding the ctrl-key).

   Right click on the first atom (with the white circle) > Swap > Bond > Single bond > OK

   Repeat for bonds N1-C6, C4-C5, C3-C4 and C4A-O4A: Press and hold the Tab or the Alt key, then the last action will be repeated: The program prompts you to click on the first, then on the second atom, and the bond will be swapped into a double bond.

3. Select atoms C2 and C3.

   Right click > Swap > Bond > Double bond > OK

   Repeat for bonds C5-C6 and C4-C4A.

4. Now, the quinoid intermediate will be built. The PLP aldehyde oxygen atom has to be changed to a nitrogen:

   Select atom O4A. Right click > Swap > Atom > Element: Nitrogen > OK

5. To further extend the molecule with additional atoms, existing hydrogen atoms are selected and swapped into carbon or other hetero atoms. After adjusting the bond order, new hydrogen atoms will appear at the introduced carbon atom, which can be further changed into carbon atoms. Note that the bond order of the carbon skeleton will determine the geometry of the newly built molecule.

   Select the hydrogen atom pointing upwards from N7 (Fig. 4).

   Right click > Swap > Atom > Element: Carbon > OK

6. Select atoms C and N7.

   Right click > Swap > Bond > Double bond > OK

7. Select atom N7.

   Right click > Add > hydrogens to: Atom > 1 hydrogen > OK

8. Select one hydrogen atom on C.

   Right click > Swap > Atom > Element: Carbon > OK

   Repeat for the other hydrogen atom on C.

9. Select one hydrogen atom on C9.

   Right click > Swap > Atom > Element: Oxygen > OK

10. Select atoms C9 and O9.

    Right click > Swap > Bond > Resonance bond of order 1.50 > OK

11. Select the remaining hydrogen atom on C9.

    Right click > Swap > Atom > Element: Oxygen > OK

    The bond should automatically swap to a resonance bond of order 1.50. If not, repeat **step 10** for atoms C9 and O10.

12. Now, the geometry of the built molecule will be optimized in a small energy minimization. As the ring of the PLP has not been changed, the atoms can be fixed and thus excluded from movements. Select C4.

    Right click > Fix > Atom

    Repeat on all atoms in the ring and on the substituents of atoms C2, C3 and C5. Tip: Hold Alt or Tab after C4 is fixed and click all other atoms that are to be fixed. This repeats the last action performed.

13. Perform energy minimization to optimize the orientation of the substrate part of the molecule.

    Options > Choose experiment > Energy minimization

14. Select any fixed atom.

    Right click > Free > Object

15. Lock the PLP molecule in a planar conformation. Select atoms C4, C4A, N7 and C.

    Right click > Geometry > Dihedral > 180 > OK
    Select C4.
    Right click > Fix > Atom

    Repeat on atoms C4A, N7 and C.

16. Save the finished ligand (Fig. 4)

    File > Save as > YASARA Object > Object: 1 PLP > Filename: PLP_Ala.yob > OK.

(b) **PLP_PEA**

1. Repeat **steps 1–8** from PLP_Ala

2. Select any hydrogen atom on C9.

   Right click > Swap > Atom > Element: Carbon > OK

   This action forms atom C10.

3. Select atoms C9 and C10.

   Right click > Swap > Bond > Resonance bond of order 1.50 > OK

4. The Alt button can be used to repeat the previous command. In case two commands are to be performed at once (atom swap and bond swap) the command needs to be repeated manually twice before it can be repeated by the use of the Alt button.

   Select any hydrogen atom on C10.

   Right click > Swap > Atom > Element: Carbon > OK

   The bond should automatically be swapped to a resonance bond of order 1.50.

5. Hold Alt and click any hydrogen atom on C11. Repeat for C12 and C13.

6. Select atoms C9 and C14.

   Right click > Add > Bond > Resonance bond of order 1.50 > OK

   This action will close the aromatic ring of the substrate.

7. Repeat **steps 12–16** from PLP_Ala.

8. Save the finished ligand (Fig. 5).

   File > Save as > YASARA Object > Object: 1 PLP > Filename: PLP_PEA.yob > OK

3.2.3 Molecular Docking Simulations

Docking simulations will be performed using the previously prepared ligands, PLP_Ala and PLP_PEA (Figs. 4 and 5), to the receptor structure.

The molecular docking simulations are performed using a pre-made script for molecular docking simulations in YASARA. The script (dock_run.mcr) [36] is found in the main YASARA folder. Here, this script (dock_run.mcr) requires a minor modification before start (*see* **Note 5**). All files involved in the docking experiment must be named as written in the script and saved in the same "Your working directory" folder. It is recommended to create a new folder for every docking experiment, e.g., create a folder named PLP_Ala_a. In this folder, the ligand and receptor of the docking experiment is saved. The filename for the ligand should end with the tail "_ligand.yob" and the receptor filename should end with the tail "_receptor.sce."

1. Open the previously saved "PLP_Ala.yob":

   File > Load > YASARA Object > Filename: PLP_Ala.yob > OK

2. Save the Object with a new filename in the correct folder:

   File > Save as > YASARA Object > Object: 1 PLP_Ala > Browse: PLP_Ala_a > Filename: PLP_Ala_a_ligand.yob > OK

3. Load the previously saved "subunit_a_receptor.sce":

   File > Load > YASARA Scene > subunit_a_receptor.sce > OK

   Loading a YASARA Scene automatically closes anything that was open before so closing the YASARA Object before loading the YASARA Scene is not necessary.

4. Save the Scene with a new filename in the correct folder:

   File > Save as > YASARA Scene > Browse: PLP_Ala_a > Filename: PLP_Ala_a_receptor.sce > OK

5. Set the receptor structure as target for the docking experiment:

   Options > Macro & Movie > Set target > Browse: PLP_Ala_a_receptor.sce > Filename: Since the YASARA Scene was selected in the previous action the name of the file should now be in the Filename box. Delete the tail "_receptor.sce" of the filename. > OK

6. Start the docking simulation.

   Options > Macro & Movie > Play macro > dock_run.mcr > OK

   As the docking simulation is completed, all generated files have been saved in "Your working directory" folder. The results of the docking simulation can be visualized interactively by running the script dock_play.mcr [37] found in the main YASARA folder.

7. The analysis of the docking simulation can be performed by various methods depending on the aim of the experiment. Here, the docking result is firstly visualized by opening the scene "PLP_Ala_a.sce":

   File > Load > YASARA Scene > Browse: PLP_Ala_a.sce > OK

   In the scene file "PLP_Ala_a.sce", all 50 docked ligands are presented as 50 separate molecules in one single object called "Ligand". The molecules are listed in the file "PLP-Ala_a.log" under "Bind.energy" by binding energies; ranging from high to low values. One could compare the binding energies calculated by YASARA for each docked ligand. A higher positive binding energy value indicates a stronger binding, while a negative energy means no binding. However, the values of the binding energy do not say anything about productive ligand positioning for catalysis (productive binding mode). Here, the four docked ligands with the lowest binding energy still have rather high binding energies compared to the highest ones, but they are docked upside down.

8. Another common method to analyze docking results is to compare predefined properties of every docked ligand or to use a productive binding mode definition. For the docked pro-(S)-quinonoid complexes, the distance between an aspartate residue (Asp261 in 3HMU.pdb) and the hydrogen atom of the pyridine nitrogen (N1) can be evaluated [38–41]. For catalysis, the proton on the N1 atom should be coordinated by a hydrogen bond (<3 Å) to the carboxylate group of an aspartic residue.

Select the hydrogen atom (atom 16040 in the ligand with the highest binding energy) bound to the N1 atom and one of the two oxygen atoms in the carboxylate of Asp261. When both atoms are selected, the distance between them will be listed under "Atom Properties" in the top left corner of the window as "Marked Distance".

As the docking simulation experiment to 3HMU_A (subunit A) is finished, all steps in this section is repeated for the second subunit as well as for the second ligand PLP_PEA.

3.2.4 Molecular Dynamics Simulations

Molecular dynamics (MD) simulations can be performed in order to visualize possible interactions between enzyme and substrate (ligand). This is a complex task, but since YASARA includes pre-made MD-scripts the task is simplified. In this protocol, MD-simulations are used to visualize the hydrogen bonding interaction between the carboxylate group of the substrate alanine and the amino acid residue Arg417.

1. First, a starting structure needs to be prepared using the previously created YASARA Scene file containing 50 docked ligands: Open the previously saved "PLP_Ala_b.sce":

File > Load > YASARA Scene > Browse: PLP_Ala_b.sce > OK

2. To run the MD-simulation, only one of the docked ligands is required. A suitable candidate should suit the criteria from the docking experiment. Here, the first of the 50 docked ligands is chosen. To join the chosen ligand to the enzyme structure, the following steps need to be performed:

Edit > Split > Object > Sequence: 3 Ligand > OK

Edit > Join > Object > Sequence: 3 LigandA > OK > Sequence: 1 Receptor > OK

The ligand is now joined to the enzyme and the complete structure can be saved as a PDB file. Make sure that the file is saved in a new folder dedicated to the MD simulation.

File > Save as > PDB file > Sequence: 1 Receptor > OK > Filename: PLP_Ala_b_MD.pdb > OK

3. Before the MD simulation can begin the script needs to be prepared. Locate the folder named mcr in the main YASARA folder and open the file md_run.mcr [42] using a text editor. In the beginning of the script there are settings that may require modification, e.g., pH, temperature, force field, and duration. In this protocol, the duration is set to 1 ns (*see* **Note 6**). The actual time that the simulation will take depends on the computer to be applied.

4. The structure does not need further preparations. The script will make a simulation cell, fill it with water and perform an energy minimization before the MD simulation starts. To start the script, open the previously saved PDB file "PLP_Ala_b_MD.pdb" in a new YASARA window:

File > Load > PDB file > Browse: PLP_Ala_b_MD.pdb > OK

Options > Macro & Movie > Set target > Browse: PLP_Ala_b_MD.pdb > OK

Options > Macro & Movie > Play macro > Browse: md_run.mcr > OK

5. During the MD simulation a series of snapshots will be saved in the folder where the target structure is located. They will be named "PLP_Ala_b_MD00000.sim" and so on. These snapshots can be visualized by running the script md_play.mcr [43]. To be able to run this script, the scene file created by the previous script (md_run.mcr) needs to be set as target. Open a new YASARA window and follow these steps:

Options > Macro & Movie > Set target > PLP_Ala_b_MD.sce > OK

Options > Macro & Movie > Play macro > md_play.mcr > OK

This script adds a user interface to the YASARA window and makes it possible to click through each snapshot to visualize how the MD simulation proceeded. As can be seen in Fig. 6, the amino acid residue Arg417 moved towards the substrate Ala and formed hydrogen bond coordinations.

Fig. 6 Screenshots of the molecular dynamics simulation after 0 ps (**a**), 250 ps (**b**), 500 ps (**c**) and 1000 ps (**d**). The atoms are colored in element color: *cyan*—carbon, *red*—oxygen, *yellow*—phosphorus, *blue*—nitrogen, and *white*—hydrogen. The protein secondary structure in *ribbon* is colored *gray*. The hydrogen bond coordinations are shown by *yellow dashes*

4   Notes

1. *Learn about homology modeling in YASARA*: Go to Help > Show Documentation, on Query write "Homology modeling" and learn how to: Build a homology model, useful homology modeling hints, how YASARA builds homology models, homology modeling the easy way and refine a homology model.

2. The script hm_build.mcr written by Elmar Krieger [27] was modified (oligostate=4 changed to oligostate=2, according to below), saved and used for the further homology modeling experiment.

   # Maximum oligomerization state, build a dimeric model
   oligostate=2

3. Validation of model structure: Analyze > Check. The following checks can be made to validate the quality of a homology structure:

   • Detect cis-peptide bonds or unusual amino acid residues.

   • Detect D-amino acid residues or mirrored side chain residues of Thr and Ile.

   • Verify naming conventions.

   • Detect flipped side chain residues of Val and Leu.

   • Detect incorrect or not normal water positions.

   • Detect the normality of bond lengths, angles, or dihedral angles (Dihedrals) according to the current force field. Ignores hydrogen atoms.

   • Detect the normality of combined Coulomb and VdW interactions, according to the current force field.

   • Detect the normality of 1D distance-dependent packing interactions (Packing1D) and 3D direction-dependent packing interactions (Packing3D), according to the YASARA2 force field.

   • Detect the model quality (ModelQuality) by the following calculation: The weighted sum of $0.145 \times$ 'Dihedrals', $0.390 \times$ 'Packing1D', and $0.465 \times$ 'Packing3D' to yield a single model quality score.

4. The correlation between the Z-score and the comment (description) used by YASARA is given in Table 5. The description of the Z-score is based on the numbers that "perfect" or "misfolded" proteins show. Perfectly folded proteins give positive values, while proteins "with serious errors" give low negative values.

5. The script dock_run.mcr written by Elmar Krieger [36] was modified (according to below), saved and used for the molecular docking simulation experiment.

   # Docking method, either AutoDockLGA or VINA
   method='AutoDockLGA'
   # Number of docking runs (maximally 999, each run can take up to an hour)
   runs=50

Table 5
Interpretation of the Z-score

Z-score[a]	Comment[b]
< -5	Disgusting
< -4	Terrible
< -3	Bad
< -2	Poor
< -1	Satisfactory
< 0	Good
> 0	Optimal

[a]The definition and calculation of the Z-score can be found at www.Applications/YASARA.app/yasara/doc/CheckAtomResObjAll.html
[b]The description of the Z-score is based on the numbers that "perfect" or "misfolded" proteins shows. Perfectly folded proteins give positive values, while proteins "with serious errors" give low negative values

6. The script md_run.mcr written by Elmar Krieger [42] was modified (according to below), saved and used for the molecular dynamics simulation experiment.

   # Duration of the simulation, alternatively use for example duration=5000 to simulate for 5000 picoseconds
   duration=1000

Acknowledgment

This work was funded by KTH Royal Institute of Technology.

References

1. Faber K (2011) Biotransformations in organic chemistry: a textbook, 6th edn. Springer, Heidelberg
2. Luetz S, Giver L, Lalonde J (2008) Engineered enzymes for chemical production. Biotechnol Bioeng 101:647–653
3. Widmann M, Pleiss J, Samland AK (2012) Computational tools for rational protein engineering of aldolases. Comput Struct Biotechnol J 2:e201209016
4. Chen R (2001) Enzyme engineering: rational redesign versus directed evolution. Trends Biotechnol 19:13–15
5. Chica RA, Doucet N, Pelletier JN (2005) Semi-rational approaches to engineering enzyme activity: combining the benefits of directed evolution and rational redesign. Curr Opin Biotechnol 16:378–384
6. Turner NJ (2009) Directed evolution drives the next generation of biocatalysts. Nat Chem Biol 5:567–573
7. Otten LG, Hollmann F, Arends IWCE (2009) Enzyme engineering for enantioselectivity: from trial-and-error to rational design? Trends Biotechnol 28:46–54
8. Lutz S (2010) Beyond directed evolution - semi-rational protein engineering and design. Curr Opin Biotechnol 6:734–743
9. Bommarius AS, Blum JK, Abrahamson MJ (2011) Status of protein engineering for biocatalysts: how to design an industrially useful biocatalyst. Curr Opin Chem Biol 15:194–200

10. Bornscheuer UT, Huisman GW, Kazlauskas RJ et al (2012) Engineering the third wave of biocatalysis. Nature 485:185–194

11. Steiner K, Schwab H (2012) Recent advances in rational approaches for enzyme engineering. Comput Struct Biotechnol J 2:e201209010

12. The Protein Data Bank http://www.rcsb.org/pdb/home/home.do

13. Bergman HM, Henrick K, Nakamura H (2003) Announcing the world-wide Protein Data Bank. Nat Struct Biol 10:980

14. Epstain CJ, Goldberger RF, Anfinsen CB (1963) The genetic control of tertiary protein structure: studies with model systems. Cold Spring Harb Symp Quant Biol 28:439–449

15. Epstain CJ (1964) Relation of protein evolution to tertiary structure. Nature 203:1350–1352

16. Chothia C, Lesk AM (1986) The relation between the divergence of sequence and structure in proteins. EMBO J 5:823–836

17. Sander C, Schneider R (1991) Database of homology-derived protein structures and the structural meaning of sequence alignment. Proteins 9:56–68

18. Rost B (1999) Twilight zone of protein sequence alignments. Protein Eng 12:85–94

19. Krieger E, Nabuurs SB, Vriend G (2003) Homology modeling. In: Bourne PE, Weissig H (eds) Structural bioinformatics. John Wiley & Sons, Hoboken, NJ

20. Tramontano A (2006) Protein structure prediction. Concepts and applications. Wiley-VCH, Weinheim

21. Venselaar H, Joosten RP, Vroling B et al (2010) Homology modeling and spectroscopy, a never ending love story. Eur Biophys J 39:551–563

22. Leach AR (2001) Molecular modelling – principles and applications, 2nd edn. Dorset Press, Dorchester

23. Information regarding the YASARA program products www.yasara.org/products.html

24. Krieger E, Vriend G (2014) YASARA View - molecular graphics for all devices - from smartphones to workstations. Bioinformatics 30:2981–2982

25. The NCBI GenBank: www.ncbi.nlm.nih.gov

26. Benson DA, Cavanaugh M, Clark K et al (2013) GeneBank. Nucleic Acids Res 41:D36–D42

27. Pre-made script in YASARA for homology modeling simulations http://www.yasara.org/hm_build.mcr

28. Altschul SF, Madden TL, Schäffer AA et al (1997) Gapped BLAST and PSI-BLAST: a new generation of protein database search programs. Nucleic Acids Res 25:3389–3402

29. Hooft RW, Vriend G, Sander C et al (1996) Errors in protein structures. Nature 381:272

30. Jones DT (1999) Protein secondary structure prediction based on position-specific scoring matrices. J Mol Biol 292:195–202

31. Konagurthu AS, Whisstock JC, Stuckey PJ et al (2006) MUSTANG: a multiple structural alignment algorithm. Proteins 64:559–574

32. Denesyuk AI, Denessiouk KA, Korpela T et al (2002) Functional attributes of the phosphate group binding cup of pyridoxal phosphate-dependent enzymes. J Mol Biol 316:155–172

33. Sayer C, Isupov MN, Westlake A et al (2013) Structural studies with *Pseudomonas* and *Chromobacterium* [omega]-aminotransferases provide insights into their differing substrate specificity. Acta Crystallogr Sect D 69:564–576

34. Baugh L, Phan I, Begley DW et al (2015) Increasing the structural coverage of tuberculosis drug targets. Tuberculosis 95:142–148

35. Silverman RB (2002) The organic chemistry of enzyme-catalysed reactions, 2nd edn. Academic Press, London, pp 388–390

36. Pre-made script in YASARA written by Elmar Krieger for molecular docking simulations http://www.yasara.org/dock_run.mcr

37. Pre-made script in YASARA written by Elmar Krieger to visualize (play up) a molecular docking simulation http://www.yasara.org/dock_play.mcr

38. Svedendahl M, Branneby C, Lindberg L et al (2010) Reversed enantiopreference of an ω-transaminase by a single-point mutation. ChemCatChem 2:976–980

39. Svedendahl Humble M, Engelmark Cassimjee K, Abedi V et al (2012) Key amino acid residues for reversed or improved enantiospecificity of an ω-transaminase. ChemCatChem 4:1167–1172

40. Steffen-Munsberg F, Vickers C, Thontowi A et al (2013) Connecting unexplored protein crystal structures to enzymatic function. ChemCatChem 5:150–153

41. Steffen-Munsberg F, Vickers C, Thontowi A et al (2013) Revealing the structural basis of promiscuous amine transaminase activity. ChemCatChem 5:154–157

42. Pre-made script in YASARA written by Elmar Krieger for molecular dynamics simulation http://www.yasara.org/md_run.mcr

43. Pre-made script in YASARA written by Elmar Krieger to visualize (play up) a molecular dynamics simulation http://www.yasara.org/md_play.mcr

Chapter 5

A Computational Library Design Protocol for Rapid Improvement of Protein Stability: FRESCO

Hein J. Wijma, Maximilian J.L.J. Fürst, and Dick B. Janssen

Abstract

The ability to stabilize enzymes and other proteins has wide-ranging applications. Most protocols for enhancing enzyme stability require multiple rounds of high-throughput screening of mutant libraries and provide only modest improvements of stability. Here, we describe a computational library design protocol that can increase enzyme stability by 20–35 °C with little experimental screening, typically fewer than 200 variants. This protocol, termed FRESCO, scans the entire protein structure to identify stabilizing disulfide bonds and point mutations, explores their effect by molecular dynamics simulations, and provides mutant libraries with variants that have a good chance (>10%) to exhibit enhanced stability. After experimental verification, the most effective mutations are combined to produce highly robust enzymes.

Key words Stabilization, Thermostability, Protein engineering, Biocatalysis, Biotransformation, Directed evolution, Mutant design strategy, Smart library, Computational design, In silico screening

1 Introduction

Thermostable enzymes are important for applications in research, analytics, diagnostics, and industry [1–4]. For many enzyme classes no thermostable variants are available from nature. With most protein engineering techniques, the reported increases in apparent melting temperature (T_M) are in the range of 2–15 °C [2]. These are small increases compared to the differences between naturally occurring thermostable enzymes ($T_M > 80$ °C) and mesostable enzymes (T_M approximately 50 °C) [5]. To obtain larger stability improvements, the FRESCO workflow was developed. FRESCO uses the computational library design approach—(sets of) mutations are prescreened in silico. The result is a small high-quality library that can be experimentally screened in a short time. The results hitherto obtained with four enzymes showed promising T_M improvements of 20–35 °C [6–9].

A challenge in thermostability engineering is related to the large size of most enzymes and their irreversible denaturation.

Uwe T. Bornscheuer and Matthias Höhne (eds.), *Protein Engineering: Methods and Protocols*, Methods in Molecular Biology, vol. 1685, DOI 10.1007/978-1-4939-7366-8_5, © Springer Science+Business Media LLC 2018

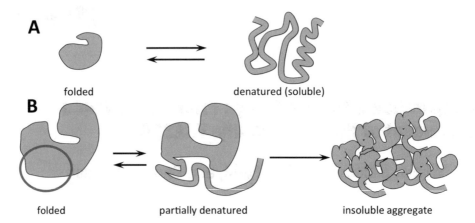

Fig. 1 Thermally induced inactivation of small versus large proteins. (**a**) For small proteins (≤20 kDa, [10]), it is most common that the entire protein unfolds in a single reversible step and that the unfolded protein remains soluble. (**b**) In large proteins, there is often a specific region (indicated with a *red circle*) that unfolds first. This partial unfolding often triggers irreversible aggregation [10]

Whereas small proteins often unfold reversibly in a single step (Fig. 1a), larger ones mostly aggregate irreversibly following the initial unfolding of certain regions (Fig. 1b) [4]. In case of reversible one-step unfolding, mutations at all positions are expected to have an effect on T_M since interactions of all amino acids change in the unfolding step. For larger proteins, mutations outside the early unfolding region have a much smaller or a negligible effect [4, 11, 12] and the spots where mutations can improve stability may be hard to find.

The FRESCO workflow addresses this challenge by in silico screening for diverse types of potentially stabilizing mutations throughout the enzyme [6, 7]. Selecting a small subset of the FRESCO generated mutations bears the risk that mutations stabilizing early unfolding regions are missed. The most stabilizing mutations generated by FRESCO were found both in flexible and in rigid (low B-factor) stretches of the protein sequence [7, 9]. If the target protein is well expressed in an easy to transform host organism like *Escherichia coli*, the complete FRESCO library can be experimentally screened in a few weeks and will produce enough stabilizing mutations to be combined into a highly robust final variant.

Below, the entire FRESCO [6, 7] protocol is described in detail for experimentalists in a way that requires no prior experience with Unix(-like) systems, which are required for running the protocol. The protocol is implemented for the user-friendly Mac OS X operating system but can be modified to be used under Linux (*see* **Note 1**). Possibly the protocol can also be implemented under the Linux bash shell that recently became available under Windows 10, or other Unix-like environments. The underlying algorithms are

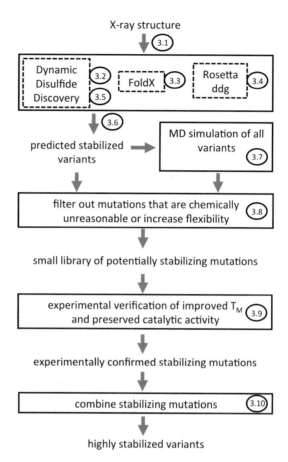

Fig. 2 Framework for rapid enzyme stabilization by computational library design (FRESCO). The numbers refer to the sections in this protocol. The protocol differs slightly from the initial approach [7], in which chemically unreasonable mutations were filtered out (Subheading 3.8) prior to the MD simulations (Subheading 3.7). The current protocol is faster as each mutant only needs to be inspected once

described elsewhere [7, 13–15]. The example that is described in the protocol, the stabilization of the enzyme limonene epoxide hydrolase (LEH) [7], enables the users to verify that all the installed software works properly. The workflow is shown in Fig. 2.

2 Materials

The FRESCO workflow consists of executables, scripts, and parameter files for running the three software packages that identify the stabilizing mutations. FoldX and Rosetta are employed to predict stabilizing point mutations [13, 14]. YASARA is used for molecular graphics, for designing disulfide bonds, and for MD simulations

[15]. The FRESCO specific software is made available via https://groups.google.com/forum/#!forum/fresco-stabilization-of-proteins. Via this forum, it is also possible to ask questions about the protocol to other FRESCO users. A short introduction to use the command line (unixIntroduction.pdf, with short exercises for people without command line experience) and instructions for obtaining and installing the other software are provided there as well. The procedure described below assumes that all software has been installed as described in "installationInstructions.pdf". YASARA Dynamics (YASARA View lacks the required functionalities), Rosetta, and FoldX require licenses. Below, differently colored layouts distinguish `UNIX command line input` and `YASARA command line input`

Hardware requirements—The only part of FRESCO that needs a large amount of calculation power is the MD simulation of the mutants (Subheading 3.7), for which a computer cluster may be needed. The calculation time of these MD simulations increases roughly with the square of protein size but also depends on protein shape. Accordingly, it depends on the target protein whether or not a computer cluster is required (*see* **Note 2**). The MD simulations for LEH (32 kDa) took <45 min per variant on a desktop computer (Intel Core i5, 4 cores, 3.2 GHz). Thus, testing 500 mutants by MD would take 10 days. On a computer cluster, this could be done in a few hours. To test the protocol on the LEH example as shown below, only a few selected MD simulations are required.

3   Methods

3.1   Setting Up a Directory Structure and Preparing the Target Protein

In this section, defects to the pdb file, such as missing hydrogen atoms, are repaired. The resulting structure, which should be representative for the protein in solution, will be used for the rest of the procedure.

1. Create a design directory and subdirectories for each step of the procedure, e.g., in your home folder (which can be abbreviated with ~):

   ```
   mkdir ~/frescoLEH
   ```

   ```
   cd ~/frescoLEH
   ```

   ```
   mkdir disulfides foldx rosetta designsMD finalVariants
   ```

2. Obtain a pdb file of the protein of interest. Best are crystal structures of high resolution and with a low R_{free} (<0.25). Structures obtained through homology modeling are probably too inaccurate. For training purposes, download the LEH structure 1NWW.pdb.

3. Move the downloaded 1NWW.pdb to ~/frescoLEH. In this directory, type `yasara1NWW.pdb&` (`&` opens the process in the background, so the console can still be used while YASARA is running).

4. In YASARA, delete buffer, ligand, and any other nonstructural molecules (*see* **Note 3**). For 1NWW, this can be done by typing `DelRes MES HPN` (MES and HPN are the names of the buffer and ligand molecules). For other proteins, possible ligands or buffer molecules have to be identified by visual inspection with YASARA. Usually, they are displayed in the amino acid sequence panel in the bottom of the YASARA window. Cofactors, such as heme or NADP, should not be deleted at this stage.

5. Use the YASARA commands `CleanAll` and `OptHydAll` to obtain reasonable protonation states for most residues. For each protein, one should carefully check by visual inspection that the protein structure is realistic (*see* **Note 4**).

6. Save the structure as a pdb file: `SavePdb OBJ 1, 1NWW_cleaned`

3.2 Running an MD Simulation for Dynamic Disulfide Discovery

To explore possible protein conformations for disulfide bond design, an MD simulation of the wild-type enzyme is carried out. The result will be a series of snapshots that provide samples of the possible protein conformations. These conformations are used under Subheading 3.5 for designing disulfide bonds (Fig. 2).

1. Within the disulfides directory, make a subdirectory: `mkdir disulfides/trajectoryMD`

2. Enter this directory (`cd disulfides/trajectoryMD/`) and copy the cleaned pdb file to this directory: `cp ../../1NWW_cleaned.pdb .`

3. To run the MD simulation, type (as a single command):

```
yasara -txt ~/frescoSoft/FRESCO/MDSimulBackboneSampl.mcr "MacroTarget = '1NWW_cleaned'" > LOG_MD&
```

This will start the macro (.mcr) file that contains all necessary specifications for the simulation.

4. Verify that the MD simulation started. The command `ls -rlt` should reveal new files being created and `top -o cpu` (*see* **Note 1**) should reveal the processes running (exit top with `q`). It should take several hours or even days before this simulation is finished. One can already start with sections 3.3 and 3.4.

3.3 Predicting Stabilizing Point Mutations with FoldX

This section starts the FoldX calculations, which predict the ΔΔGfold for individual mutations. These results will be used to select stabilizing point mutations (Fig. 2).

1. In the frescoLEH directory (`cd ~/frescoLEH`), create a table file (.tab) that lists the protein residues that are allowed to mutate by executing the following command, type:

```
yasara -txt ~/frescoSoft/FRESCO/FarEnoughZone.mcr "MacroTarget = '1N
WW'" "AvoidResidue = 'HPN'" "AvoidDistance = 5"
```

For other proteins, replace "HPN" with the PDB abbreviation of either an active site ligand, or cofactor. The "AvoidDistance" is the minimal distance that residues should have from the "AvoidResidue" to be allowed to mutate (*see* **Note 5**). If the entire protein should be allowed to mutate, use 'XXX'.

2. This should result in a new file (`ls -rtl`) named 1NWW_MoreThan5AngstromFromHPN.tab.

3. Go to the FoldX directory (`cd foldx`) and copy the rotabase.txt file (`cp ~/frescoSoft/FoldX_2017/rotabase.txt .`).

4. Set up the FoldX calculations using (the text in between <X> should be replaced, including the brackets themselves):

```
~/frescoSoft/FRESCO/DistributeFoldX Phase1 1NWW_cleaned 2 A B ../1NW
W_MoreThan<X>AngstromFrom<AvoidedResidue>.tab 1000
~/frescoSoft/FoldX_2017/foldx<version>
```

A short explanation will appear on the command line describing what DistributeFoldX does. This explanation will also provide guidance when setting up the calculation for one's own protein of interest. Write down the number of mutations that will be analyzed by FoldX.

5. Start the calculations by running the todolist file: `./todolist&`.

6. Verify that no error messages appear and check whether the calculations are indeed running (`top -o cpu`). Type `tail */LOG` to verify that no problems were encountered (*see* **Note 6**). It may take a day for the calculations to finish. One can estimate how much time the calculations will take, based on the information provided by the command `ls -rlt Subdirectory*`.

3.4 Predicting Stabilizing Point Mutations Through Rosetta

This is essentially the same procedure as described for FoldX in Subheading 3.3.

1. Enter the Rosetta directory (`cd ~/frescoLEH/rosetta`) and copy the necessary parameter file FLAGrow3 into this directory (`cp ~/frescoSoft/FRESCO/FLAGrow3 .`).

2. Open this file FLAGrow3 using a plain text editor (`open -e FlAGrow3`) and adapt the Rosetta database location, behind "–database", to match that in your own computer (*see* **Note 1**). Alternatively, for manual editing you might use Perl:

```
perl -pi -e "s,-database.*,-database ~/frescoSoft/rosetta_bin_<versi
on>_bundle/main/database/,g" FLAGrow3
```

3. The Rosetta_ddg application is parameterized for implicit water and has not (yet) been programmed to accept multimeric proteins. Therefore, explicit water molecules have to be

deleted. If the pdb file of the protein contains more than one chain, residues and chain IDs have to be renamed. For example if there are two chains—A and B having 400 amino acids each—residues of chain B have to be renumbered to 401–800, and Rosetta will accept this as a "monomeric" protein. Use YASARA to adapt the earlier cleaned pdb file: For LEH type: `yasara ../1NWW_cleaned.pdb&`. Then, remove all water molecules (`DelRes HOH`) (*see* **Note 3**), remove an amino acid that occurs only in one of the LEH subunits (`DelRes 4`), rename subunit B to A (`RenameMol B, A`) and ensure the software forgets that the protein are two different chains with (`JoinRes protein`). Ensure consecutive residue numbering, without shifting the original positions in the first subunit, with `RenumberRes protein, 5` and save the file in the current directory (`SavePdb OBJ 1, 1NWW_forRosetta`).

4. Set up the calculations by typing (use Tab-completion):

```
~/frescoSoft/FRESCO/DistributeRosettaddg Phase1 ../1NWW_MoreThan5Ang
stromFromHPN.tab 2 A 5 B 150 1NWW_forRosetta.pdb 4000 FLAGrow3 ~/fre
scoSoft/rosetta_bin_<version>_bundle/main/source/bin/ddg_monomer.mac
osclangrelease
```

5. Start the calculations (`./todolist&`)

6. Again, verify that these calculations are running correctly (`top -o cpu`, `ls -rlt` , `Subdirectory*`, `tail */LOG`, *see* **Note 6**).

3.5 Predicting Stabilizing Disulfide Bonds Through Dynamic Disulfide Discovery

The snapshots created under Subheading 3.2 are now used to design disulfide bonds.

1. Verify the snapshot files exist in the disulfides directory:
`cd ~/frescoLEH/disulfides` with
`ls trajectoryMD/*pdb`
The files have names ending with for example 1000ps.pdb, where ps stands for picoseconds.

2. In the disulfides directory, make a new subdirectory `mkdir all_designs` and copy both the cleaned pdb file (`cp ../1NWW_cleaned.pdb all_designs/`) and the snapshots from the MD trajectory in there `cp trajectoryMD/*ps.pdb all_designs/`

3. Go into this new directory (`cd all_designs/`)

4. If desired, the minimum number of residues spanning between a disulfide bond can be increased by editing ~/frescoSoft/FRESCO/DisulfideDiscovery.mcr. However, this is unnecessary for thermostability engineering [16].

5. Type

```
chmod +x ~/frescoSoft/FRESCO/commandRunningDisulfideDesign
```

and `~/frescoSoft/FRESCO/commandRunningDisulfideDesign` to generate a todolist file. Inspect the todolist: `less todolist`.

6. Start the calculations with `./todolist&`. Verify with `top` that YASARA started. The calculations may take several hours to finish on a desktop computer.

7. Type `tail LOG` to verify no errors were encountered. Type

```
ls disulfideBonds_1NWW_cleaned__1NWW_MoreThan<X>AngstromFrom<avoidRe
sidue>/*pdb | wc -l
```

This command counts the number of pdb files in the directory with disulfide bonds. For the LEH example, this should be 27 once the calculation has finished. This includes multiple conformations of the same disulfide bond.

8. Use the appropriate script to create an overview of all the disulfide bonds (`~/frescoSoft/FRESCO/OverviewDisulfides`) and inspect the result (`less BestEnergyUniqueDisulfideBonds.tab`). The UniqueDisulfides should now contain the pdb files of the disulfide bonds structures with the best energy, as well as their templates.

3.6 Selecting Computationally Designed Variants for MD Simulation

In this section, mutations that are predicted to be stabilizing by FoldX and Rosetta are identified. The predicted 3D structures of the mutants, and those of the disulfide bond mutants, are collected to carry out MD simulations (*see* section 3.7).

1. Go to the all_designs directory and copy the UniqueDisulfides folder to designsMD (`cp -r UniqueDisulfides ../../designsMD/`). Also copy the list with structures there

   (`cp BestEnergyUniqueDisulfideBonds.tab ../../designsMD`)

2. Go to the Rosetta folder (`cd ../../rosetta`). Select all mutations that are predicted to have a more than 5 kJ/mol improvement of $\Delta\Delta G\text{fold}^{\text{predicted}}$ (*see* **Note 7**):

```
~/frescoSoft/FRESCO/DistributeRosettaddg Phase2 ../1NWW_MoreThan5Ang
stromFromHPN.tab 2 A 5 B 150 1NWW_forRosetta 4000 -5
```

   Use the resulting command line output to verify that indeed all targeted mutations were screened.

3. In the FoldX folder (`cd ../foldx`), type:

```
~/frescoSoft/FRESCO/DistributeFoldX Phase2 1NWW_cleaned 2 A B ../1NW
W_MoreThan5AngstromFromHPN.tab 1000 -5
```

   Again, verify that the planned number of mutations has indeed been screened.

4. Make a list of all the mutations that are predicted to be stabilizing, by either FoldX or Rosetta, by entering (using Tab-completion):

```
cat ../rosetta/MutationsEnergies_BelowCutOff.tab > list_SelectedMuta
tions.tab && tail -n +2 MutationsEnergies_BelowCutOff.tab >> list_Se
lectedMutations.tab
```

5. Before doing MD simulations, re-add the water molecules (and possibly the cofactors) to the pdb files of the designs. Do this by running the HydrateDesigns script:

```
yasara -txt ~/frescoSoft/FRESCO/HydrateDesigns.mcr > log_conversion &
```

A few lines of this short script need to be altered if any other protein than LEH is targeted, as indicated in the script itself.

6. Look at the resulting directory (`ls -rlt NamedPdbFiles/`). Verify that there are indeed pdb files in the generated sub-directories. With `top` and `tail -f log_conversion` one can check whether YASARA has already finished.

7. For a selected target protein, use YASARA to open one of the pdb files with waters added to verify that the structure is realistic, e.g., with all cofactors present (*see* **Note 8**).

8. Once finished, copy the subdirectory with pdb files of the hydrated structures to the designsMD folder (`cp -r NamedPdbFiles ../designsMD/`; there should not be a / behind `NamedPdbFiles`) as well as the list of selected mutations (`cp list_SelectedMutations.tab ../designsMD/`).

3.7 MD Simulations of Mutants

For each of the mutants, MD simulations are carried out. This is done to predict their flexibility.

1. Go to the MD directory `cd ~/frescoLEH/designsMD`. One should see (`ls`) two directories named UniqueDisulfides and NamedPdbFiles.

2. Run the script to set up the MD simulations `~/frescoSoft/FRESCO/commandRunningMDsimulations` (if one is targeting another protein than the LEH example, this file should be modified using a text editor according to the instructions in the file itself). After that, run the resulting todolist `./todolist&`. Verify that YASARA is running with `top -o cpu`.

3. For the LEH example, it will probably take several hours for this step to finish, as only a few selected designs will be subjected to an MD simulation. For any other protein than LEH, after a few MD simulations have finished on a desktop computer do a visual inspection (section 3.8) and verify there are still no problems with the protein structure (such as missing cofactors, *see* **Note 8**). After careful inspection of a few structures, the remaining MD simulations can be done without risking that a large amount of CPU time is wasted.

4. Also determine how much time it takes for the MD simulations of a single mutant to finish (`ls -rlt */*/*yob`). If the pace of MD simulations is too slow for all selected variants to finish in a reasonable time, obtain an account at a computer cluster and carry out the calculations there (*see* **Note 9**).

3.8 Visual Inspection

Those mutations that are computationally predicted to be stabilizing will often have one or more identifiable biophysical errors due to simplifications in the energy functions and incomplete conformational sampling [7, 17]. With visual inspection, such variants are eliminated (*see* **Notes 10 and 11**). This further improves the quality of the library that will be screened experimentally, and thus reduces the screening effort that is required.

1. Copy the YASARA plugin file in the appropriate folder:

```
cp ~/frescoSoft/FRESCO/MutantInspectPlugin.py ~/frescoSoft/YASARA.app/yasara/plg/
```

2. Enter the MD directory (`cd ~/frescoLEH/designsMD`) and run YASARA with `yasara 1NWW_cleaned.pdb&`. In the menu bar, go to Analyze → FRESCO → Prepare Excel file from Mutations list. Wait until the file is created, open it and copy/paste the text into a blank sheet of your favorite spreadsheet application. To this list of mutations, the user should add his or her own observations and a final judgment whether to keep or discard the mutation. Start the visual inspection in YASARA by clicking Analyze → FRESCO → Start Inspection of Mutants. The plugin will load the mutations showing the static structures of wild type and mutant in a panel called main and the MD simulations of mutant and wild type in two other panels (*see* **Note 12**).

3. Optionally, set YASARA to stereo vision (`Stereo CrossEyed` or `Stereo Parallel`) (*see* **Notes 3** and **13**).

Carry out visual inspections for all mutations in the sequence of **steps 4–10** (*see* **Note 14**). Variants can be eliminated as soon as they fail an inspection step. Usually, about 40–50% of the mutations will be eliminated both during inspection for biophysical credibility (**steps 4–6**) and during inspection for conserved rigidity (**steps 7–10**). Thus, about 25–35% of the mutations usually survive the visual inspection. Some information for practicing visual inspection skills on example mutations of LEH are provided in **Note 10**.

4. *Eliminate mutations that result in unusual solvent exposure of hydrophobic side chains.* Inspect the structure of both wild type and mutant around the mutated residue to see whether the introduced (hydrophobic) side chain atoms becomes unusually water exposed (Fig. 3a). The inspection can be done both for the static structure and for the structures from the MD simulation.

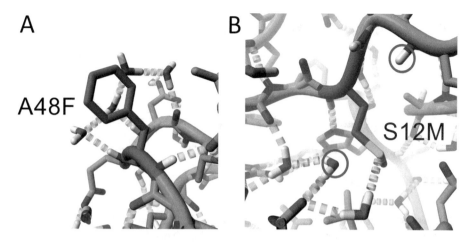

Fig. 3 Examples of the most common structural errors encountered amongst top-ranked point mutations. Both mutations belong to the example set of LEH (*see* **Note 10**). The visualization is as provided by the YASARA FRESCO plugin. (**a**) Introduction of a hydrophobic residue that is solvent exposed. F48 is surrounded by water molecules. (**b**) A mutation that results in an unsatisfied H-bond donor (the backbone amide) and an unsatisfied H-bond acceptor (the water oxygen). In the native structure, S12 makes an H-bond to the backbone amide while there is room for an additional water molecule to make an H-bond to the now unsatisfied water

Visually inspect how many water molecules can contact hydrophobic atoms in the side-chain and evaluate whether this is still normal. In case of doubt, make a comparison by looking at the same type of residue elsewhere in the enzyme (e.g., for phenylalanines, type `ShowRes Phe`, `ColorAtom Res Phe element C, red`, `ShowRes res with distance < 5 from res Phe`). For trained eyes, the identification of this common problem is very fast, leading to elimination of mutants within seconds.

5. *If the number of unsatisfied H-bond donors/acceptors increases due to the mutation, eliminate the mutant.* This is the second most common reason for elimination. Count the number of unsatisfied H-bond donors and acceptors around the mutation (Figs. 3b and 4, *see* **Note 15**). H-bond acceptors and donors that are only involved in one three-centered H-bond interaction (Fig. 4c) are counted as half unsaturated [18].

6. *Verify that the mutations do not violate other biophysical criteria.* With most proteins, for one or a few positions almost all substitutions are predicted to be stabilizing, which probably reflects a systematic error in the energy calculations (*see* **Note 16**). In such cases, only accept the mutations if the wild-type protein features structural problems (unsatisfied H-bonds, cavities) that are repaired by the proposed mutations. Further, no prolines should be introduced in an α-helix. In case of a disulfide bond mutation, eliminate the proposed mutations if these create a large cavity in the protein interior.

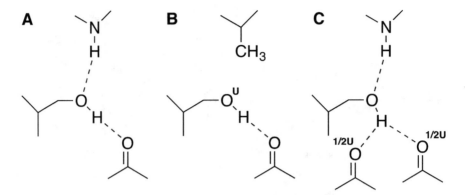

Fig. 4 Schematic examples of saturated, unsaturated, and partially unsaturated H-bond networks. (**a**) All H-bond donors and acceptors are saturated. (**b**) The hydroxyl oxygen, a good H-bond acceptor, is unsaturated. (**c**) Both carbonyl oxygens are half-unsaturated since they share a single H-bond donor, forming a 3-center H-bond

7. *Make the most different MD structures invisible for both wild type and mutant.* It is often found that one of the MD simulations samples a different conformation than all the others (Fig. 5) and thus behaves as an outlier. If the results of these MD simulations were evaluated in an identical manner, the differences between mutant and wild type would be randomly exaggerated. To prevent this, always remove the most different structures. Click on the visibility button in the HUD display one by one for the structures while watching the screen. The picture will change most when hiding the most different structure.

8. *Eliminate the mutant if the introduced side chain is unusually flexible.* Flexibility depends on the nature of the side chain. For example, high flexibility would be normal for a lysine but not for a tryptophan. When in doubt, compare with similar wild-type residues (for example, type `ShowRes Trp`).

9. *Eliminate the mutation if the backbone at the mutation site, or in the flanking regions, becomes significantly more flexible* (Fig. 6).

10. *Eliminate the mutant if the overall structure becomes significantly more flexible.* This rarely occurs by introduction of single point mutations.

3.9 Experimental Verification of the Selected Variants

The variants that survive visual inspection should be screened experimentally. The protocols for genetic engineering and thermostability assays are widely used and are therefore only briefly summarized here. Genetic engineering can be done rapidly and inexpensively using 15 μL scale QuikChange reactions (Agilent Technologies) in 96-well plates. The reactions should be very reliable, we find mostly only a single clone needs to be sequenced. The T_M of the variants can be determined with the Thermofluor

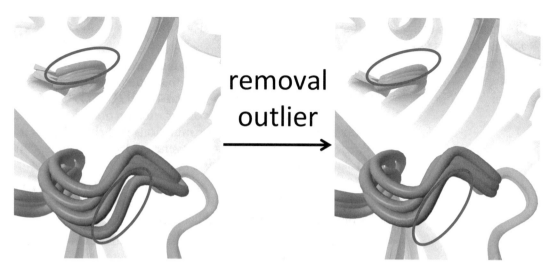

Fig. 5 Example of identifying an outlier. From the averaged structures of five independent MD simulations, the structure that differs most from the other four is removed. For clarity, only the part of the protein with the largest differences is shown

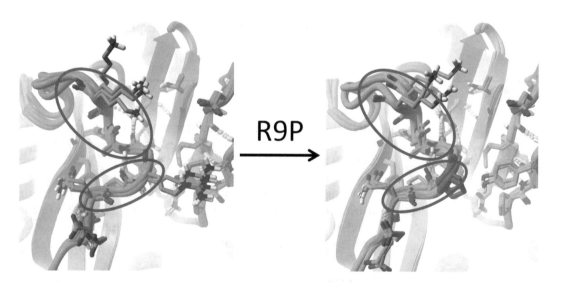

Fig. 6 Example of a mutation that is predicted to increase local backbone flexibility. R9P is one of the LEH example mutations (*see* **Note 10**). Parts of the backbone that show significant increase of flexibility are marked with *red*. MD-averaged structures (*see* caption Fig. 5) of the wild type are shown in *sea green* while the corresponding structures of the mutant are shown in *orange*. The mutated residue is in *magenta*

method [19, 20] after a small-scale purification (from 1 to 5 mL of culture). The mutants with improved thermostability should also be tested for preserved catalytic activity. Additional details have been described elsewhere [6, 7].

3.10 Combination of Stabilizing Mutations to a Hyperstable Final Variant

1. Enter the finalCombinations directory: `cd frescoLEH/finalCombinations`.

2. Combine all compatible stabilizing mutations that do not decrease catalytic activity. Predict the structure of the proposed final variant(s) using the ~/frescoSoft/FRESCO/Combine-Mutations.mcr. This script contains instructions for how to generate a table file listing the mutations that should be combined. The protein structure as generated under Subheading 3.1, **step 6** should be used as a template.

3. The generated pdb file(s), which already contain(s) the crystal waters, should then be used as starting point(s) for MD simulation as described under Subheading 3.7.

4. The resulting structures should be inspected as described under Subheading 3.8.

5. If the combination fails these inspection steps, identify possible (combinations of) mutations that cause problems and repeat **steps 2–4** while omitting these mutations.

6. Prepare the final variant(s) using consecutive QuikChange reactions (*see* Subheading 3.9). Determine the T_M of all intermediate mutants as well. This allows experimental identification of incompatible mutations.

---

4   Notes

1. This protocol could be adapted for any Unix(-like) system on which one can install the required software. Typical challenges then are finding the correct graphic drivers that YASARA needs to run smooth molecular graphics, compiling Rosetta if the precompiled binaries do not work for that particular Linux distribution, and Linux distribution specific differences in the commands that need to be given. For example, `top -o cpu` does not work under Linux but `top` does. It is possible to run the visual inspection under Windows.

2. The increase of calculation time with the square of the protein length is due both to the increasing number of mutants that will be screened as well as the increase in computation time per MD simulation.

3. For more information about the YASARA commands and their syntax, use the SearchDoc function within YASARA. For example `SearchDoc AddBond`.

4. For other proteins, common problems encountered are: unusual numbering of the amino acids in the pdb file, gaps in the protein sequence, unusual residues, unusual protonation states, cofactors that need to be manually adapted to ensure the

simulated state is physically relevant, etc. Careful inspection and editing solves such problems.

5. Both FoldX and Rosetta_ddg were not parameterized for use with cofactors. Thus, the predicted $\Delta\Delta G^{Fold}$ for residues close to cofactors would be unreliable.

6. Input files or commands often cause problems due to (small) abnormalities in the formatting or errors in spelling and punctuation. If Rosetta, FoldX, or YASARA do not work as expected, it is best to first examine whether the log files contain error messages. This can be done by entering `tail log` if the output of the failing program was redirected to a file called log.

7. The cutoff of -5 kJ/mol can be made less strict (e.g., -2.5 kJ/mol) to increase the number of stabilizing mutations that can be discovered.

8. The HydrateDesigns.mcr script ought to put cofactors back together with the crystallographic water molecules. This script has been tested for several cofactors but may fail for others. If cofactors or covalent bonds are missing, or other errors occur, the user needs to adapt the HydrateDesigns.mcr script (*see* **Note 3**) and rerun it.

9. To log in to a cluster, `ssh` can be used after one obtains a user name. Only YASARA will need to be installed at the cluster. The most useful command to transfer a large number of files to and from a computer (cluster) is `rsync –avu <origin> <destiny>`. To start the calculations, cluster specific scripts will be needed that can normally be obtained via the cluster's website.

10. For the LEH example, Q7M, E68L, A48F, and S111M introduce highly surface-exposed hydrophobic side chains. Mutations S12M, T22D, and G129S introduce unsatisfied H-bond donors or acceptors while E49P, Y96W, and R9P cause local flexibility which is larger than that of the template structure. All other variants, both those that solve structural problems (T85V) and those that merely lack clear biophysical errors (E45K, E124D) should be selected for experimental testing.

11. Visual inspection is a standard step in computational design and molecular modeling and is therefore often not mentioned in the materials and methods sections of publications.

12. The plugin automatically finds the files used for the visual inspection but it requires the above provided standard file and folder names to function properly. In a directory called designsMD there have to be two subdirectories: NamedPdbFiles and UniqueDisulfides. The pdb and yob files in subdirectories of these directories should bear a name of the type <anything>_cleaned<name of the mutations><furtherExtensions>.

13. Cross-eyed or parallel stereo needs to be learned by the user. This will take a few hours. See the YASARA website for the available other forms of stereo. Some users prefer to manually rotate the structures for 3D depth perception.

14. Experienced protein designers can inspect more than 120 variants per day while beginners should aim at 30–50 variants per day. The fastest method for inspecting is to follow the described sequence of steps, in which mutants are initially eliminated based on common and fast to analyze problems.

15. The algorithms used in molecular modeling are poor at assessing whether an H-bond is made. They typically use some kind of distance cutoff and use surface accessibility to predict whether water can form an H-bond. For this reason, also distrust the H-bonds as displayed by YASARA. There could be additional H-bonds that are not visualized. Visual inspection is needed to eliminate cases where the computer overestimates the feasibility of water H-bonds or fails to identify the three-center H-bonds, which are energetically unfavorable [18].

16. The calculated energy of the wild-type structure is subtracted from those of the mutants to predict $\Delta\Delta G^{Fold}$. If almost all mutations are predicted to be stabilizing at a particular position, this suggests an error in the energy calculation of the wild-type structure.

Acknowledgments

This research was supported by the European Union Seventh framework project Kyrobio (KBBE-2011-5, 289646), by the European Union Horizon 2020 program (project LEIT-BIO-2014-1, 635734) by NWO (Netherlands Organization for Scientific Research) through an ECHO grant, and by the Dutch Ministry of Economic Affairs through BE-Basic (www.be-basic.org).

References

1. Tokuriki N, Tawfik DS (2009) Stability effects of mutations and protein evolvability. Curr Opin Struct Biol 19:596–604

2. Wijma HJ, Floor RJ, Janssen DB (2013) Structure- and sequence-analysis inspired engineering of proteins for enhanced thermostability. Curr Opin Struct Biol 23:588–594

3. Bommarius AS, Paye MF (2013) Stabilizing biocatalysts. Chem Soc Rev 42:6534–6565

4. Eijsink V, Bjork A, Gaseidnes S et al (2004) Rational engineering of enzyme stability. J Biotechnol 113:105–120

5. Haki GD, Rakshit SK (2003) Developments in industrially important thermostable enzymes: a review. Bioresour Technol 89:17–34

6. Floor RJ, Wijma HJ, Colpa DI et al (2014) Computational library design for increasing haloalkane dehalogenase stability. ChemBioChem 15:1659–1671

7. Wijma HJ, Floor RJ, Jekel PA et al (2014) Computationally designed libraries for rapid enzyme stabilization. Protein Eng Des Sel 27:49–58

8. Wu B, Wijma HJ, Song L et al (2016) Versatile peptide C-terminal functionalization via a computationally engineered peptide amidase. ACS Catal 6:5405–5414

9. Arabnejad H, Dal Lago M, Jekel PA et al (2017) A robust cosolvent-compatible halohydrin dehalogenase by computational library design. Protein Eng Des Sel 30:173. doi:10.1093/protein/gzw068

10. Wijma HJ (2016) In silico screening of enzyme variants by molecular dynamics simulation. In: Svendsen AS (ed) Understanding enzymes; function, design, engineering and analysis. Pan Stanford, Singapore, pp 805–833

11. Eijsink V, Gaseidnes S, Borchert T et al (2005) Directed evolution of enzyme stability. Biomol Eng 22:21–30

12. Veltman O, Vriend G, Hardy F et al (1997) Mutational analysis of a surface area that is critical for the thermal stability of thermolysin-like proteases. Eur J Biochem 248:433–440

13. Kellogg EH, Leaver-Fay A, Baker D (2011) Role of conformational sampling in computing mutation-induced changes in protein structure and stability. Protein Struct Funct Bioinform 79:830–838

14. Guerois R, Nielsen J, Serrano L (2002) Predicting changes in the stability of proteins and protein complexes: a study of more than 1000 mutations. J Mol Biol 320:369–387

15. Krieger E, Vriend G (2015) New ways to boost molecular dynamics simulations. J Comput Chem 36:996–1007

16. van Beek HL, Wijma HJ, Fromont L et al (2014) Stabilization of cyclohexanone monooxygenase by a computationally designed disulfide bond spanning only one residue. FEBS Open Bio 4:168–174

17. Wijma HJ, Janssen DB (2013) Computational design gains momentum in enzyme catalysis engineering. FEBS J 280:2948–2960

18. Feldblum ES, Arkin IT (2014) Strength of a bifurcated H bond. Proc Natl Acad Sci U S A 111:4085–4090

19. Lavinder JJ, Hari SB, Sullivan BJ et al (2009) High-throughput thermal scanning: a general, rapid dye-binding thermal shift screen for protein engineering. J Am Chem Soc 131:3794–3795

20. Ericsson UB, Hallberg BM, DeTitta GT et al (2006) Thermofluor-based high-throughput stability optimization of proteins for structural studies. Anal Biochem 357:289–298

Chapter 6

Directed Evolution of Proteins Based on Mutational Scanning

Carlos G. Acevedo-Rocha, Matteo Ferla, and Manfred T. Reetz

Abstract

Directed evolution has emerged as one of the most effective protein engineering methods in basic research as well as in applications in synthetic organic chemistry and biotechnology. The successful engineering of protein activity, allostery, binding affinity, expression, folding, fluorescence, solubility, substrate scope, selectivity (enantio-, stereo-, and regioselectivity), and/or stability (temperature, organic solvents, pH) is just limited by the throughput of the genetic selection, display, or screening system that is available for a given protein. Sometimes it is possible to analyze millions of protein variants from combinatorial libraries per day. In other cases, however, only a few hundred variants can be screened in a single day, and thus the creation of smaller yet smarter libraries is needed. Different strategies have been developed to create these libraries. One approach is to perform mutational scanning or to construct "mutability landscapes" in order to understand sequence–function relationships that can guide the actual directed evolution process. Herein we provide a protocol for economically constructing scanning mutagenesis libraries using a cytochrome P450 enzyme in a high-throughput manner. The goal is to engineer activity, regioselectivity, and stereoselectivity in the oxidative hydroxylation of a steroid, a challenging reaction in synthetic organic chemistry. Libraries based on mutability landscapes can be used to engineer any fitness trait of interest. The protocol is also useful for constructing gene libraries for deep mutational scanning experiments.

Key words Synthetic biology, Directed evolution, Protein engineering, Site-directed mutagenesis, Saturation mutagenesis, Scanning mutagenesis, Mutability landscapes, Deep mutational scanning, Cytochrome P450 monooxygenase, Stereoselectivity

1 Introduction

Directed evolution has emerged as one of the most effective protein engineering methods in basic research [1] as well as in applications in synthetic organic chemistry and biotechnology [2–10]. Traits that can be improved by this method include, *inter alia*, protein activity, allostery, binding affinity, expression, folding, fluorescence, solubility, substrate scope, selectivity (enantio-, stereo-, and regio-selectivity) as well as stability (temperature, organic solvents, and pH). Directed evolution consists of repetitive cycles of random or focused gene mutagenesis, expression, screening or selection, and

Uwe T. Bornscheuer and Matthias Höhne (eds.), *Protein Engineering: Methods and Protocols*, Methods in Molecular Biology, vol. 1685, DOI 10.1007/978-1-4939-7366-8_6, © Springer Science+Business Media LLC 2018

amplification of the fittest variant. To create diversity, various methods can be used, the most common ones being error-prone PCR (epPCR), homologous recombination-based methods (e.g., DNA shuffling) and saturation mutagenesis (SM). The former ones usually target partial or complete gene sequences randomly, whereas the latter can be directed to practically any gene position or region.

Depending on the setup, combinatorial libraries bearing large numbers of enzyme variants in the range of 10^8–10^{12} can be handled in a single experiment provided high-throughput settings are available such as genetic selection systems [11, 12], display systems (plasmid, SNAP, phage, mRNA, ribosome, liposome, bacteria, and yeast) [1, 11–13], fluorescence-based systems (e.g., FACS) [14], or in vitro compartmentalization (e.g., water-in-oil emulsions) [15] systems. The implementation of these systems, however, is hampered by the nature of the protein trait that has to be optimized. For example, it is very difficult to devise reliable genetic selection or display systems for engineering stereoselective enzymes [16], which are important biocatalysts in the synthesis of stereoisomers and/or regioisomers in organic chemistry [5, 8]. In consequence, medium (10^4–10^5) to low (10^2–10^3) throughput screening of enzyme libraries becomes the only option available. To increase the speed of analysis, screening can be automatized using chromatographic systems based on HPLC, GC, NMR, or UV-Vis absorbance, with the most common format being a microtiter plate (MTP) of 96 or 384 wells for library assaying. Single colonies from a library are inoculated manually or automatically using a liquid-handling robot into each MTP well. The cells are grown, followed by protein expression and functional analysis. Thereafter, the best mutant is chosen and the process is iterated until the desired biocatalyst has been evolved. The faster that an enzyme can be engineered, the better it is for both practical and economical reasons.

Although the most common methods in directed evolution are random epPCR and DNA shuffling [17], semirational approaches have emerged as superior ones because the same or better results can be achieved with less effort, for example, when engineering selectivity and stability [5, 8]. One of these methods relies on Iterative Saturation Mutagenesis (ISM), in which careful library design is crucial for success: To increase the activity, substrate scope as well as stereoselectivity and/or regioselectivity, amino acids lining or near the substrate binding pocket are grouped and randomized using the Combinatorial Active-site Saturation Test (CAST) [18, 19]. If the goal is to increase the tolerance against chemical (organic solvents) and thermal denaturation, the most flexible residues analyzed from crystallographic data are chosen for ISM via the B-Factor Iterative Test (B-FIT) [18, 19]. For example, using CAST-based ISM, our group reported the evolution of the cytochrome monooxygenase $P450_{BM3}$ for the efficient regioselective and stereoselective hydroxylation of steroids [20]. It is also possible to combine CAST and B-FIT to engineer multiple protein fitness traits simultaneously [21].

An unresolved challenge in ISM, however, is the selection and grouping of residues involved in phenotype (or fitness) and the choice of the optimal amino acid alphabet (AAA) for randomization. For example, in small binding pockets of many hydrolases (e.g., esterases, lipases, or epoxide hydrolases) there can be up to ten first-sphere residues that can serve as targets for mutagenesis, but in larger binding pockets such as P450s, there can be more than 40 potential target residues. Obviously, this difference increases the library complexity and difficulty for improving one or more enzyme traits depending on the protein family. Furthermore, targeting second-sphere residues [22] as well as distant residues from the active site [22] can help in controlling allostery and selectivity. For these reasons, it is often difficult to predict what are the best residues for mutagenesis. Fortunately, strategies have been developed for selecting the optimal AAAs:

1.1 Bioinformatics and Computational Methods

Multiple-sequence alignments (MSAs) can be used to identify consensus-like sequences [23] based on the assumption that naturally occurring amino acids are less likely to cause deleterious effects [24–26]. This strategy is particular helpful in enzymes that share a high degree of sequence homology because fewer residues will change across protein family members, thus allowing to have smaller libraries depending on the number of sequence alignments. Computational methods that also gain information from MSAs, gene or protein 3D structure databases have emerged more recently as valuable tools to guide the selection of suitable AAAs [27].

1.2 Random Methods

In early protein engineering studies, the degenerate codon NNN (64 codons) was used for SM, but soon it was realized that NNK/S degeneracy is better to reduce redundancy because 20 amino acids are encoded by 32 codons. To further reduce the genetic code redundancy, the methodology evolved to make use of the degenerate codon NDT (12 codons), which encodes 12 amino acids bearing a balanced mixture of physicochemical properties [28]. This codon was been used successfully in many different settings [29]. More recently, smaller AAAs have been successfully used including five [30], four [31, 32] or two [33, 34] members of different physicochemical properties. Theoretically, small randomization alphabets are better than large ones because the total number of variants using small AAAs is still low compared to 12 codons (Table 1). Practically, however, overly small randomization alphabets may prove to be useless if these fail to contain sequences that are needed to code for a properly folded and improved biocatalyst.

Table 1
The numbers problem in directed evolution focusing on the number of codons as a function of the reduced amino acid alphabet (AAA)

Residues	NDT	6-AAA	4-AAA	3-AAA	2-AAA
1	12	6	4	3	2
2	144	36	16	9	4
3	1728	216	64	27	8
4	20,736	1296	256	81	16
5	248,832	7776	1024	243	32
6	2.9×10^6	46,656	4096	729	64
7	3.5×10^7	279,936	16,384	2187	128
8	4.3×10^8	1.6×10^6	65,536	6561	256
9	5.1×10^9	1.0×10^7	262,144	19,683	512
10	6.2×10^{10}	6.0×10^7	1.0×10^6	59,049	1024
11	7.4×10^{11}	3.6×10^8	4.2×10^6	1.8×10^5	2048
12	8.9×10^{12}	2.2×10^9	1.6×10^7	5.3×10^5	4096

The number of variants increases exponentially as the number of target residues and amino acid alphabet increases. The screening effort for a given library coverage needs to be computed as well, which goes from ~3 to ~10 oversampling for 95–100%, respectively, without library bias. The library sizes were computed with TopLib [35]

This phenomenon has been seen even in the simplest case (binary alphabet), in which one of two residues (the wild type and another amino acid) was used randomly as building block [33].

1.3 Rational Methods   These rational design methods are based on sequence–function relationships from previous mutagenesis studies to derive beneficial mutations that can increase fitness while avoiding deleterious ones. Several non-combinatorial scanning mutagenesis techniques have emerged to explore genotype-to-phenotype relationships by introducing all 20 canonical amino acids at a handful sites, partial regions or even complete proteins. In vitro scanning mutagenesis [36], codon-based mutagenesis [37], and shotgun mutagenesis [38] were essentially developed to scan antibody binding affinity, which can be applied in a high-throughput manner thanks to phage-display [39]. More recently, all-codon scanning [40], deep-mutational scanning [41], and massively parallel single-amino-acid mutagenesis [42] have emerged as powerful high-throughput scanning approaches, yet their potential is limited to genetic selection, display, and high-throughput colorimetric or fluorescent systems [43] that can handle large protein library sizes (10^6–10^{12} variants). Since the function of many proteins, for example stereoselective enzymes [16], can only be analyzed using screening-based systems in a small to medium-throughput manner (10^3–10^5), other scanning strategies are required. Three scanning techniques with different names but same principle are site-saturation mutagenesis (SSM) [44], gene-site saturation mutagenesis (GSSM) [45], and single-site saturation mutagenesis (SSSM) [46], all employing conventional degenerate, doped, spiked, or contaminated oligonucleotides for the randomization of individual codons based on SM. The potential of these technologies has been shown for complete enzymes including nitrilase [45], DNA polymerase [46], haloalkane dehydrogenase [47], phytase [48], lipase [49], and two xylanases [50, 51]. In all these studies, degenerate nucleotides such as NNN [44, 46, 50, 51] (N = A/T/G/C) or NNK [45, 47–49] (K = G/T) have been used, assuming a uniform randomization of target sequences that usually are oversampled with 95% library coverage or more, resulting in the screening of one MTP of 96 wells or up to four, which corresponds to 96–384 samples per amino acid position. This setup represents a formidable screening effort.

However, there is accumulating evidence showing that the randomization of target codons does not occur uniformly [52, 53]. For example, strong hybridization of the parental codon to the wild-type (WT) sequence can occur, regardless of whether NNK or nonredundant combinations of codons are used [52]. Likewise, recent work by our lab also indicates that library bias and thus library yield and quality also depends on primer purity and source, with oligos from some suppliers being excellent and others poor regardless of purity [52]. To ensure that the yield of the library is maximal, it is usually recommended to perform oversampling, according to the Patrick and Firth [54, 55] algorithm. For example, an oversampling factor of ~3 is needed to ensure 95% library coverage, which corresponds to about 96, 66 and 60 transformants, respectively, when NNK (32 codons), 22-Trick (22 codons) [56] or Tang (20 codons) [57] randomization is used as AAA at one single site (Fig. 1). Typically, it is assumed that library yield is 100%, i.e., that there is no bias during library creation and that each variant has the same probability to be created. However, this is not always the case as shown in different enzyme systems by standard DNA sequencing [52, 53, 58, 59]. Thus, amino acids important for the fitness trait under quest (e.g., activity or selectivity) can be missed because these were either not sampled, or not created during the mutagenesis process regardless of library oversampling.

To reduce library bias, library creation at each target codon should be optimized in terms of primer length, melting temperature, overlapping region (if using Quik-Change), PCR conditions including annealing temperature and number of cycles, DNA template amount, and transformation conditions. Another way to reduce bias is to reduce library redundancy by using specific

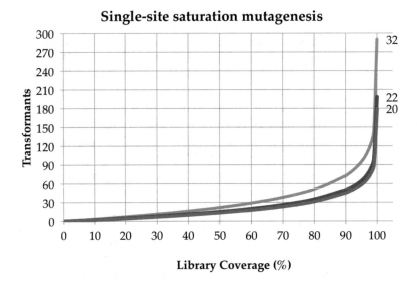

Fig. 1 Oversampling factor for screening single-site saturation mutagenesis libraries. For 95% library coverage, traditional NNK/S (32 codons) requires the screening of ca. 90 transformants, whereas 66 and 60 samples are needed when the Trick (22 codons) or Tang (20 codons) methods are employed

degenerate/nondegenerate mixtures of three [56] or four [57] primers specific for 22 (i.e., Trick-22c) or 20 (i.e., Tang-20c) codons encoding the 20 canonical amino acids, respectively. However, we have observed that bias can be also introduced by using these special primer mixtures [52]. Furthermore, the disadvantage of using the Trick-22c or Tang-20c method is that six or eight primers, respectively, are needed per randomization site if a Quik-Change-like protocol is used, as it is done in most SM experiments [17]. On the other hand, only two primers are needed in the case of NNK/S-based libraries; thus, the Trick22c and Tang-20c strategy is three and four times more expensive than using NNK/S, especially when large regions or complete proteins are targets for mutagenesis.

For this reason, statistically, the most economical strategy to obtain all 19 point mutants per target site is to perform first a round of SM using NNK/S degeneracy, then sequence a portion of the resulting library and finally create in a second round the remaining mutants using site-directed mutagenesis (SDM) [60]. Although it might be too expensive to sequence many variants at each amino acid site of a large region or full protein, the costs of library creation versus library screening have to be balanced. For example, in some cases HPLC or GS screening might be too expensive (<100 samples per day). Thus, if the step of library screening is more expensive than library creation, it might be even cheaper to create each of the 19 variants [52] using emergent SDM approaches [61]. As indicated above, performing NNK/S at each residue of a target codon is nothing new regardless of the technique used for that purpose

Table 2
Mutability landscapes studies of proteins and enzymes

Protein/enzyme	Target residues	Randomization codon	Number of variants created	Reference
26–10 monoclonal antibody	6	Nondegenerate codons	114	Burks et al. [36]
Abscisic acid receptor	39	NNN	741	Mosquna et al. [65]
Abscisic acid receptor	25	NNK	475	Park et al. [66]
Lipase	181	NNS	3439	Frauenkron-Machedjou et al. [58] and Fulton et al. [59]
Tautomerase	62	Unknown	At least 930	van der Meer et al. [64]
P450$_{BM3}$	34	Nondegenerate codons	646	Acevedo-Rocha et al. (in preparation)

The target variants have been created and sequence-verified for binding or screening assays (excluding genetic selection and fluorescence-based studies)

(i.e., SSM, GSSM, or SSSM). In most of the studies using these techniques, however, the creation of diversity was not assessed, with a few exceptions of a limited number of sampled residues [45, 51]. Verification by sequencing that all mutants are actually created would be ideal. This strategy, also dubbed "mutability landscapes" [62–64], has been reported for various protein classes in which some regions or the full-length protein has been entirely mutated (Table 2).

Once all mutants have been created and sequenced, each possible variant is known and can be placed in a unique MTP well after DNA sequencing. This information is valuable because it can be used to learn structure–function relationships of multiple substrates with the lowest screening effort since no variant duplicates are generated (for five residues, each with 19 variants, a 96-well MTP suffices). Importantly, this strategy replaces completely other different noncombinatorial methods including alanine [67], cysteine [68], tryptophan [69], glycine [70], serine [71], proline [72], tyrosine [73], and lysine [74] scanning. Furthermore, these "smart libraries" differ from typical libraries in directed evolution (SSM, GSSM, and SSSM), which are stored and then used to screen new substrates because these procedures increase expenses and production of waste, especially when most of the wells contain redundant variants.

We envision that gene sequencing will be replaced by ultrahigh-throughput gene synthesis in the near future, but the cost of such libraries (in which all variants are synthesized) will remain too high

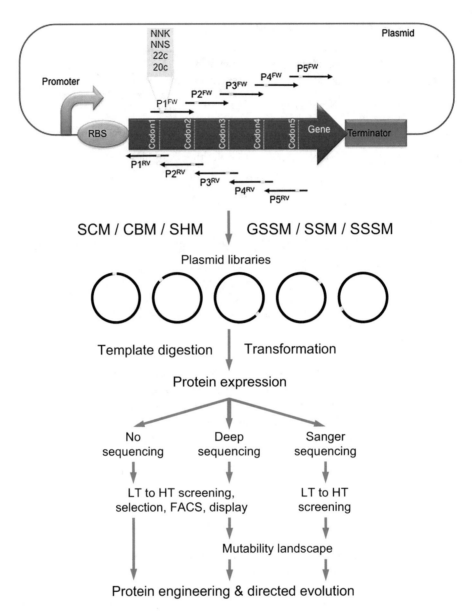

Fig. 2 Rational approaches based on saturation mutagenesis commonly used in protein engineering and directed evolution. The region of a protein or a complete protein can be scanned using PCR. The promoter and ribosome binding site (RBS) are shown upstream of the target gene in which five codons (*yellow*) are targets of mutagenesis. A downstream terminator is also shown. These sequences are important to consider when randomizing the initial and terminal codons of a whole gene. All PCR methods are based on the QuikChange protocol that uses completely overlapping oligonucleotides. Other protocols use a pair of partially overlapping nucleotides, as shown for five sets of sense (FW) and antisense (RV) primers P1–5. In each primer, it is possible to introduce degenerate (e.g., NNK or NNS), nondegenerate codons or combinations thereof (e.g., 22c and 20c refer to a mixture of three and four codons, respectively). To create diversity at each codon, different methodologies with the same principle but different names can be recalled: Scanning Mutagenesis (SCM), codon-based mutagenesis (CBM), shotgun mutagenesis (SHM), Site-Saturation Mutagenesis (SSM), Gene-Site Saturation Mutagenesis (GSSM), and Single-Site Saturation Mutagenesis (SSSM). The resulting libraries are

for many labs with modest budgets. For this reason, practical and economical alternatives are necessary. In this chapter, we provide a general protocol for scanning mutagenesis in which all point mutants of a protein region (or complete proteins) can be obtained in an economical and efficient manner using PCR as well as suitable equipment and reagents. This means that all 19 amino acid variants per residue site have to be generated and sequence-verified (Fig. 2). The resulting mutants are arranged in a standard MTP of 96-wells (or 384), which are ready for screening by, for instance, GC or HPLC. There are several advantages of sequencing and obtaining all mutants for constructing mutability landscapes vs. typically doing NNK randomization: (1) *Detailed knowledge about sequence–function relationships.* In NNK randomization, only the "winners" are sequenced and nothing else can be learned from the study, for example, the identity of neutral and deleterious mutations. Gaining insights about the protein system is important for subsequent iterative rounds of mutagenesis and for optimizing machine learning systems. (2) *Reliable and bias-free library design.* NNK randomization assumes a uniform creation of the 32 variants; however, this might not be the case, as it has been shown in several cases [52, 53, 58, 59]. Thus, rare beneficial mutations that are important for the desired fitness trait might not be created during mutagenesis due to bias. Mutability landscapes do not have assumptions and, thus, are reliable. (3) *Economics and versatility.* If the screening step is too expensive (e.g., a few dozens of samples can be screened per day per fitness trait per substrate), in the long-term the cost of creating the mutability landscape could be cheaper than using NNK randomization. Because the mutability landscape can be screened relatively fast and with the minimal screening effort, in principle, any fitness trait can be explored with various substrates and in different environments, thus accelerating the engineering process.

It is also possible to use the current protocol for creating libraries that do not require low-throughput screening like in deep mutational scanning coupled to genetic selection [41] or fluorescence-readouts [75]. In contrast to new approaches like nicking mutagenesis [76], we believe that our method can decrease

Fig. 2 (Continued) transformed into a suitable host upon eliminating the template and protein expression is induced. Typically, most libraries are usually not sequenced and used directly for blind screening in a low (LT) to high-throughput (HT) setting (*left panel*). If the libraries are sequenced using deep or next generation (*middle panel*) and Sanger (*right panel*) sequencing, it is possible to construct "mutability landscapes" in order to learn insightful sequence–function relationships that can guide more rationally the protein engineering or directed evolution process. Deep Mutational Scanning (DMS) libraries are deep-sequenced prior and after genetic selection or Fluorescence-Activated Cell Sorting (FACS). For proteins where genetic selection systems, display or fluorescence-based methods are difficult or not possible to devise, screening is the only method available, which can be in low- (LT) or high-throughput (HT) mode

significantly the costs for scanning even whole proteins because it does not require ligase, phosphorylated oligos as well as many different restriction sites and nucleases. Last but not least, our method is exemplified for protein-coding sequences, but any gene sequence or element (promoters, RBSs, linkers, etc.) can be studied as well.

2   Materials

2.1   Databases

1. The nucleotide sequence of interest obtained from a suitable database, e.g., GenBank https://www.ncbi.nlm.nih.gov/genbank/ [77].

2.2   Reagents

1. Target gene cloned into or assembled to a suitable vector (*see* **Note 1**).

2. Oligonucleotides of good quality preferentially resuspended at 2–5 μM in ultra-pure water or suitable buffer (*see* **Note 2**).

3. DNA polymerase for PCR (*see* **Note 3**).

4. PCR Master mix including buffer, dNTPs, and polymerase (*see* **Note 4**).

5. *Dpn*I restriction enzyme (*see* **Note 5**).

6. *E. coli* BL21(DE3)Gold, or another suitable expression host (*see* **Note 6**).

7. SOC medium: 20 g/L tryptone, 5 g/L yeast extract, 10 mM NaCl, 2.5 mM KCl, 10 mM $MgCl_2$, and 10 mM $MgSO_4$. Sterilize by autoclaving, then add glucose to 20 mM using a 2 M sterile stock solution (*see* **Note 7**).

8. LB medium and LB agar plates.

9. Suitable antibiotic stock solution(s) (1000×) according to the used plasmid.

10. Plasmid miniprep, midiprep, and/or maxiprep kit (*see* **Note 8**).

2.3   Equipment

1. Gradient thermocycler suitable for 96-well microtiter plates (MTPs).

2. Equipment for high-throughput (HT) DNA electrophoresis (*see* **Note 9**).

3. Incubator with shaker for MTPs.

4. Incubator with shaker for Eppendorf tubes.

5. Centrifuge for MTPs and Eppendorf tubes.

6. Spectrophotometer for determining DNA concentration.

7. Multichannel pipettes (8 or 12 channels).

2.4 Materials

1. Standard plastic ware (PCR tubes, 1.5 and 2.0 mL plastic reaction tubes, 15 and 50 mL falcon tubes, petri dishes of 10 and 15 cm diameter).

2. 96-Well MTPs for PCR (*see* **Note 10**).

3. Agarose gels for HT DNA electrophoresis (*see* **Note 11**).

4. Standard 14 mL round-bottom tubes for cell culturing.

5. 2.2 mL deep-well 96-well MTPs with paper or metal lids.

6. 6-/12-Well multidish plates with flat-bottom well design (*see* **Note 12**).

7. Glass beads of ca. 3 mm diameter.

8. Wood toothpicks.

2.5 Services

1. A company providing service for HT DNA plasmid extraction and sequencing in MTP format (*see* **Note 13**).

2.6 Software and Servers

1. Programs for designing primers in HT format:

    (a) **Deepscan** is a program currently in development that allows introducing degenerate or non-degenerate codons into partially overlapping primers alike to QuikChange mutagenesis for gene segments or complete genes (*see* **Note 14**).

    (b) **AAscan** [78] was primarily created to find optimal primers for alanine scanning using a QuikChange-like protocol [79].

    (c) **MegaWHOP** [80] is based on the MegaPrimer [81] method.

2. Software for aligning and processing multiple DNA sequencing files (*see* **Note 15**).

3. Software to measure base peak heights from DNA chromatograms (*see* **Note 16**).

4. An online program to calculate library screening effort (*see* **Note 17**):

    (a) **TopLib** [35]: http://stat.haifa.ac.il/~yuval/toplib/

    (b) **GLUE-IT** [55]: http://guinevere.otago.ac.nz/cgi-bin/aef/glue-IT.pl

3 Methods

The stages of this protocol are summarized in Fig. 3. It has been applied to a cytochrome P450 monooxygenase as described below in the case study. However, it can be applied to any other protein and even non-coding regions that are part of a plasmid.

3.1 Library Design In Silico

1. *Choose a suitable system as starting point*: It is advisable to know the performance of the gene/enzyme in the cloned/assembled plasmid before starting high-throughput mutagenesis experiments (*see* **Note 1**). Also, it may not be necessary to start mutating the WT gene as model system: a single, double or triple mutant that is sufficiently active and stable or exhibits, for instance, high activity, selectivity, or fluorescence can be a good starting point. This means that some interesting mutations for optimizing the target fitness trait could be tested and added to the WT gene before embarking on a more thorough mutational scanning project.

2. *Choose target region*: Before starting primer design, it is very important to define the region of interest to be scanned depending on the budget allocated to the project. Of course, scanning a whole gene is a great scientific endeavor for learning insightful sequence–function relationships, but it is extremely costly, especially if a bottleneck in the screening procedure exists (i.e., no genetic selection, display, or fluorescent system is available). If this is the case and the goal is to engineer substrate scope and selectivity, for instance, the binding pocket of a protein could be scanned firstly. If the goal is to increase thermal and chemical stability, residues exhibiting the highest B-Factors based on X-ray crystallography (or homology models) can be the targets for mutagenesis. The selection of either set of residues via CAST-/B-FIT-based ISM can be done following a suitable protocol reported

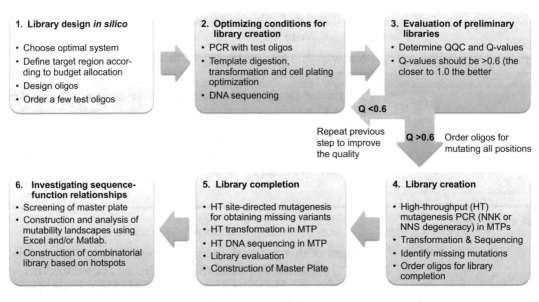

Fig. 3 Protocol overview. The flowchart describes the most critical steps for the generation and construction of mutability landscapes. *QQC* quick quality control, *MTP microtiter* plate, *HT* high throughput

elsewhere [18]. If the purpose of the study is to optimize other fitness traits such as protein activity, allostery, fluorescence, or product inhibition, there are still no general rules as to what residues to mutate. Nevertheless, some statistical approaches based on amino acid coevolution can provide some insight in some protein families [63], which are worth considering when no guideline is available and the budget is a limitation.

3. *Design of oligonucleotides*: Upon choosing targets for mutagenesis, oligonucleotides have to be designed either manually or automatically, for example, using NNK degeneracy per site. In previous studies, we designed oligos manually until we found the program AAscan, which seems to be very convenient as it takes into account primer melting temperature, overall length, length of overlap regions, and presence of GC clamps at the 3' end [78]. Other researchers have used this software successfully to scan large protein regions to a limited set of residues [82, 83]. Still, since there is no program dedicated to provide oligos with degenerate codons at any gene sequence, we are developing a server for this purpose (Ferla et al. in preparation). Regardless of the design approach, we prefer to include partially overlapping oligonucleotides for saturation (SM) and site-directed mutagenesis (SDM) to reduce the formation of dsDNA homoduplexes. With this setting we have observed lesser primer misinsertions but the primers should have an optimal overlapping region. Xia et al. found that three out of seven high-fidelity polymerases were not able to amplify a target plasmid using completely overlapping oligos, including Phusion polymerase [79]. Nevertheless, with Phusion it is possible to amplify the target vector when the primer-overlapping region entails 12–24 nucleotides (but not more). Based on these findings, we have used Phusion successfully to amplify vectors with an optimal overlapping region of 16–20 nucleotides (*see* **Note 3**). It is also important to mention that the quality of the oligos has to be good otherwise the library quality will be compromised and the project will become more expensive (*see* **Note 2**).

3.2 Optimizing Conditions for Library Creation

1. Transform your target vector in a Dam/Dcm⁺ *E. coli* strain to obtain methylated template using chemical-based transformation according to standard procedures (*see* **Note 18**).

2. Purify the vector using the miniprep kit and dissolve it in ultrapure water.

3. Determine concentration of template.

4. Find optimal conditions for heat-shock transformation (*see* **Note 19**) as well as parental template digestion (*see* **Note 20**).

5. Choose optimal DNA template concentration for PCR.

6. Order a small set of oligo pairs with NNK degeneracy and dilute them to 5 μM for testing PCR efficiency at different regions of the target gene (*see* **Note 21**).

7. Mix 2 μL of each oligo with 1 μL template and 5 μL 2× PCR Master Mix. The final reaction volume of the PCR is 10 μL.

8. Perform gradient PCR in tubes (*see* **Note 22**).

9. Run 1–2 μL of samples in appropriate agarose gel.

10. Determine DNA concentrations and yields using 1–2 μL.

11. Digest parent template with *Dpn*I at the optimal conditions previously found by adding the enzyme directly to the same PCR buffer.

12. Have ready a 14 mL round-bottom polypropylene round-bottom tube with 1 mL SOC medium or equivalent pre-warmed at 37 °C.

13. Add 1–5 μL of DNA at the optimal amount to 10–50 μL host strain and transform according to the optimal conditions previously found (*see* **Note 19**). The transformation can be done in a standard PCR tube using the thermocycler or water bath. After heat-shock, place samples on ice for 1–2 min.

14. Transfer the cells to the 14 mL prewarmed tube containing SOC broth.

15. Incubate samples at 37 °C for 40–60 min (or longer) and vigorous shaking.

16. Plate for example 1/100th (10 μL) and 1/20th (50 μL) of the 1-mL cell suspension onto selective LB agar plates.

17. Incubate the plates in the oven at 37 °C for 16–20 h.

18. Add 1 mL of LB + 1 mL of LB having 2× the antibiotics concentration to the tube containing the remaining cells and incubate it in a shaker at 37 °C for 16–20 h. The final volume is 2 mL but this volume can be scaled up depending on the plasmid yield.

19. The next day count the colonies on the agar plates and determine transformation efficiency in CFU per microgram DNA.

20. Transfer 2 mL from the 14 mL tubes to 2 mL Eppendorf tubes for extracting and sequencing the pooled plasmids using the miniprep kit (*see* **Note 13**).

21. Send libraries for sequencing with a suitable primer for the target codon(s).

3.3 Evaluation of Preliminary Libraries

1. Open a suitable program for processing abl files (*see* **Note 15**).

2. Measure the height of each peak (*see* **Note 16**).

3. Perform a Quick Quality Control (QQC) and calculate the Q_{pool}-values [52, 53] (*see* **Note 23**).

4. If the libraries show the desired properties (i.e., no or little bias that correlates with a high Q_{pool}-value), proceed to order oligos of good quality dissolved at an appropriate concentration (e.g., 5 µM), with an optimal melting temperature (e.g., 55–60 °C) and length (e.g., 18–60 bases) for HT library creation (*see* **Note 2**).

3.4 Library Creation

1. Transform your target vector in a Dam/Dcm$^+$ *E. coli* strain to obtain methylated template by chemical transformation according to standard procedures (*see* **Note 18**) and in an appropriate amount according to the total number of reactions expected to be performed (*see below*).

2. Purify vector using miniprep, midiprep or maxiprep kit depending on the expected plasmid yield and dissolve it in ultrapure water that is sufficient for the number of reactions that you need (*see* **Note 24**).

3. Determine concentration of template.

4. Using a multichannel pipette, mix in a new MTP 2 µL of each oligo, 1 µL template and 5 µL 2× PCR Master Mix. Close the plate with a lid or several PCR lid-stripes, vortex the samples and spin down.

5. Perform PCR in 96-well MTP using the optimal conditions determined previously.

6. Run 1–2 µL of samples in 96-well agarose gel.

7. Determine if there is a DNA band at the expected plasmid size (*see* **Note 25**).

8. Digest parent template with *Dpn*I at the optimal conditions previously found by adding 1 µL enzyme directly to the PCR buffer (*see* **Note 5**).

9. Have ready a deep-well MTP with 450 mL SOC medium or equivalent broth incubated at 37 °C.

10. Add 1–5 µL of PCR mixture previously digested to 50 µL freshly cultured host strain and transform according to the optimal conditions previously found (*see* **Note 18**). The transformation can be done using the thermocycler or a water bath. After heat-shock, place samples on ice for 1–2 min.

11. Transfer the cells to the prewarmed deep-well MTP containing 450 mL recovery broth and close it with a lid (e.g., adhesive paper foil with pores for oxygen transfer).

12. Incubate samples at 37 °C for 40–60 min or longer with vigorous shaking, for example, 200–250 rpm of a standard MTP shaker.

13. Plate optimal amount of culture previously found onto selective LB agar plates in standard petri dishes so that colonies can be easily picked.

14. Incubate plates in the oven at 37 °C for 16–20 h.

15. The next day, count the colonies on the agar plates and determine transformation efficiency in CFU per microgram DNA.

16. Pick up 32–48 colonies per randomization site into 200 μL of LB broth in a deep-well MTP and incubate the plate in a shaker at 37 °C and 220 rpm for 16–20 h (*see* **Note 26**).

17. The next day, transfer 100 μL of preculture to MTPs containing 50 μL of 70% glycerol (final conc. ca. 20% glycerol) and 5–10 μL preculture to a well of a 96-well MTPs containing selective LB agar provided by the sequencing provider.

18. Dry the plates under sterile conditions for 20–30 min.

19. Send the plates for sequencing with a suitable primer that targets the randomized codon (*see* **Note 27**).

20. Store the glycerol plates at −80 °C until further use.

3.5 Library Evaluation

1. Open a suitable program for processing ab1 files (*see* **Note 15**).

2. Align sequencing reads using a suitable algorithm (*see below case study*) and discard sequences containing primer misinsertions and unexpected gene mutations.

3. If gene variants are missed, order the corresponding primer pairs and repeat the procedure above until obtaining the desired mutants. In other cases, you can also increase oversampling to 48–96 samples per MTP but this strategy is also expensive if many codons are mutated.

4. Once all variants have been created, take a deep-well MTP and place each unique variant in a well (*see* **Note 28**) to prepare a "Master plate." Take out all those glycerol stock plates containing 95 mutants, for example, 19 mutants of five target sites. Scratch the surface of the plates with a toothpick and transfer it to 200 μL of LB broth in a deep-well MTP. Incubate the plate in a shaker at 37 °C and 220 rpm for 16–20 h.

5. The next day, transfer 100 μL of preculture to MTPs containing 50 μL of 70% glycerol (final conc. ca. 20% glycerol) and store the glycerol plates at −80 °C until further use.

3.6 Investigating Sequence–Function Relationships

1. Transfer an aliquot of your master plate to an expression plate. Grow and harvest cells and assay your enzyme using appropriate protocols.

2. Export the data to a suitable format. The data is normally obtained in a table matrix, but the best way to analyze it is by making visual representations of it. A way to do this is by arranging all amino acids according to their physicochemical properties, as shown in Fig. 4. This can be done relatively quickly in programs such as Excel.

3. The arranged data can now be depicted using bar charts or heat-maps, which helps to quickly identified patterns of the fitness trait in quest (e.g., activity, selectivity, or stability) when it increases or decreases across a large set of data (*see below case study*).

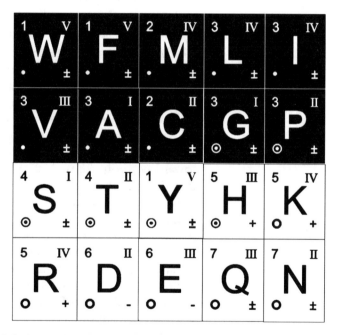

Fig. 4 Amino acid side chains sorted by physicochemical properties. The 20 nonpolar and polar amino acids are respectively highlighted in *black* and *white* background. Each residue is classified upon four different properties of the side-chains according to Pommie et al. [84]: *Top left.* Chemical type: (*1*) Aromatic; (*2*) Sulfur-containing; (*3*) Aliphatic; (*4*) Hydroxyl; (*5*) Basic; (*6*) Acidic; (*7*) Amide. *Top right.* Volume ($Å^3$): (*I*) Very small (60–90); (*II*) Small (108–117); (*III*) Medium (138–154); (*IV*) Large (162–174); (*V*) Very large (189–228). *Bottom left.* Hydropathy: Hydrophobic (•), neutral (⊙), hydrophilic (□). *Bottom right.* Polarity: Positively charged (+), uncharged (±), and negatively charged (−). Amino acids are abbreviated in *one-letter code*

4. Alternatively, the data can be analyzed a more advanced software such as Matlab. In either case, analyze the data and identify what are the most important residues involved in the fitness trait under study by comparing the activity of the wild-type protein to that of the mutants.

5. After the hotspot mutations are identified, it is possible to combine these by recombination methods. The experimental protocols for employing these methods have been published elsewhere recently including DNA Shuffling [85], Assembly of Designed Oligonucleotides [18] and gene assembly [86].

3.7 Case Study: Controlling Regioselectivity and Stereoselectivity of Steroids Using P450$_{BM3}$

1. *Choose a suitable system as starting point*: We chose the cytochrome P450 monooxygenase P450$_{BM3}$ from *Bacilllus megaterium* because it is perhaps the most active P450 known to date [87]. However, the WT enzyme does not display significant activity toward large nonnatural substrates. This is the reason why we started with the mutant F87A as template. The gene length of P450$_{BM3}$ is 3147 bp, which is translated in 1049 amino acid residues. The gene coding for BM3 was previously cloned in the pMET11 vector, a low-copy number plasmid of ca. 9 kb [20].

2. *Choose target region*: The active site of P450$_{BM3}$ is very large, with more than 20 residues surrounding the heme prosthetic group, which could serve as potential targets for mutagenesis (Fig. 5). We previously found that mutating residues R47, T49, Y51, V78, and A82 enables the discovery of two highly active mutants capable of hydroxylating the steroidal substrate testosterone (**1**) with high stereoselectivity and regioselectivity toward positions 2β or 15β, which are very challenging reactions in synthetic organic chemistry [20].

   In a separate study, we scanned the same five residues of mutant F87A to construct a mutability landscape, finding that some hotspot-mutants are regioselective toward 2β- (**2**), 15β- (**3**), or 16β- (**4**) hydroxytestosterone (Fig. 6). Interestingly, variant A82M/F87A is the most active and **3**-selective one exhibiting 78% conversion of **1** and 78% **3**-regioselectivity, respectively (Acevedo-Rocha et al., in preparation). To further study the evolution of the fitness landscape of **3**, we chose mutant A82M/F87A as template for scanning the remaining four residues R47, T49, Y41, and V78 and construct a mutability landscape with $19 \times 4 = 76$ newly created mutants.

3. *Design of oligonucleotides*: For PCR, we used at that time a QuikChange-like protocol based on the KOD polymerase in which the primers overlap completely. The oligos were designed manually including nondegenerate codons encoding all 19 amino acids (Table 3) according to *E. coli* codon usage. It is important to order high-quality primers in resuspended form (e.g., water) to increase speed. We usually order oligos at a final

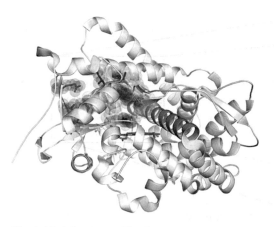 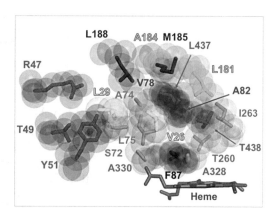

Fig. 5 Model enzyme. The P450_{BM3} heme domain is seen in the *left panel*. The *right panel* zooms into the active site. The double mutant parental enzyme is F87A/A82M. Mutational scanning was performed at residues R47, T49, Y51, and V78. The image was created using PYMOL based on PDB 1JPZ [88]. Many other residues are potential targets for mutagenesis

Fig. 6 Model reaction. Testosterone (**1**) can be hydroxylated at position 2-beta (**2**), 15-beta (**3**) or 16-beta (**4**) by P450_{BM3}F87A-derived mutants

Table 3
Oligos used for scanning four active site residues of P450_{BM3}

Name	Sequence (5′ → 3′)
R47_FW	CGCCTGGT**XXX**GTAACGCG
R47_RV	CGCGTTAC**YYY**ACCAGGCG
T49_FW	CTGGTCGTGTA**XXX**CGCTACTTATC
T49_RV	GATAAGTAGCG**YYY**TACACGACCAG
Y51_FW	GTAACGCGC**XXX**TTATCAAGTCAGC
Y51_RV	GCTGACTTGATAA**YYY**GCGCGTTAC
V78_FW	CAAGCGCTTAAATTT**XXX**CGTGATTTTATGG
V78_RV	CCATAAAATCACG**YYY**AAATTTAAGCGCTTG

XXX/YYY = FW/RV codon used for each amino acid: ATG/CAT (M), ATC/GAT (I), ACA/TGT (T), AAA/TTT (K), AAC/GTT (N), TCC/GGA (S), CGT/ACG (R), GGA/TCC (G), GAC/GTC (D), GAA/TTC (E), GCA/TGC (A), GTA/TAC (V), CAC/GTG (H), CAA/TTG (Q), CCA/TGG (P), CTC/GAG (L), TTC/GAA (F), TAC/GTA (Y), TGC/GCA (C), TGG/CCA (W)

concentration of 2–5 μM in desalted form from one of the best suppliers (*see* **Note 2**).

4. *Library creation*. We did not perform PCR optimization tests for creating the four gene libraries using the P450$_{BM3}$ template A82M/F87A, as protocols for this gene are well established in our lab. We set up four PCR using 0.1 μL parental plasmid (20 ng), 5 μL of 2 μM of each primer pair (Table 3) and 10 μL KOD 2× PCR Master Mix in a total volume of 20 μL. The PCR program entails 95 °C for 3 min, 30 cycles of 95 °C for 30 s, 55 °C for 20 s, 72 °C for 6 min; 72 °C for 10 min, followed by sample cooling. Upon checking the presence of bands in DNA gel electrophoresis, the reactions were purified using the Qiagen miniprep kit with elution in 45 μL water and treated with 1 μL *Dpn*I for 16–20 h in the supplier's buffer. Thereafter, samples were dialyzed in 50 mL water and 2 μL were used for transforming electrocompetent *E. coli* BL21 (DE3)-Gold cells (*see* **Note 19**). The cells were recovered with LB broth at 37 °C for 60 min and 100 × *g* in an Eppendorf thermo-shaker. Afterward, 100 μL of cells was plated onto LB agar Kan^R plates. The plates were placed in an incubator at 37 °C for 16–20 h. The next day, 200–300 colonies appeared onto each agar plate, of which 96 were picked up in 200 μL of LB Kan^R broth in a 2.2 mL MTP with 96 wells. The cells were cultured for 16–20 h. The next day, using a liquid-handling robot (Tecan, Maennedorf, Switzerland), 10 μL were inoculated on 96-well MTPs containing LB agar Kan^R provided by the sequencing provider. The plates were left to dry inside a laminar flow hood for 30 min and where sent for sequencing with a primer specific for the T7 promoter. Using the same tips, 100 μL of the overnight culture were mixed with 50 μL of 70% glycerol (final conc. ca. 20% glycerol) and the plates were stored at −80 °C until further use.

5. *Library evaluation*. The first step is to download the data files from the sequence provider. Some providers provide a good overview of the sequencing results (Fig. 7). Interestingly, not all 96 samples sent to the provider came back with a legible sequence. In some cases, the sequencing may have been poorly performed, but the supplier usually includes quality controls and if the service is not performed properly, the analysis is usually repeated one more time. If a large number of samples are not sequenced, it is likely related to a bad quality library preparation. In these cases, additional unexpected sequences upstream or downstream of the target codon can be introduced (e.g., primer misinsertions), which would impair a correct ligation of the plasmid. Furthermore, not all obtained sequences are of good quality. Some of them will have shorter or longer reads, which increases the chances of getting lesser

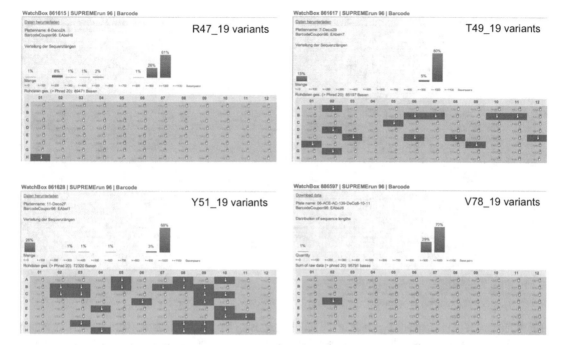

Fig. 7 Exemplary overview of 96-well sequencing results of four target residues. *Green* fields indicate successful sequencing runs, whereas *red* fields code for nonsuccessful sequencings. Clearly, there is more bias during library construction at residues T49 and Y51 than R47 or V78. This demonstrates the need for checking quality of the different libraries. *Top left*: Residue R47 yielded 95 out of 96 sequencing reads (99%) of which 85 have lengths of >700 bp (88%). *Top right*: Residue T49 yielded 82 out of 96 sequencing reads (86%) of which 82 have lengths of >700 bp (85%). *Bottom left*: Residue Y51 yielded 71 out of 96 sequencing reads (74%) of which 68 have lengths of >700 bp (71%). *Bottom right*: Residue V78 yielded 50 out of 51 sequencing reads (99%) of which 50 have lengths of >700 bp (99%). In all cases 96 samples were sent for sequencing, except for residue V78 in which 51 samples were delivered (wells A1 to E3)

sequences correctly aligned. For example, in library A82/F87A targeting residue R47, even though 1 out of 95 sequences were not delivered, about 11 samples have lengths of <500 bp whereas the remaining 85 samples are >800 bp long (Fig. 7a).

All sequences were downloaded and stored in a suitable folder. Files with ab1 extension are required for the alignment and peak assessment. For illustrative purposes, we show the alignment of files from the first library (R47). The 95 files were opened in MegAlign (enter sequences option) and aligned using the Clustal V algorithm. A screenshot of the resulting alignment is shown in Fig. 8.

In total, we found 26 primer misinsertions, which decrease the number of useful sequences from 95 to 69. All those 'bad' sequences having primer insertions were removed from the alignment to proceed. The remaining files are aligned again using the Clustal V algorithm. A screenshot of the alignment is shown in Fig. 9.

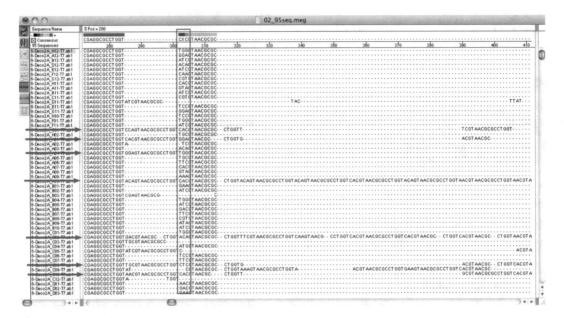

Fig. 8 Alignment of 95 sequencing reads of the R47 library using MegAlign software. There are several primer misinsertions of different length upstream and downstream of the target codon (marked by the *red arrows*). The target codon CGT is represented with the *blue* bases CXC (in the *red rectangle*)

Subsequently, 14 sequences showing additional mutations were removed from the alignment and the remaining 55 files were realigned using the Clustal V algorithm. Overall, 55 sequences were translated in the corresponding amino acids. A summary of the results for the R47 and the other libraries is shown in Table 4. In general, the success or yield of sample randomization lies between 57% and 71%, which means that for 95% library coverage, 79–99 colonies had to be sequenced. It is generally assumed that randomization efficiency is 100%, but this is not the case in the present examples.

Table 5 shows amino acid distributions for the libraries of residues R47, T49, Y51, and V78. As discussed above, some libraries show bias: the template should not be present, but it appears several times in the library R47 (10) and Y51 (6). This means that *Dpn*I overnight digestion was suboptimal in these cases, but not in the libraries T49 and Y78 that showed parental template in just one case. Also, it is unexpected to find a stop codon in R47 library, as only 19 sense and 19 antisense primers were mixed for creating these libraries. Still, all target amino acids were found in library T49, but 1 (Ala), 2 (Cys, Asp) and 3 (Met, Gly, Lys) amino acids were missed in libraries R47, Y51 and V78, respectively. In the latter cases, the 6 remaining residues were created using SDM as described in the experimental section. For each case, four colonies were harvested after PCR and transformation, followed by plasmid extraction

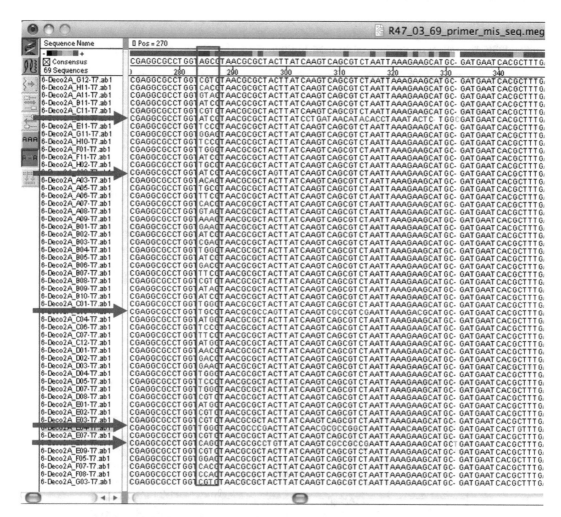

Fig. 9 Alignment of 69 sequencing reads of R47 library using MegAlign software. There are mutations in *red* in several cases downstream of the target codon CGT, which has bases AGC as consensus sequence (*blue*) within the *red rectangle*. The screenshot shows about 53 aligned sequences. Interestingly, even though the sequencing reads are of good quality and sequences showing primer misinsertions have been removed, there are still several point mutations and frameshift mutations that affect the open reading frame (indicated by the *red arrows*). These effects are usually present in libraries that have not been optimized. However, they can also be found in optimized libraries. This example shows how difficult it is to randomize a given gene at a certain location without bias. The success is affected by many factors, such as GC content, secondary structures, primer melting temperature as well as annealing temperature, polymerase, number of cycles of the PCR

and sequencing (*see* **Note 28**). In this way, all target 76 mutants were created, but it should be noticed that, contrary to expectations, amino acid distributions are not uniform, with residues being represented as many as 8 (R47 and T49 libraries) or 14 (W in Y51 library) times versus once (N in libraries R47, Y51, and V78). It is important to consider these findings when

Table 4
Sequencing results of single-site saturation mutagenesis libraries at four residues

Samples	R47	T49	Y51	V78
Sent for sequencing	96 (100%)	96 (100%)	96 (100%)	51 (100%)
Failed sequencing	1 (1%)	14 (15%)	24 (25%)	1 (2%)
Primer misinsertions	26 (27%)	15 (16%)	6 (6%)	8 (16%)
Other mutations	14 (15%)	1 (1%)	9 (10%)	6 (12%)
In-frame sequences	55 (57%)	66 (69%)	57 (59%)	36 (71%)
Oversampling for 95% library coverage based on real yield (correct sequences)	99	85	95	79
Oversampling for 95% library coverage based on 100% theoretical yield	56	56	56	56

The oversampling factor was calculated using the TopLib server [35]

Table 5
Amino acid distributions determined in site-saturation mutagenesis libraries of the four target residues

AA (codon)	R47 (CGT)	T49 (ACG)	Y51 (TAC)	V78 (GTA)
W (TGG)	4	1	14	2
F (TTC)	3	2	2	3
M (ATG)	2	6	1	0
L (CTC)	2	4	6	2
I (ATC)	8	5	1	4
V (GTA)	2	8	5	1
A (GCA)	0	4	3	2
C (TGC)	2	1	0	1
G (GGA)	2	2	2	0
P (CCA)	1	3	3	2
S (TCC)	3	4	0	4
T (ACA)	3	1	3	4
Y (TAC)	1	3	6	0
H (CAC)	2	3	2	2
K (AAA)	1	4	2	0
R (CGT)	10	2	1	1
D (GAC)	3	1	0	4
E (GAA)	3	2	1	2
Q (CAA)	1	3	4	1
N (AAC)	1	7	1	1
Stop (TAA)	1	0	0	0
Total	**55**	**66**	**57**	**36**

WT residue is highlighted in black

building mutability landscapes that required the careful control of regioselectivity and stereoselectivity, at least in the case of steroid hydroxylation (*see* **Note 29**).

6. *Investigating sequence–function relationships.* Upon identification of correct mutants and creation of remaining ones, a toothpick was used to scratch the surface of the frozen glycerol and placed into a deep-well MTPs containing 800 µL of LB Kan^R broth. Cells were placed in the incubator at 37 °C for 16–20 h. The next day, 100 µL preculture were transferred with the liquid handling robot to a deep-well MTP containing TB Kan^R with additives including 100 µM IPTG for inducing protein expression, as reported elsewhere [20]. After 24 h, cells were centrifuged at 4000 rpm for 10 min at 4 °C and washed once with 100 mM potassium phosphate buffer at pH 8.0. The cells were frozen twice with liquid nitrogen and suspended in 600 µL reaction mixture containing potassium phosphate buffer, 100 mM NADH with a recycling system based on 10% glucose and 1 U/mL glucose dehydrogenase and 1 mM testosterone (dissolved in 1% dimethylformamide as cosolvent). The reaction was stopped after 24 h by adding 350 µL ethyl acetate (EA). The organic phase was extracted and transferred to a 96-well MTP of 250 µL suitable for HPLC analysis. The EA was left to dry and the steroid was dissolved in 150 µL acetonitrile. The samples were processed in the HPLC as described elsewhere (Acevedo-Rocha et al. submitted). Screening data of steroid hydroxylation was processed in triplicates and analyzed with Microsoft Excel. Table 6 shows the screening data of **1** with the created mutants. However, the data arranged in this way lacks visual power, so it should be represented visually to understand sequence–function relationships, which can be displayed in different ways. One strategy is to use bar charts where the genotype and phenotype or fitness are linked in a 3D matrix. This can be done fairly easy in Microsoft Excel (Fig. 10).

Interestingly, half of the mutations at position R47 confer an increase of ca. 10% conversion without losing selectivity (Fig. 10a/c). At position T49, a smaller amount of mutants exhibited similar patterns. On the other hand, mutations at position Y51 lose **3**- in favor of **2**-selectivity, which can be nicely seen in Fig. 10b. Finally, residue V78 shows the most active and **3**-selective mutants when M, L or I are introduced as well as some mutants that are **4**-selective (Fig. 10c/d).

The disadvantage of bar charts, however, is that they occupy large space so that these can be used for analyzing a small number of variants. If a large dataset is generated, it is more suitable to use heat-maps (Fig. 11). This visual representation can be also generated in Excel.

Table 6
Fitness analysis of all possible single steps of parent mutant A82M/F87A at residues R47, T49, Y51, and V78 using resting cells

Entry	Mutant	1 Conversion (%)	Selectivity (%) 2	3	4	Others [a]
1	----M	78.7 ± 0.7	14.5 ± 0.8	78.4 ± 0.1	2.9 ± 0.2	4.2 ± 0.3
2	W---M	80.8 ± 1.4	14.4 ± 0.2	76.7 ± 1.5	3.7 ± 1.8	5.3 ± 1.8
3	F---M	91.0 ± 1.2	5.6 ± 0.3	81.5 ± 1.2	0.7 ± 0.7	12.2 ± 1.1
4	M---M	87.5 ± 2.6	7.5 ± 0.5	81.1 ± 2.4	0.8 ± 0.6	10.7 ± 1.7
5	L---M	89.7 ± 0.7	7.4 ± 0.7	80.8 ± 0.7	0.8 ± 0.7	11.0 ± 1.1
6	I---M	88.2 ± 1.5	7.3 ± 0.6	80.8 ± 1.1	0.8 ± 0.7	11.1 ± 0.8
7	V---M	85.2 ± 4.7	6.6 ± 0.9	79.9 ± 4.1	1.1 ± 0.9	12.3 ± 1.1
8	A---M	87.0 ± 1.5	7.1 ± 0.6	80.7 ± 1.6	0.9 ± 0.9	11.2 ± 1.3
9	C---M	91.2 ± 2.3	8.0 ± 0.4	80.2 ± 2.1	0.9 ± 0.6	10.9 ± 1.0
10	G---M	87.9 ± 3.0	7.8 ± 1.1	80.9 ± 2.1	1.7 ± 0.9	9.6 ± 0.8
11	P---M	68.4 ± 3.3	14.9 ± 0.4	78.3 ± 4.0	0.9 ± 0.6	5.9 ± 1.0
12	S---M	88.3 ± 2.5	7.6 ± 0.8	81.3 ± 2.1	1.1 ± 1.0	9.9 ± 1.2
13	T---M	84.8 ± 1.6	8.0 ± 0.4	80.8 ± 1.9	1.2 ± 1.0	10.1 ± 1.5
14	Y---M	88.3 ± 1.2	5.5 ± 0.7	80.5 ± 1.1	0.8 ± 0.7	13.2 ± 1.9
15	H---M	88.0 ± 1.2	8.8 ± 0.7	81.9 ± 0.9	0.9 ± 0.8	8.4 ± 0.8
16	K---M	78.1 ± 2.4	13.0 ± 0.5	80.1 ± 3.3	0.7 ± 0.6	6.2 ± 0.9
17	D---M	88.3 ± 2.1	8.6 ± 0.4	80.9 ± 1.9	0.8 ± 0.8	9.7 ± 1.5
18	E---M	86.5 ± 2.4	8.2 ± 0.3	79.2 ± 2.6	0.8 ± 0.6	11.8 ± 1.1
19	Q---M	82.9 ± 2.3	9.9 ± 0.8	81.0 ± 2.2	1.1 ± 0.9	8.0 ± 0.9
20	N---M	17.8 ± 1.7	13.8 ± 1.8	81.9 ± 8.2	0.4 ± 0.1	3.9 ± 0.7
21	-W--M	73.7 ± 4.5	11.7 ± 0.6	82.0 ± 5.6	0.7 ± 0.4	5.6 ± 0.5
22	-F--M	69.9 ± 1.1	13.0 ± 0.2	80.9 ± 1.1	0.8 ± 0.5	5.3 ± 0.4
23	-M--M	86.9 ± 2.7	11.0 ± 0.5	79.8 ± 2.8	2.5 ± 1.0	6.7 ± 0.9
24	-L--M	76.6 ± 0.1	12.0 ± 0.4	82.5 ± 0.0	1.6 ± 1.3	4.0 ± 0.3
25	-I--M	78.2 ± 2.9	11.2 ± 0.5	80.3 ± 3.7	2.0 ± 0.9	6.5 ± 1.1
26	-V--M	82.2 ± 0.6	10.0 ± 0.3	80.6 ± 0.7	2.4 ± 0.9	7.0 ± 1.4
27	-A--M	69.6 ± 1.7	11.9 ± 1.0	80.4 ± 1.4	2.2 ± 0.8	5.5 ± 1.2
28	-C--M	85.0 ± 1.8	16.5 ± 0.7	74.6 ± 2.2	2.1 ± 0.8	6.8 ± 0.7
29	-G--M	80.7 ± 3.0	14.2 ± 0.2	78.9 ± 3.3	1.4 ± 0.5	5.5 ± 0.7
30	-P--M	45.6 ± 1.9	12.8 ± 0.2	82.4 ± 3.9	1.4 ± 0.5	3.5 ± 0.5
31	-S--M	84.6 ± 2.3	12.6 ± 0.3	78.4 ± 2.1	1.1 ± 0.9	7.9 ± 0.8
32	-Y--M	86.1 ± 1.2	9.2 ± 0.4	81.7 ± 1.4	2.6 ± 1.0	6.4 ± 1.1
33	-H--M	78.1 ± 1.8	13.0 ± 0.8	80.3 ± 1.8	1.3 ± 0.4	5.4 ± 0.2
34	-K--M	72.1 ± 2.1	19.1 ± 1.0	72.3 ± 2.1	1.4 ± 0.7	7.2 ± 0.7
35	-R--M	46.5 ± 2.4	14.7 ± 0.7	80.4 ± 4.1	0.6 ± 0.4	4.2 ± 0.2
36	-D--M	87.0 ± 1.7	9.7 ± 0.4	79.6 ± 1.9	0.5 ± 0.4	10.2 ± 0.8
37	-E--M	86.6 ± 1.8	11.4 ± 0.2	72.2 ± 2.3	0.5 ± 0.2	15.9 ± 0.9
38	-Q--M	85.6 ± 2.1	6.7 ± 0.3	84.5 ± 2.3	2.1 ± 0.8	6.7 ± 0.6
39	-N--M	72.5 ± 0.7	11.9 ± 0.5	81.2 ± 0.8	3.0 ± 0.1	4.0 ± 0.4
40	--W-M	89.1 ± 3.0	10.2 ± 0.3	82.5 ± 2.4	1.6 ± 0.2	5.6 ± 0.8
41	--F-M	89.9 ± 2.1	17.4 ± 0.4	74.7 ± 1.6	1.1 ± 0.5	6.8 ± 0.3
42	--M-M	87.6 ± 1.9	25.2 ± 0.5	68.2 ± 1.8	0.8 ± 0.2	5.8 ± 0.5
43	--L-M	84.4 ± 3.0	25.8 ± 1.4	66.9 ± 2.0	0.7 ± 0.2	6.6 ± 0.4
44	--I-M	86.7 ± 1.6	33.5 ± 0.4	59.5 ± 2.0	0.6 ± 0.2	6.4 ± 0.8
45	--V-M	71.7 ± 0.7	36.8 ± 0.3	56.5 ± 0.5	0.5 ± 0.2	6.2 ± 0.4
46	--A-M	34.7 ± 1.7	34.9 ± 1.7	60.6 ± 3.1	0.7 ± 0.2	3.8 ± 1.0
47	--C-M	52.3 ± 0.2	34.5 ± 1.1	59.9 ± 0.9	0.7 ± 0.2	5.0 ± 0.1
48	--G-M	40.0 ± 4.1	25.3 ± 2.6	70.2 ± 7.9	0.8 ± 0.3	3.6 ± 0.5
49	--P-M	52.1 ± 2.8	36.8 ± 1.7	58.1 ± 3.2	0.9 ± 0.5	4.3 ± 0.5
50	--S-M	54.5 ± 4.0	35.4 ± 2.6	59.7 ± 4.2	0.6 ± 0.2	4.2 ± 0.4
51	--T-M	76.6 ± 2.2	34.9 ± 0.7	58.8 ± 1.3	0.3 ± 0.2	6.0 ± 0.8
52	--H-M	85.9 ± 2.9	16.2 ± 0.2	76.2 ± 2.4	1.5 ± 0.7	6.1 ± 0.7
53	--K-M	77.6 ± 3.1	25.8 ± 0.8	67.8 ± 2.8	1.1 ± 0.6	5.2 ± 0.8
54	--R-M	12.9 ± 0.7	17.4 ± 1.9	72.4 ± 5.3	2.7 ± 0.6	7.6 ± 1.4
55	--D-M	21.4 ± 3.6	23.0 ± 3.6	72.8 ± 11.8	0.7 ± 0.7	3.6 ± 0.9
56	--E-M	51.9 ± 5.4	18.0 ± 1.9	75.5 ± 7.6	1.2 ± 0.7	5.3 ± 0.3
57	--Q-M	63.7 ± 3.1	20.2 ± 1.2	74.1 ± 3.8	1.6 ± 0.6	4.1 ± 0.8
58	--N-M	70.7 ± 4.3	22.2 ± 1.0	70.9 ± 4.0	0.9 ± 0.5	6.1 ± 1.3
59	---WM	10.7 ± 0.8	8.8 ± 1.7	69.9 ± 4.3	6.9 ± 2.6	14.4 ± 4.9
60	---FM	19.5 ± 2.3	13.2 ± 1.5	80.3 ± 9.5	1.0 ± 0.9	5.4 ± 2.1
61	---MM	83.0 ± 2.6	5.2 ± 0.4	83.9 ± 2.6	1.7 ± 1.7	9.3 ± 1.9
62	---LM	91.5 ± 1.7	7.7 ± 0.3	81.0 ± 1.4	0.6 ± 0.2	10.7 ± 0.3
63	---IM	88.4 ± 1.4	6.4 ± 0.7	84.2 ± 1.5	0.5 ± 0.4	8.9 ± 0.9
64	---AM	58.1 ± 2.4	6.9 ± 0.3	82.4 ± 4.2	4.2 ± 2.1	6.5 ± 2.1
65	---CM	69.9 ± 2.7	15.3 ± 0.8	77.6 ± 2.9	1.8 ± 0.8	5.3 ± 1.0
66	---GM	19.5 ± 1.1	9.4 ± 0.8	85.9 ± 5.4	1.1 ± 0.8	3.6 ± 0.8
67	---PM	15.5 ± 0.4	9.3 ± 0.8	83.5 ± 1.9	1.8 ± 0.6	5.3 ± 1.3
68	---SM	31.3 ± 1.7	8.6 ± 0.2	85.5 ± 4.8	1.1 ± 1.1	4.9 ± 1.4
69	---TM	56.3 ± 1.9	10.5 ± 0.7	82.1 ± 2.1	1.2 ± 1.1	6.3 ± 1.0
70	---YM	12.5 ± 2.2	15.7 ± 2.7	77.0 ± 12.4	1.1 ± 0.2	6.3 ± 2.2
71	---HM	2.0 ± 0.1	6.5 ± 6.5	52.8 ± 1.2	3.3 ± 1.7	37.5 ± 12.1
72	---KM	2.1 ± 0.4	4.1 ± 4.1	50.2 ± 9.8	10.0 ± 1.2	35.7 ± 8.6
73	---RM	1.6 ± 0.1	0.0 ± 0.0	5.9 ± 5.9	14.2 ± 2.7	79.9 ± 24.5
74	---DM	6.8 ± 0.8	12.5 ± 1.9	72.3 ± 9.5	3.4 ± 0.5	11.8 ± 1.6
75	---EM	1.5 ± 0.0	0.0 ± 0.0	42.6 ± 5.1	17.5 ± 2.3	39.9 ± 8.4
76	---QM	1.5 ± 0.0	3.4 ± 3.4	47.9 ± 3.1	10.0 ± 1.0	38.6 ± 12.8
77	---NM	13.4 ± 0.8	14.9 ± 1.3	76.5 ± 5.5	2.5 ± 0.8	6.0 ± 2.2

Mutant A82M/F87A (——M) was used as template to create all single mutants at positions R47 (_—M), T49 (-_—M), Y51 (–_-M), and V78 (——_M). Values shown in *green/red* are above/below those of the parent mutant
[a]Other oxidation sites occur at positions 19, 1β and unknown products
Screening values are the average of three experiments displaying the standard error mean ($n = 3$)

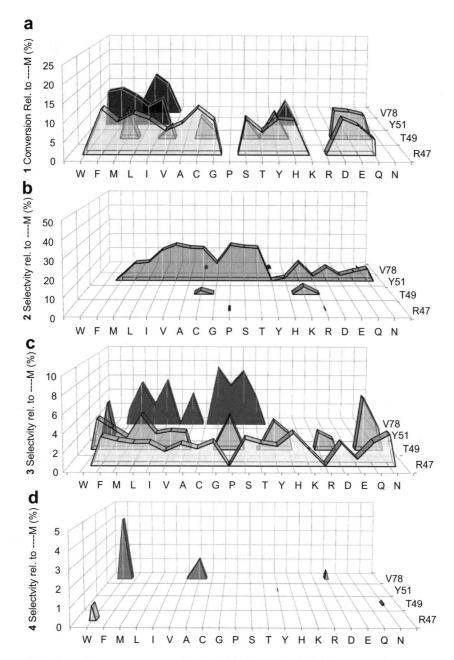

Fig. 10 Mutability landscapes of P450$_{BM3}$ mutant A82M/F87A (M) in 3D bars format created with Microsoft Excel. The traits are (**a**) conversion of **1** toward (**b**) **2**, (**c**) **3**, and (**d**) **4**. Data represent the %HPLC average values normalized to the respective parent mutant based on three independent experiments ($n = 3$). Amino acids are arranged according to their physicochemical parameters (Fig. 3). *M* means that the parent mutant A82M/F87A is used to create the mutability landscape at four residues represented by the *hyphens*

Microsoft Excel is useful for representing scanning data, but it is limited in statistical analyses that can be performed. Alternatively, other programs such as Matlab or R can be used. For

AA	W	F	M	L	I	V	A	C	G	P	S	T	Y	H	K	R	D	E	Q	N	%	
R47	81	91	88	90	88	85	87	91	88	68	88	85	88	88	78	72	88	87	83	23	100	
T49	87	87	81	78	82	86	85	70	86	73	70	72	20	85	78	82	72	85	46	87	75	Conversion of 1
Y51	89	90	88	84	87	76	35	52	40	52	55	77	72	86	78	13	26	52	64	71	50	
V78	11	20	83	92	88	72	58	70	19	15	31	56	13	2	2	2	7	2	1	13	25	
R47	14	6	8	7	7	7	7	8	8	15	8	8	5	9	13	15	9	8	10	14	100	
T49	10	11	14	11	10	7	16	13	9	12	12	15	79	16	13	12	19	13	13	11	75	2-Regioselectivity
Y51	10	17	25	26	33	27	35	34	25	37	35	35	15	16	26	17	23	18	20	22	50	
V78	9	13	5	8	6	15	7	15	9	9	9	10	16	6	4	0	13	0	3	15	25	
R47	77	82	81	81	81	80	81	80	81	78	81	81	81	82	80	79	81	79	81	81	100	
T49	80	72	79	80	81	85	75	81	82	81	80	79	17	75	80	65	72	78	82	80	75	3-Regioselectivity
Y51	83	75	68	67	60	66	61	60	70	58	60	59	79	76	68	72	73	75	74	71	50	
V78	70	80	84	81	84	79	82	78	86	84	85	82	77	53	50	6	72	43	48	76	25	
R47	4	1	1	1	1	1	1	1	1	2	1	1	1	1	1	2	1	1	1	2	100	
T49	1	0	1	2	2	2	2	1	3	3	2	2	1	2	1	4	1	1	1	3	75	4-Regioselectivity
Y51	2	1	1	1	1	0	1	1	1	1	1	0	2	2	1	3	1	1	2	1	50	
V78	7	1	2	1	0	2	4	2	1	2	1	1	1	3	10	14	3	17	10	3	25	

Fig. 11 Mutability landscapes of P450$_{BM3}$ mutant A82M/F87A (M) in heat-map format. The color code was generated using Microsoft Excel. The traits shown are (**a**) conversion of **1** (*first panel*) toward (**b**) **2** (*second panel*), (**c**) **3** (*third panel*), and (**d**) **4** (*fourth panel*). The data represent the %HPLC average values based on three independent experiments (*n* = 3). Amino acids are arranged according to their physicochemical parameters (Fig. 4)

example, heatmaps can be even combined in one image (Fig. 12) and it is also possible to build histograms (Fig. 13) and ternary plots (Fig. 14) to identify the critical mutations for a determined fitness trait (*see* **Notes 30–32**).

This case study was developed to study how evolution proceeds at single mutation steps, a fascinating topic related to fitness landscapes, epistasis, protein evolution, directed evolution and protein engineering in general [89].

4   Notes

1. As any other well-planned scientific project, it is important to test what is the optimal plasmid for overexpressing a target protein. Plasmids with different inducible and non-inducible promoters, ribosomal binding sites (RBSs) sequences, terminators, replication origins, and markers are available from many suppliers (DNA 2.0 provides a series of custom vectors for

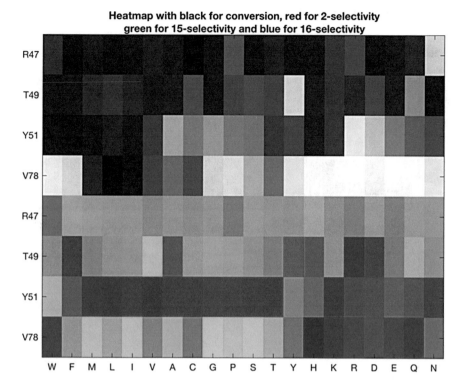

Fig. 12 Mutability landscapes of P450$_{BM3}$ mutant A82M/F87A (———M) in heatmap format with colors corrected to show diversity. The figure was generated by using Matlab (*see* **Note 30**). In the upper four lanes, larger conversions are indicated by a *black color*. The lower four lanes code for selectivity: the degree of *red*, *green*, and *blue* color codes for hydroxylation in position 2 (formation of **2**), position 15 (formation of **3**), and position 16 (formation of **4**)

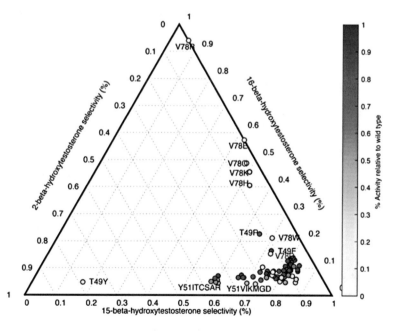

Fig. 13 Mutability landscapes of P450$_{BM3}$ mutant A82M/F87A (M) in ternary plot format generated with Matlab. The datapoints are color-coded based on the activity relative to wild type (*see* **Note 31**)

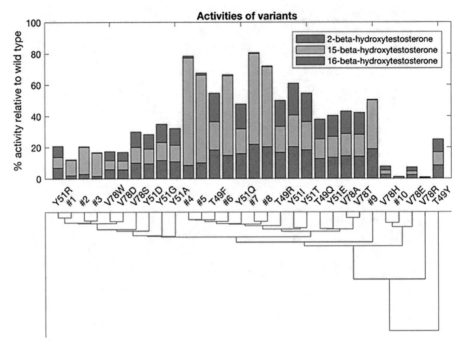

Fig. 14 Clustered mutability landscapes of P450$_{BM3}$ mutant A82M/F87A (M) in bar chart format with a partially collapsed dendrogram. The figure was generated by using Matlab (*see* **Note 32**). Some nodes were clustered together, namely, cluster #1 (V78Y, V78N), #2 (R47N, V78F), #3 (V78G, V78P), #4 (R47W/F/M/L/I/V/A/C/G/S/T/Y/H/K/D/E/Q, T49W/M/L/I/V/A/G/H/K/E/N, Y51W/F/H, V78M/L/I), #5 (R47P/R, T49C/P/S/T, Y51Y, V78V/C), #6 (T49D, Y51N), #7 (Y51M/L), #8 (Y51V/K), #9 (Y51C/P/S), #10 (V78K/Q)

various hosts including *E. coli*). If the target codons for mutagenesis are not located close upstream or downstream of the open reading frame (ORF), the oligos can be used for mutagenesis regardless of the plasmid. Also consider the possibility of optimizing the codon usage according to the host if it is intended to maximize protein expression and solubility. Importantly, the gene chosen for mutagenesis should display a reasonable output (e.g., measurable activity or fluorescence) because the distribution of hotpots mutations is usually very low (probably below 0.1–1%), which means that most mutations will be neutral or deleterious and just a few beneficial. Thus, there is a risk of losing some important hotspot-mutations if the activity of the gene is too low. Sometimes the WT gene may not exhibit the desired activity toward a nonnatural substrate. If this is the case, a set of variants should be tested first and the best mutant can be chosen as the parental gene.

2. Although there are many different suppliers of oligonucleotides, it is advisable to get the best quality oligos for the lowest cost. That means that HPLC-purified is prohibited and only column-purified (desalted) oligos had to be considered in the

present case. We have tested three different suppliers finding that oligos from IDT (Leuven, Belgium) show the same quality regardless of purification (desalted vs. HPLC) [52]. Furthermore, in IDT the cost per base can go as low as USD 0.10 by ordering more than 288 oligos of 15–60 bases in MTP format of 96 or 384 wells ("500 picomole DNA Plate Oligo"), which represents the lowest cost compared to any other provider. IDT thus fulfills oligos without purification of highest quality, lowest price, and fastest turnaround times. In addition, the oligos can be normalized to a specific concentration in water or suitable buffer without additional or minor costs. In our lab a concentration of 2 μM works fine but 5 μM is ideal to decrease volume and hence minimize PCR costs. Also, it is advisable to bear in mind that buying two oligos with NNK degeneracy is cheaper than 38 oligos per site (if using QuikChange) from the economical perspective. This is why it is recommended to first perform NNK/S mutagenesis, followed by the creation of the missing mutants (60). Lastly, it is possible to ask the provider to mix the oligos without additional costs.

3. It is important to choose a polymerase whose PCR buffer is compatible with the buffer of the *Dpn*I restriction enzyme to ensure complete removal of the methylated and hemimethylated parental plasmid when obtained from a strain containing methyl transferases (Dam or Dcm) without additional purification steps, which would otherwise increase the cost of library preparation. The polymerase that fulfils this requirement is *Phusion* (New England Biolabs, Ipswick, USA). It generates blunt-ended PCR products and its buffer is compatible with *Dpn*I enzyme from the same supplier. However, there might be other enzymes with buffers that should be compatible. In this case, testing is essential.

4. It is more practical and convenient to mix the template DNA and primers with a PCR master mix (generally 2×) so that the process can be automatized or done manually in a high-throughput fashion and with minimal errors. Obviously, the choice of the PCR master mix depends on the polymerase chosen (*see* **Note 3**). In the case of our preferred enzyme, *Phusion*, there are two types of PCR Master Mixes (HG vs. GC buffers according to high fidelity vs. high GC content in the parental plasmid). Either buffer works, but if your target gene contains a high GC content, it is advisable to use the latter buffer. Either buffer can be obtained in two versions allowing for 100 or 500 reactions of 50 μL each (25 μL of 2× Master Mix are needed per reaction), however, the PCR volume can be minimized to 10 μL to decrease costs by using 5 μL of PCR Master Mix (see below).

5. It is advisable to choose a *Dpn*I restriction enzyme that is fast and whose buffer is compatible with the polymerase (*see* **Note 4**). We have routinely used the enzyme from New England Biolabs, but others should work as well.

6. Creating a large number of libraries is not only expensive, but also time-consuming. Thus, the less number of steps performed, the more efficient the procedure becomes. If subcloning or retransformation steps can be avoided, it is certainly advantageous. Instead of starting with a common cloning strain (*E. coli* Top10, D5H-alpha, XL1-Blue, etc.), it is advisable to use directly the expression host, if possible. It is of course important to test its transformation efficiency before creating all the libraries. We have successfully used *E. coli* BL21(DE3)Gold (Novagen, Merck-Millipore, Darmstadt, Germany), which allows for both convenient DNA preparation and protein expression.

7. It is also possible to use typical LB or $2 \times$YT broth, but SOC may be more suitable for some host strains. If necessary, it is advisable to test first different media for cell recovery before performing high-throughput experiments.

8. There are many miniprep plasmid DNA kits available from different suppliers (Qiagen, New England Biolabs, Sigma, Promega, Thermo Fisher Scientific, Macherey Nagel, etc.). We have found that "Zyppy™ Plasmid Miniprep Kit" from Zymo Research (Irvine, USA) offers the fastest protocol and lowest cost for isolation good-quality plasmid DNA preparations. In the case of midiprep or maxiprep kits, we have only experience with Qiagen, but there should be no big difference compared to other suppliers. If the plasmid copy number is low, lower yields will be obtained. Therefore, it is more convenient in such cases to choose a midiprep or maxiprep kit. The midiprep kit can be used to prepare several reactions in parallel (e.g., two columns), whereas the maxiprep can be used to do at once a big reaction. If the plasmid copy number is high, it may suffice to prepare several miniprep tubes in parallel (e.g., 6–8). Regardless of the plasmid copy number, it is advisable to use a "fresh" plasmid stock (maximum 2 weeks after preparation) to increase mutagenesis performance and parental template elimination.

9. It is advisable to have a DNA electrophoresis system suitable for high-throughput (HT) sample processing in which gels of 96 wells can be used to apply samples from MTPs upon PCR using a multichannel pipette. Companies such as Thermo Fisher Scientific (Waltham, USA), Labnet International (Edison, USA), and Embi Tec (San Diego, USA) offer such kind of systems. The "E-Base™ Integrated Power System" from the former company works very well with precasted gels (*see* **Note 11**).

10. Be sure to have PCR MTPs in which the lids can seal properly the wells. If small volumes are used, relatively volatile sample will evaporate quickly if the seal is suboptimal.

11. Precasted gels for HTP electrophoresis are commercially available, but it should be also possible to prepare homemade gels [90]. We have used successfully the "E-Gel® 96 1% Agarose Gels" from Thermo Fisher Scientific, as 96 samples can be processed and analyzed in <20 min.

12. Multidish plates of 6 or 12 wells are excellent tools when aiming to a small number of colonies. In these plates, LB agar can be economized because the volume needed for a 6- or 12-well plate is ca. 4 mL or 2 mL, respectively which corresponds in either case to a total volume of 24 mL. In the 6-well format, 20–40 colonies can be easily isolated. This format could be even useful for plating the NNK libraries if a low number of samples are chosen for DNA sequencing. In the 12-well format, about 10–20 colonies can be isolated. This format may be more suitable for SDM of a few gene variants. In either case, the volume for cell plating has to be optimized depending on the transformation efficiency and amount of cells. In either format, the fastest way to culture the cells is by adding 2–3 glass beads using a sterile metal spoon of small size. We have used successfully plates from Corning, Thermo Fischer Scientific and Merck-Millipore.

13. If many samples are processed in parallel for sequencing, it is cheaper to send directly an inoculum of the cultures in 96-well LB agar plates with the right antibiotic provided by a sequencing company. The service provider will perform plasmid extraction and sequencing. It is advisable to negotiate with the company for a reasonable price if many MTPS are planned to be sequenced, as a discount of even 50% on the overall cost may well be granted.

14. Deepscan generates a list of partially overlapping QuikChange-based primers aimed at constructing mutability landscapes using PCR. It iterates across a DNA sequence, codon by codon and generates a primer pair that employs the partially overlapping primer pair strategy [79]. Namely, the primers in a pair will have an user-defined overlap length (e.g., 22 bp) centred around the codon to mutate and will have a 3′ overhand long enough to allow the region beyond the mutagenized codon to anneal with the template above a given melting temperature, while taking into account terminal GC clamp. It is available as both a Python 3 script (https://github.com/matteoferla/mutational_scanning) and as a web app (http://deepscan.matteoferla.com/).

15. We have used successfully the MegAlign program from DNA-star (Madison, Wisconsin, US) to quickly align DNA sequences (using the Clustal V algorithm) and translate them directly into amino acid sequences. This program requires a one-time license that is not so expensive compared to other software for analyzing DNA sequences. Other suitable software that can be used is CLC Main Workbench (Qiagen), which is an excellent program but the license is considerably more expensive. Both programs are compatible with Windows and Mac operating systems.

16. There are various software programmes available but we find four peaks (http://nucleobytes.com/4peaks/) very useful. This is freeware software for measuring quickly the peak heights, which is needed to calculate the Q_{pool}-values for determining the quick quality control (QQC) at each codon. The program can be installed in both Windows and Mac operating systems.

17. TopLib is a newer server that can calculate oversampling in libraries in which up to 12 codons are randomized, whereas GLUE-IT is limited to six codons.

18. The following protocol for preparing chemically competent cells works in our hands: (a) Cool centrifuges and all solutions. (b) Inoculate 100 mL LB/2×YT medium with 1 mL of fresh overnight culture. (c) Add 10 mM $MgSO_4$ and 10 mM $MgCl_2$. (d) Check OD_{600} every hour and harvest cells at $OD_{600} = 0.4$. (e) Place the cells on ice for 15 min. (f) Centrifuge at $3500 \times g$ for 8 min at 4 °C and remove the supernatant. (g) Resuspend cell pellet in 50 mL of RF1 solution at pH 5.8 (100 mM RbCl, 50 mM $MnCl_2 \cdot 4H_2O$, 30 mM CH_3CO_2K, 10 mM $CaCl_2 \cdot 2H_2O$, and 15% glycerol) and incubate 15 min on ice. (h) Centrifuge at $3500 \times g$ for 8 min at 4 °C and remove the supernatant. (i) Resuspend cell pellet in 5 mL of RF2 solution at pH 5.8 (10 mM RbCl, 10 mM MOPS, 75 mM $CaCl_2 \cdot 2H_2O$, and 15% glycerol) and incubate 15 min on ice. (j) Using a multichannel pipette, transfer 25 µL aliquots per well in two MTPs. (k) Transform immediately or store at −80 °C until further use. Always check transformation efficiency.

19. To find optimal transformation conditions, it is best to harvest cells when $OD_{600} = 0.3$–0.4. The cells have to be cooled all times and treated gently (no vortex and gentle pipetting). If combinatorial libraries are needed, electroporation is recommended. In this case, cells are harvested and washed carefully two or three times with either ultra-pure water or 10% glycerol (each wash requires $3500 \times g$ at 4 °C for 8 min). The final OD_{600} has to be determined as well. A final $OD_{600} = 100$/mL should be concentrated in 50 or 100 µL cells for optimal

efficiency. There should be also an optimal amount of DNA used, which is 1–100 ng depending on the vector. In the case of transformations in which a few colonies or small libraries of <10,000 members are needed, the heat-shock protocol is recommended. The final OD_{600} has to be adjusted depending on the strain by diluting the final cell suspension to different ODs such as 1, 10 or more ODs/mL concentrated in 25 or 50 μL so that material and effort can be saved. DNA amounts of 1–100 ng could be also tested. It is advisable to test optimal conditions according to the host and protocol for transformation.

20. Old *Dpn*I enzymes were not so efficient as modern ones. Suppliers now typically recommend digesting template DNA in <60 min. While this protocol depends on the nature of the DNA and optimal buffer conditions, it is advisable to test first a series of digestions using the target vector only. For example, 10 and 100 ng of vector can be *Dpn*I digested in 10 μL for 1, 4, and 16 h. The reactions can then be dialyzed using filter disks of 0.022 μM pore diameter (Merck-Millipore): Add 30 mL deionized water in a typical petri dish, place the filter on the water surface using tweezers and pipette carefully the digestion reactions. It is possible to use one filter for up to six reactions each of 50 μL or up to ten reactions of 10 μL. Dialyze samples for 15–30 min (this time can be optimized as well). Use 1–5 μL of digested samples to host strain via electroporation. After cell recovery in SOC broth as usually performed, centrifuge cells for 3 min at 6000 rpm in a Benchtop centrifuge. Remove supernatant and suspend cells in 100 μL of suitable medium and plate all the suspension on selective agar plates. Incubate the plates at 37 °C in the oven for 16–20 h. The next day count the colonies and determine optimal digestion conditions. Do not forget to include positive controls where samples are also dialyzed but without *Dpn*I enzyme treatment.

21. For example, if your scanning region consists of 35 amino acid residues, which corresponds to 105 bases, you can choose arbitrarily a codon at the first (bases 1–3), second (bases 30–33), third (bases 60–63), and fourth (bases 90–93) quarter of the gene. Ideally, you target four codons every 100 nucleotides. For each target codon order a primer pair (if using QuikChange format).

22. To reduce the workload, two annealing temperatures with a difference of 5–10 °C can be tested. If the Tm of the primers = 60 °C, you can test 55 and 60 °C. At this point, it is also advisable to optimize the number of cycles needed for your PCR. It may suffice with 20 cycles, so you could try both 20 and 30 cycles of template denaturation, primer annealing and polymerase extension. Thus there can be up to four different

conditions for each target site. If four sites are chosen for testing, you will end up with 16 samples. Also, depending on the polymerase chosen (*see* **Note 3**), the extension temperature and speed (e.g., kb/s) can differ. Use the settings provided by the supplier to reduce time and thus increase efficiency.

23. The QQC is a single sequencing reaction from pooled plasmids that can be obtained from a solid-agar plate or from liquid culture [52]. The QQC tells whether diversity has been introduced at the target codon, but it does not tell if all amino acids have been found. The QQC is qualitative, but the Q_{pool}-values were introduced by Stewart and colleagues to convert the QQC in a quantitative score [53]. The closer the Q_{pool}-value is to 1.0, the more efficient the randomization is. The calculation of the Q-value is cumbersome if done for many different codons at many different PCR conditions. To avoid this, we have implemented a server that allows obtaining Q_{pool}-values from DNA electropherograms (Ferla et al., in preparation). This server allows the automation of the Q_{pool}-values calculations. We have provided two MatLab scripts to do so, in QQC_automated.m (https://raw.githubusercontent.com/matteoferla/mutational_scanning/master/QQC_fully_automated.m) a whole folder of alignments is processed, in QQC_manual.m (https://raw.githubusercontent.com/matteoferla/mutational_scanning/master/QQC_manual.m) only one alignment, but with extra detail. These operate thanks to the inbuilt function scfread, which reads the alignment files. Once the span corresponding to the mutated codon is identified on the chromatogram, the maximum height of each of the four channels can be obtained. In the case of sequences mutated with multiple codons a rough estimate of the contribution of each can be obtained by optimizing the function in Eq. 1.

$$\operatorname*{argmin}_{x\in[0,\infty]} \sum_{i=1;j=1}^{4,3} \left| p_a x_{i,j,a} a_{i,j} + p_b x_{i,j,b} b_{i,j} + p_c x_{i,j,c} c_{i,j} - m_{i,j} \right| \quad (1)$$

where p is a vector of the proportions of the three primers (A, B, C), a, b, c, and d are matrices of dimensions 4×3 of the proportion of bases in each position (columns) either expected for the three primers (a, b, and c) or empirical values of the mixture (m), the tensor x of dimensions $4 \times 3 \times 3$ to be optimised is a scaling factor for each of the bases at each position for each codon. The component-wise multiplication of x_A with a gives the predicted proportion of the bases for primer A. The sum of the predicted frequency of the bases in each position may differ from one as the proportion of that primer may differ from the expected proportion (p_A); however, the sums of the predicted base frequencies for each position in a primer mix should be the same, therefore making it a good indicator of the accuracy of the fit. Finally the predicted proportion of amino acids can be also predicted. This is done first by calculating the frequency of each codon assuming no covariance. Namely, the codon proportion is found by multiplying the base frequencies at each position with each other, or more technically two tensor products between the three vectors of base frequencies (e.g., a_1 in the deconvolution problem above) arranged as horizontal, vertical, and stacked, resulting in a $4 \times 4 \times 4$ tensor. Lastly, the frequencies of codons encoding the same amino acid are summed together. One caveat is that this calculation assumes no covariance, that is, the neighboring positions are independent on each other, which may not be overly correct.

24. For example, if four MTPs are used for HT PCR and 1 μL template (e.g., 10 ng) is used per well, it means that you need 400 μL (4 μg) of template.

25. If bands are detected, proceed with template digestion. If not, PCR may require optimization by changing annealing temperature and/or number of cycles, among other potential factors.

26. It is important to optimize PCR conditions (at least from a set of samples) to reduce bias during library creation. The less bias introduced, the more variants will be found. The most expensive part of creating mutability landscapes is to sequence many variants until finding 19 unique sequences per randomized codon. In one MTP, libraries from two to three codons could be included for DNA sequencing. In the former case, 48 colonies could be analyzed, whereas 32 would suffice in the latter format. About 10–19 mutants could be likely found in either scenario. While this might not be desired in some projects, in others it might be sufficient. Experience tells that at least ten variants can be found in the first 30–48 samples analyzed.

27. If the gene is too long to cover it with a single primer, another sequencing primer could be included if it is necessary. If the parameter in quest increases significantly, e.g., tenfold activity, the result could be the introduction of an additional mutation at a distal site within or even outside the gene. In such cases, it is important to confirm that only a single mutation is responsible of the effect observed to avoid making wrong conclusions.

28. Once all the target variants are found, each one can be placed in a unique well to explore sequence–function relationships with the lowest screening effort. If 19 variants are created for each of five codons, all can be arranged in 95 wells, living one for the control template without mutations if using 96-well MTPs.

29. We have found that for controlling regio- and diastereoselective hydroxylation of steroids, it is important to find the right

amino acid at the right position. In some cases, having equivalent amino acids is useful (e.g., M vs. L, F vs. W, E vs. D, and S vs. T). However, in other cases it appears that only a particular side chain is necessary to improve the fitness trait under study. For this reason, we find important to construct mutability landscapes when subtle changes in sequence have a profound change in selectivity and/or activity.

30. The selectivity toward the predominant product in the wild-type enzyme (15-β-hydroxytestosterone) dominates over the others in most variants, and therefore in order to accentuate the differences between the mutants the data was scaled, shifted, and gamma-corrected so that the minimum value, the median, and the maximum were converted to 0, 0.5, and 1, respectively, using Eq. 2. The Matlab script for Figs. 12, 13, and 14 can be found at: https://raw.githubusercontent.com/matteoferla/mutational_scanning/master/P450_analysis.m

$$\xi = \left(\frac{x - \min(x)}{\max(x) - \min(x)} \right)^{\gamma} \quad \text{where } \gamma = \frac{\log(0.5)}{\log(\text{median}(x))} \quad (2)$$

31. A ternary plot is a barycentric plot where each sample is represented by three variables which sum to 1. If there were only two different enzymatic products, a 1D plot could be used (akin to a timeline), but generally a 2D plot would be used, such as a scatter plot with selectivity against activity. When there are three possible different products a ternary plot can be used. If there were four different products, it could either be plotted on a tetrahedral plot, which is the 3D extension of a ternary plot, or on a 2D plot with square-shaped axes. Beyond that, collapsing dimensions is the only alternative. Additional dimensions can be encoded with data-point size or color (as done in this example).

32. Many possible clustering schemes are commonly used to reduce the size of large datasets. k-means clustering groups the data into a specified number (k) of different groups based on the distance from the centroid of each group, while a dendrogram is based on the pairwise distance between each element and in this network the lower branches can be grouped ("collapsed").

References

1. Molina-Espeja P, Vina-Gonzalez J, Gomez-Fernandez BJ et al (2016) Beyond the outer limits of nature by directed evolution. Biotechnol Adv 34:754–767

2. Bommarius AS (2015) Biocatalysis: a status report. Annu Rev Chem Biomol Eng 6:319–345

3. Denard CA, Ren H, Zhao H (2015) Improving and repurposing biocatalysts via directed evolution. Curr Opin Chem Biol 25:55–64

4. Nestl BM, Hammer SC, Nebel BA et al (2014) New generation of biocatalysts for organic synthesis. Angew Chem Int Ed 53:3070–3095

5. Reetz MT (2013) Biocatalysis in organic chemistry and biotechnology: past, present, and future. J Am Chem Soc 135:12480–12496

6. Goldsmith M, Tawfik DS (2012) Directed enzyme evolution: beyond the low-hanging fruit. Curr Opin Struct Biol 22:406–412

7. Bornscheuer UT, Huisman GW, Kazlauskas RJ et al (2012) Engineering the third wave of biocatalysis. Nature 485:185–194

8. Reetz MT (2011) Laboratory evolution of stereoselective enzymes: a prolific source of catalysts for asymmetric reactions. Angew Chem Int Ed 50:138–174

9. Turner NJ (2009) Directed evolution drives the next generation of biocatalysts. Nat Chem Biol 5:567–573

10. Jackel C, Kast P, Hilvert D (2008) Protein design by directed evolution. Annu Rev Biophys 37:153–173

11. Xiao H, Bao Z, Zhao H (2015) High throughput screening and selection methods for directed enzyme evolution. Ind Eng Chem Res 54:4011–4020

12. Tizei PA, Csibra E, Torres L et al (2016) Selection platforms for directed evolution in synthetic biology. Biochem Soc Trans 44:1165–1175

13. Baker M (2011) Protein engineering: navigating between chance and reason. Nat Methods 8:623–626

14. Yang G, Withers SG (2009) Ultrahigh-throughput FACS-based screening for directed enzyme evolution. ChemBioChem 10:2704–2715

15. Aharoni A, Griffiths AD, Tawfik DS (2005) High-throughput screens and selections of enzyme-encoding genes. Curr Opin Chem Biol 9:210–216

16. Acevedo-Rocha CG, Agudo R, Reetz MT (2014) Directed evolution of stereoselective enzymes based on genetic selection as opposed to screening systems. J Biotechnol 191:3–10

17. Tee KL, Wong TS (2013) Polishing the craft of genetic diversity creation in directed evolution. Biotechnol Adv 31:1707–1721

18. Acevedo-Rocha CG, Hoebenreich S, Reetz MT (2014) Iterative saturation mutagenesis: a powerful approach to engineer proteins by systematically simulating Darwinian evolution. Methods Mol Biol 1179:103–128

19. Reetz MT, Carballeira JD (2007) Iterative saturation mutagenesis (ISM) for rapid directed evolution of functional enzymes. Nat Protoc 2:891–903

20. Kille S, Zilly FE, Acevedo JP et al (2011) Regio- and stereoselectivity of P450-catalysed hydroxylation of steroids controlled by laboratory evolution. Nat Chem 3:738–743

21. Li G, Zhang H, Sun Z et al (2016) Multiparameter optimization in directed evolution: engineering thermostability, enantioselectivity, and activity of an epoxide hydrolase. ACS Catal 6:3679–3687

22. Wu S, Acevedo JP, Reetz MT (2010) Induced allostery in the directed evolution of an enantioselective Baeyer-Villiger monooxygenase. Proc Natl Acad Sci U S A 107:2775–2780

23. Steipe B, Schiller B, Pluckthun A et al (1994) Sequence statistics reliably predict stabilizing mutations in a protein domain. J Mol Biol 240:188–192

24. Jochens H, Bornscheuer UT (2010) Natural diversity to guide focused directed evolution. ChemBioChem 11:1861–1866

25. Steffen-Munsberg F, Vickers C, Kohls H et al (2015) Bioinformatic analysis of a PLP-dependent enzyme superfamily suitable for biocatalytic applications. Biotechnol Adv 33:566–604

26. Höhne M, Schätzle S, Jochens H et al (2010) Rational assignment of key motifs for function guides in silico enzyme identification. Nat Chem Biol 6:807–813

27. Sebestova E, Bendl J, Brezovsky J et al (2014) Computational tools for designing smart libraries. Methods Mol Biol 1179:291–314

28. Reetz MT, Kahakeaw D, Lohmer R (2008) Addressing the numbers problem in directed evolution. ChemBioChem 9:1797–1804

29. Acevedo-Rocha CG, Reetz MT (2014) Assembly of designed oligonucleotides: a useful tool in synthetic biology for creating high-quality combinatorial DNA libraries. Methods Mol Biol 1179:189–206

30. Parra LP, Agudo R, Reetz MT (2013) Directed evolution by using iterative saturation mutagenesis based on multiresidue sites. ChemBioChem 14:2301–2309

31. Sun Z, Lonsdale R, Ilie A et al (2016) Catalytic asymmetric reduction of difficult-to-reduce ketones: triple-code saturation mutagenesis of an alcohol dehydrogenase. ACS Catal 6:1598–1605

32. Sun Z, Lonsdale R, Wu L et al (2016) Structure-guided triple-code saturation mutagenesis: efficient tuning of the stereoselectivity

of an epoxide hydrolase. ACS Catal 6:1590–1597

33. Sun Z, Lonsdale R, Kong XD et al (2015) Reshaping an enzyme binding pocket for enhanced and inverted stereoselectivity: use of smallest amino acid alphabets in directed evolution. Angew Chem Int Ed 54:12410–12415

34. Sun Z, Wikmark Y, Backvall JE et al (2016) New concepts for increasing the efficiency in directed evolution of stereoselective enzymes. Chemistry 22:5046–5054

35. Nov Y (2012) When second best is good enough: another probabilistic look at saturation mutagenesis. Appl Environ Microbiol 78:258–262

36. Burks EA, Chen G, Georgiou G et al (1997) In vitro scanning saturation mutagenesis of an antibody binding pocket. Proc Natl Acad Sci U S A 94:412–417

37. Wu H, Beuerlein G, Nie Y et al (1998) Stepwise in vitro affinity maturation of Vitaxin, an $\alpha_v\beta_3$-specific humanized mAb. Proc Natl Acad Sci U S A 95:6037–6042

38. Davidson E, Doranz BJ (2014) A high-throughput shotgun mutagenesis approach to mapping B-cell antibody epitopes. Immunology 143:13–20

39. Rojas G, Tundidor Y, Infante YC (2014) High throughput functional epitope mapping: revisiting phage display platform to scan target antigen surface. MAbs 6:1368–1376

40. Baronio R, Danziger SA, Hall LV et al (2010) All-codon scanning identifies p53 cancer rescue mutations. Nucleic Acids Res 38:7079–7088

41. Fowler DM, Fields S (2014) Deep mutational scanning: a new style of protein science. Nat Methods 11:801–807

42. Kitzman JO, Starita LM, Lo RS et al (2015) Massively parallel single-amino-acid mutagenesis. Nat Methods 12:203–206

43. Dietrich JA, McKee AE, Keasling JD (2010) High-throughput metabolic engineering: advances in small-molecule screening and selection. Annu Rev Biochem 79:563–590

44. O'Donohue MJ, Kneale GG (1996) A method for introducing site-specific mutations using oligonucleotide primers and its application to site-saturation mutagenesis. Mol Biotechnol 6:179–189

45. DeSantis G, Wong K, Farwell B et al (2003) Creation of a productive, highly enantioselective nitrilase through gene site saturation mutagenesis (GSSM). J Am Chem Soc 125:11476–11477

46. Steffens DL, Williams JG (2007) Efficient site-directed saturation mutagenesis using degenerate oligonucleotides. J Biomol Tech 18:147–149

47. Gray KA, Richardson TH, Kretz K et al (2001) Rapid evolution of reversible denaturation and elevated melting temperature in a microbial haloalkane dehalogenase. Adv Synth Catal 343:607–617

48. Garrett JB, Kretz KA, O'Donoghue E et al (2004) Enhancing the thermal tolerance and gastric performance of a microbial phytase for use as a phosphate-mobilizing monogastric-feed supplement. Appl Environ Microbiol 70:3041–3046

49. Eggert T, Funke SA, Andexer JN et al (2008) Evolution of enantioselective Bacillus subtilis lipase. In: Lutz S, Bornscheuer UT (eds) Protein engineering handbook. Wiley-VCH, Weinheim

50. Palackal N, Brennan Y, Callen WN et al (2004) An evolutionary route to xylanase process fitness. Protein Sci 13:494–503

51. Dumon C, Varvak A, Wall MA et al (2008) Engineering hyperthermostability into a GH11 xylanase is mediated by subtle changes to protein structure. J Biol Chem 283:22557–22564

52. Acevedo-Rocha CG, Reetz MT, Nov Y (2015) Economical analysis of saturation mutagenesis experiments. Sci Rep 5:10654

53. Sullivan B, Walton AZ, Stewart JD (2013) Library construction and evaluation for site saturation mutagenesis. Enzyme Microb Technol 53:70–77

54. Patrick WM, Firth AE (2005) Strategies and computational tools for improving randomized protein libraries. Biomol Eng 22:105–112

55. Firth AE, Patrick WM (2008) GLUE-IT and PEDEL-AA: new programmes for analyzing protein diversity in randomized libraries. Nucleic Acids Res 36:W281–W285

56. Kille S, Acevedo-Rocha CG, Parra LP et al (2013) Reducing codon redundancy and screening effort of combinatorial protein libraries created by saturation mutagenesis. ACS Synth Biol 2:83–92

57. Tang L, Gao H, Zhu X et al (2012) Construction of "small-intelligent" focused mutagenesis libraries using well-designed combinatorial degenerate primers. Biotechniques 52:149–158

58. Frauenkron-Machedjou VJ, Fulton A, Zhu L et al (2015) Towards understanding directed evolution: more than half of all amino acid positions contribute to ionic liquid resistance of Bacillus subtilis lipase A. ChemBioChem 16:937–945

59. Fulton A, Frauenkron-Machedjou VJ, Skoczinski P et al (2015) Exploring the protein stability landscape: *Bacillus subtilis* lipase A as a model for detergent tolerance. ChemBioChem 16:930–936

60. Nov Y, Fulton A, Jaeger KE (2013) Optimal scanning of all single-point mutants of a protein. J Comput Biol 20:990–997

61. Mingo J, Erramuzpe A, Luna S et al (2016) One-tube-only standardized site-directed mutagenesis: an alternative approach to generate amino acid substitution collections. PLoS One 11:e0160972

62. van der Meer JY, Biewenga L, Poelarends GJ (2016) The generation and exploitation of protein mutability landscapes for enzyme engineering. ChemBioChem 17:1792–1799

63. Hecht M, Bromberg Y, Rost B (2013) News from the protein mutability landscape. J Mol Biol 425:3937–3948

64. van der Meer JY, Poddar H, Baas BJ et al (2016) Using mutability landscapes of a promiscuous tautomerase to guide the engineering of enantioselective Michaelases. Nat Commun 7:10911

65. Mosquna A, Peterson FC, Park SY et al (2011) Potent and selective activation of abscisic acid receptors in vivo by mutational stabilization of their agonist-bound conformation. Proc Natl Acad Sci U S A 108:20838–20843

66. Park SY, Peterson FC, Mosquna A et al (2015) Agrochemical control of plant water use using engineered abscisic acid receptors. Nature 520:545–548

67. Cunningham BC, Wells JA (1989) High-resolution epitope mapping of hGH-receptor interactions by alanine-scanning mutagenesis. Science 244:1081–1085

68. Kanaya E, Kanaya S, Kikuchi M (1990) Introduction of a non-native disulfide bridge to human lysozyme by cysteine scanning mutagenesis. Biochem Biophys Res Commun 173:1194–1199

69. Sharp LL, Zhou J, Blair DF (1995) Features of MotA proton channel structure revealed by tryptophan-scanning mutagenesis. Proc Natl Acad Sci U S A 92:7946–7950

70. Weinglass AB, Smirnova IN, Kaback HR (2001) Engineering conformational flexibility in the lactose permease of *Escherichia coli*: use of glycine-scanning mutagenesis to rescue mutant Glu325–>Asp. Biochemistry 40:769–776

71. Weis R, Gaisberger R, Gruber K et al (2007) Serine scanning: a tool to prove the consequences of *N*-glycosylation of proteins. J Biotechnol 129:50–61

72. Schulman BA, Kim PS (1996) Proline scanning mutagenesis of a molten globule reveals non-cooperative formation of a protein's overall topology. Nat Struct Biol 3:682–687

73. Kuhn DM, Sadidi M, Liu X et al (2002) Peroxynitrite-induced nitration of tyrosine hydroxylase: identification of tyrosines 423, 428, and 432 as sites of modification by matrix-assisted laser desorption ionization time-of-flight mass spectrometry and tyrosine-scanning mutagenesis. J Biol Chem 277:14336–14342

74. Leitolf H, Tong KP, Grossmann M et al (2000) Bioengineering of human thyrotropin superactive analogs by site-directed "lysine-scanning" mutagenesis. Cooperative effects between peripheral loops. J Biol Chem 275:27457–27465

75. Romero PA, Tran TM, Abate AR (2015) Dissecting enzyme function with microfluidic-based deep mutational scanning. Proc Natl Acad Sci U S A 112:7159–7164

76. Wrenbeck EE, Klesmith JR, Stapleton JA et al (2016) Plasmid-based one-pot saturation mutagenesis. Nat Methods 13:928–930

77. Benson DA, Karsch-Mizrachi I, Lipman DJ et al (2008) GenBank. Nucleic Acids Res 36: D25–D30

78. Sun D, Ostermaier MK, Heydenreich FM et al (2013) AAscan, PCRdesign and MutantChecker: a suite of programs for primer design and sequence analysis for high-throughput scanning mutagenesis. PLoS One 8:e78878

79. Xia Y, Chu W, Qi Q et al (2015) New insights into the QuikChange process guide the use of Phusion DNA polymerase for site-directed mutagenesis. Nucleic Acids Res 43:e12

80. Krauss U, Jaeger KE, Eggert T (2010) Rapid sequence scanning mutagenesis using in silico oligo design and the Megaprimer PCR of whole plasmid method (MegaWHOP). Methods Mol Biol 634:127–135

81. Sanchis J, Fernandez L, Carballeira JD et al (2008) Improved PCR method for the creation of saturation mutagenesis libraries in directed evolution: application to difficult-to-amplify templates. Appl Microbiol Biotechnol 81:387–397

82. Ostermaier MK, Peterhans C, Jaussi R et al (2014) Functional map of arrestin-1 at single amino acid resolution. Proc Natl Acad Sci U S A 111:1825–1830

83. Sun D, Flock T, Deupi X et al (2015) Probing Galphai1 protein activation at single-amino acid resolution. Nat Struct Mol Biol 22:686–694

84. Pommie C, Levadoux S, Sabatier R et al (2004) IMGT standardized criteria for statistical analysis of immunoglobulin V-REGION amino acid properties. J Mol Recognit 17:17–32

85. Behrendorff JB, Johnston WA, Gillam EM (2014) Restriction enzyme-mediated DNA family shuffling. Methods Mol Biol 1179:175–187

86. Currin A, Swainston N, Day PJ et al (2017) SpeedyGenes: exploiting an improved gene synthesis method for the efficient production of synthetic protein libraries for directed evolution. Methods Mol Biol 1472:63–78

87. Whitehouse CJ, Bell SG, Wong LL (2012) P450(BM3) (CYP102A1): connecting the dots. Chem Soc Rev 41:1218–1260

88. Haines DC, Hegde A, Chen B et al (2011) A single active-site mutation of P450BM-3 dramatically enhances substrate binding and rate of product formation. Biochemistry 50:8333–8341

89. Reetz MT (2013) The importance of additive and non-additive mutational effects in protein engineering. Angew Chem Int Ed 52:2658–2666

90. Gaunt TR, Hinks LJ, Rassoulian H et al (2003) Manual 768 or 384 well microplate gel 'dry' electrophoresis for PCR checking and SNP genotyping. Nucleic Acids Res 31:e48

Part II

Protein Library Expression

Chapter 7

A Brief Guide to the High-Throughput Expression of Directed Evolution Libraries

Ana Luísa Ribeiro, Mario Mencía, and Aurelio Hidalgo

Abstract

The process of protein production optimization requires time and labor, constituting one of the main bottlenecks for the downstream utilization of the proteins. However, once through this bottleneck, the protein production process can be easily standardized and multiplexed to find the fittest variants in large libraries created by random mutagenesis. In this chapter, we present an overview of the most important choices to achieve homogeneous and functional expression of directed evolution libraries in microplate format: (1) choice of induction system and host strain, (2) choice of media and growth conditions, and (3) modifications to the genetic sequence.

Key words High-throughput recombinant expression, *E. coli*, Host strains, Autoinduction medium, Solubilization tags, Affinity purification

1 Introduction

Directed evolution is often described as an iterative algorithm consisting of alternating phases of genetic diversification and selection or screening of the fittest individuals. However, the most carefully designed library and directed evolution campaign may fail to find the fittest individual if the library members are not functionally expressed in soluble form. Consequently, time and labor must be dedicated to the process of protein production optimization for the parental enzyme, to ensure an easy standardization and implementation in high-throughput format.

This chapter does not pretend to give an infallible and universal method for functional expression, as there is none. However, it is intended as an overview of the most important factors that may hamper functional expression and their corresponding fixes. Furthermore, we focus on the practical case of achieving homogeneous expression in microplate (MTP) format of the maximal number of variants instead of traditional flask cultivation, which may also condition some of our suggestions.

Uwe T. Bornscheuer and Matthias Höhne (eds.), *Protein Engineering: Methods and Protocols*, Methods in Molecular Biology, vol. 1685, DOI 10.1007/978-1-4939-7366-8_7, © Springer Science+Business Media LLC 2018

For practical purposes, we cover three possible strategies to improve functional expression of proteins using *Escherichia coli* as host, ranging from the macroscopic to the genetic level: (1) choice of induction system and host strain, (2) choice of media and growth conditions, and (3) modifications to the genetic sequence.

1.1 Choice of Induction System and Host Strain

While, in principle, a protein should be ideally expressed in its natural host organism for maximal activity, the gold standard from economical and practical points of view is the expression in the bacteria *E. coli* [1, 2]. This does not imply that the expression in many other systems is not possible, but the facts simply remain that *E. coli* is the best known organism, easy, fast and cheap to grow, easy to lyse, and with streamlined genetic manipulation methods. However, despite, all these advantages, the major drawbacks of *E. coli* as an expression system include the inability to perform many of the post-translational modifications found in eukaryotic proteins, the lack of a secretion mechanism for the efficient release of protein into the culture medium, and the limited ability to facilitate extensive disulfide bond formation [3].

Expression vectors should incorporate specific features for protein production in the selected host, i.e., transcriptional promoters, replicon, or antibiotic-resistance markers. The protein purification strategy should also be kept in mind when selecting the vector since solubility and/or affinity fusion tags, and sequences to direct the protein synthesis to the cytoplasm or periplasm might be necessary [4, 5]. From the many *E. coli* promoters tested for induced protein expression, the $\Phi10$ promoter of the T7 bacteriophage is the most widely used, because it can be tightly controlled as it is solely read by the RNA polymerase of that bacteriophage and not by the *E. coli* RNA polymerase. This system requires that the T7 RNA polymerase gene is expressed in turn from an IPTG inducible cellular promoter, normally inserted in the genome. Expression systems based on pET vectors [6, 7] are by far the most commonly used for recombinant expression in *E. coli*. Aiming to provide a parallel cloning of the target gene and an easy protein purification method, all the generated vectors should contain the same T7 promotor sequence, insertion site and antibiotic resistance. Affinity tags with a protease cleavage site are fused to all proteins to facilitate their purification or capture.

Considering the above-mentioned shortcomings of *E. coli* and assuming the use of the T7 promoter for recombinant expression, Table 1 compiles possible fixes for almost every situation using variant strains of *E. coli* developed to optimize the expression or folding of different types of protein of interest.

Table 1
Overview of the most frequently used *E. coli* expression strains

E. coli strain	Use when...	Characteristics	Ref
BL21(DE3) (Novagen)	General use	• T7 RNA polymerase is encoded in the genome • Deletion mutant of the Lon and Omp proteases to increase the yield of recombinant protein. However, producing high levels of expressed protein can lead to toxicity and/or insolubility	[8]
BL21(DE3) pLysS/ pLysE (Novagen) Lemo (NEB) KRX (Promega)	Recombinant protein is toxic	• Plasmids pLysS or pLysE express the phage T7 lysozyme upon induction, which inhibits the basally expressed T7 RNA polymerase, so the protein of interest is only obtained after full induction of the system • Two independent regulations: arabinose for the T7 RNA polymerase and IPTG for the promoter of the protein of interest • Two independent regulations: rhamnose for the T7 RNA polymerase and IPTG for the promoter of the protein of interest	[9] [10]
BL21(DE3) Tuner (Novagen)	The amount of induced protein needs to be finely adjusted	• Entrance of IPTG into the cell has been enhanced to prove proportional induction to the IPTG added within a given range	
BL21(DE3) C41, C43	Recombinant protein is a membrane protein or has low solubility	• The cytoplasmic membrane shows involutions and foldings to accommodate membrane proteins produced with the T7 system	[11]
BL21(DE3) Rosetta (Novagen)	Codon usage of the original source is very different from that of *E. coli*	• Plasmid pRARE carrying tRNA genes to read 10 codons has been engineered and included in this strain	[12]
BL21(DE3) Origami (Novagen) SHuffle	Recombinant protein contains disulfide bridges important for protein folding or function	• Thioredoxin reductase (*trxB*) and glutathione reductase (*gor*) genes are mutated to achieve oxidizing cytoplasm • Also available with pRARE as Rosetta-gami • DsbC is overexpressed	[13]
SoluBL21 (DE3) (Gelantis) Arctic Express DE3 (Agilent).	Recombinant protein needs to be expressed at temperatures below 37 °C	• Specifically selected, after random mutagenesis, for high expression of low solubility proteins • Growth in M9 minimal medium is recommended to maximize the solubility of proteins • Produces the Cpn10 and Cpn60 chaperonins from *Oleispira antarctica* • Growth in M9 minimal medium is recommended to maximize the solubility of proteins	[14]

*1.2 Choice of
Medium and Growth
Conditions*

Once the issues of post-translational modifications, codon usage and disulfide bonds have been addressed (Table 1), "only" the issue of proper folding remains. However, whereas all other difficulties can be rationalized and subsequently countered, protein folding largely remains a black box, subject to empirical test. The physiological parameters of native protein production are exacerbated when the cell machinery is "hijacked" and a recombinant protein is expressed under the control of a strong promoter, producing proteins at a much faster rate than the native. As a result, this excess of protein may not be able to fold quickly enough, and it will either be degraded or it will aggregate to reach a more thermodynamically favored conformation (inclusion bodies), which shields hydrophobic areas from being exposed to a hydrophilic medium [15], and neither of the two outcomes is desired.

Misfolding of recombinantly expressed proteins will be mitigated if the rate of synthesis is decreased, and this is mostly achieved through lowering the cultivation temperature to 12–25 °C. Under these conditions, the rate of protein synthesis will decrease notably while the rate of protein folding is hypothesized to be affected only slightly. This will provide sufficient time for protein folding, yielding active proteins and avoiding the formation of inactive protein aggregates. One further advantage of cold production is the concomitant synthesis of a cold-shock protein, which is usually better than heat-shock proteins to promote folding. However, one disadvantage of cold induction is that longer times will be needed in order to achieve yields similar to those obtained at 37 °C.

Differences in lag time or growth rate, such as those seen in cultivation in microplate format, typically generate a situation where different cultures will be ready for induction at different times. Even if cultures could be read simultaneously in a plate reader, considerable effort would be required to follow growth and add inducer to each culture at the proper time. If all of the cultures were collected at once, choosing a collection time when all had been induced to optimal levels and none had suffered overgrowth by cells incapable of expressing target protein might be difficult or impossible [16]. Autoinduction media solve this problem because the uptake of the inducer depends on the metabolic state of bacteria.

Autoinduction media take advantage of diauxic growth phenomena: this is the ability to prioritize one carbon source over another, due to the fact that utilization of the preferred carbon source represses transcription of the machinery needed to utilize the secondary source. Two of the most popular promoters, P*lac* and P*ara*, are based on utilization of secondary carbon sources and can be successfully repressed by the presence of glucose, while other carbon sources, e.g., glycerol have no effect [16]. Autoinduction media are supplemented with sugars to achieve higher cell densities, Mg^{2+} and phosphate and sometimes succinate is included to buffer the acid metabolites produced. In further detail, autoinduction medium contains glucose, lactose and glycerol as carbon sources, which will be used sequentially. Glucose is used as a preferential carbon source during the initial growth period, keeping recombinant expression low due to catabolite repression of the *lac* or *ara* operons. When glucose is depleted, catabolite repression is lifted and a small amount of lactose may enter through constitutively expressed transporter molecules [17], converted into allolactose, which in turn induces the lac operon, while growth proceeds on lactose and glycerol.

From the perspective of high-throughput protein production for structural genomics or directed evolution, autoinduction is more convenient than IPTG induction because the expression strain is simply inoculated into autoinducing medium and grown to saturation without the need to follow culture growth and add inducer at the proper time. Furthermore, the culture density and concentration of target protein per volume of culture are typically considerably higher concentrations of target protein per mL of culture. This stability of autoinduced cultures at or near saturation, together with the relative uniformity of the inoculating cultures grown to saturation in non-inducing media, makes autoinducing media very convenient for enhancing solubility and uniformity in high-throughput expression setups.

1.3 Modifications in the Genetic Sequence

When all of the above strategies fail to achieve soluble expression, the gene sequence itself may be modified to improve yield and solubility through three different strategies: adjusting the codon usage, improving stability of the transcript and appending additional sequences to promote solubilization (and streamline purification).

With decreasing prices, custom gene synthesis has become affordable for many laboratories. Among the services available, companies offer sequence optimization to address codon bias by using synonymous codons, easily adapting the codon usage of the original organism to the codon usage of the recombinant host of choice, usually *E. coli*. Thus, this strategy is equivalent if not better than the use of rare tRNA-encoding strains, such as Rosetta or CodonPlus (Table 1).

At the same time that codon usage is optimized, the resulting sequence aims to improve the secondary structure of the mRNA, which contributes to stability and half-life of the messenger, and therefore increases the yield of recombinant protein. A number of other important factors, such as splicing sites, mRNA destabilizing motifs, and repeats play a crucial role in sequence optimization as well. Therefore, the companies providing this kind of service use advanced multiobjective optimization algorithms to improve all relevant features of a given sequence, thus maximizing protein expression.

The expression of a stretch of amino acids (peptide tag) or a large polypeptide (fusion partner) in tandem with the desired protein can be exploited to improve protein solubility and folding, facilitating protein purification or downstream processing and to increase production yields [18, 19]. Table 2 provides an overview on the most frequently used and interesting tag-protein fusion systems.

The exact mechanism(s) by which solubility enhancing fusion tags transfer that property to their partners remain unclear. Many researchers hypothesize that fusion proteins attract chaperones or have an intrinsic chaperone-like activity. It was also proposed that they could form micelle-like structures or inhibit protein aggregation by electrostatic repulsion [25, 44, 48–52]. When selecting which tag to use, one is faced with an enormous number of possibilities [18]. In most cases, the identification of an optimal tag for a given experiment requires significant trial and error [20], as fusion partners do not function equally with all target proteins, and each target protein can be differentially affected by several fusion tags [53]. Furthermore, when these tags are removed, the final solubility of the desired product is unpredictable [53].

Affinity tags display different size ranges from a single amino acid to entire proteins, and allow their partner to be selectively captured and purified through association with a tag-specific affinity resin [20]. They allow for the purification of virtually any protein without any prior knowledge of its biochemical properties. In fact, HTS would not be feasible if specific purification procedures were to be developed for each individual protein. The use of affinity tags enables different proteins to be purifed using a common method as opposed to highly customized procedures used in conventional chromatographic purification [54].

Affinity and solubility enhancing tags may represent a disadvantage for the projected application of the protein [54, 55]. The removal of fusion partners is usually made by specific protease sites included between the fusion tag and the target protein. By placing an affinity tag at the N terminus of the fusion partner, cleavage can be conducted either in solution, following purification, or immediately after enzyme capture, on the chromatography resin itself (in situ) [56]. Many different protease cleavage enzymes are commercially available, such as enterokinase, factor Xa, thrombin, PreScission™ protease, and tobacco etch virus (TEV) protease, among others [57]. The main limitations in the use of proteases are related to the tag incompatibility under certain cleavage conditions (e.g., buffers and temperature), thus influencing its operation, and accessibility to the cleavage site [58]. For many groups PreScission™ and TEV appear to offer the best solutions with regard to minimal non-specific secondary effects [56].

Table 2
Overview of the most frequently used tag-protein fusion systems

Tag	Use for...	Characteristics	Ref
His-Tag	Affinity purification	• Consists of 6–10 histidines in tandem and can reversibly interact with positively charged metal ions (most commonly Ni^{2+} or Co^{2+}) immobilized in a metal chelate matrix • Tendency for contaminant proteins with external histidine residues to co-purify with His-tagged targets • Tolerance to N- or C-terminal insertion must be determined experimentally	[20–22]
Strep-TagII (Strep-Tactin)	Affinity purification	• Formed by eight amino acids (WSHPQFEK) • Exhibits intrinsic affinity towards an engineered variant of streptavidin (streptactin) • Optimal for the purification of intact protein complexes, even if only one subunit carries the tag • More efficient with two tandem copies of the tag	[23, 24]
MBP (maltose-binding protein)	Solubility enhancement and affinity purification	• MBP is a 42 kDa periplasmic and highly soluble protein of *E. coli* engaged in the transport of maltose and maltodextrins across the cytoplasmic membrane • Promotes target protein solubility by showing intrinsic chaperone activity • MBP fusion proteins can bind strongly to cross-linked amylose resins, and elution can be carried out under mild conditions with free maltose	[25–28]
GST (glutathione-S-transferase)	Solubility enhancement and affinity purification	• GST is a 26 kDa protein from *Schistosoma japonicum* and catalyzes the reaction between a nucleophile, reduced glutathione, and electrophilic compounds • GST tag can bind tightly to glutathione resins, and reduced glutathione can be employed as a competitive agent for elution • Reducing conditions must be guaranteed to avoid oxidative aggregation due to the existence of four cysteine residues exposed at the surface of the GST tag • In many cases, the target protein precipitates after cleavage of the fusion	[29–34]
Trx (Thioredoxin A)	Solubility enhancement	• Trx is a small 12 kDa *E. coli* thermostable oxidoreductase that facilitates the reduction of other proteins • *trxA* translates very efficiently and contributes to high yields of soluble protein • The fusion partner Trx is more effective at the N-terminus	[35–39]

(continued)

Table 2
(continued)

Tag	Use for...	Characteristics	Ref
NusA (N-utilization substance A)	Solubility enhancement	• NusA is a 55 kDa transcription elongation and anti-termination factor of *E. coli* • It is intrinsically soluble and slows down translation at the transcriptional pauses, offering more time for protein folding	[40–43]
SUMO (small ubiquitin-like modifier protein)	Solubility enhancement	• SUMO is a yeast protein involved in post-translational modification • It promotes the proper folding and solubility of its target proteins possibly by exerting chaperoning effects in a similar mechanism to the described for its structural homolog ubiquitin • It generates a native N-terminus for the target protein after cleavage with the SUMO protease	[41, 44–47]

2   Materials

2.1   Biological Agents, Chemicals, and Labware

- Competent *E. coli* BL21(DE3) or any other suitable strain from Table 1.

- Directed evolution library cloned in vector under the control of the T7 RNA polymerase promoter.

- SOC medium: 0.5% yeast extract, 2% tryptone, 10 mM NaCl, 2.5 mM KCl, 10 mM $MgSO_4$, 10 mM $MgCl_2$ and 20 mM glucose. Prepare a solution containing the first four reagents, autoclave and let it cool down. Add $MgCl_2$ and glucose previously sterilized by passing it through a 0.2 μm filter.

- Sterile toothpicks.

- 96-Well microplates, sterile (ranging from 300 to 2000 μL capacity/well).

- Breathable adhesive film, e.g., Breathable Sealing Tape (Corning).

- Luria–Bertani lysogeny broth (LB): dissolve 10 g tryptone, 5 g yeast extract, and 10 g NaCl in ~950 mL ultrapure water. Adjust the pH of the medium to 7.0 using 2 M NaOH. Set the volume to 1 L with water. Dispense into bottles and autoclave.

- LB agar: Add 15 g of agar to 1 L of LB medium before autoclaving. Preparation of agar plates: after autoclaving allow the LB agar bottles to cool to approximately 45 °C. Add the required antibiotics and mix well. Pour onto plates immediately.

- Autoinduction medium: 928 mL ZY medium, 500 μL 2 M MgSO$_4$, 20 mL 50× 5052, 50 mL 20× NPS, in that order.
- ZY medium: dissolve 10 g of N-Z-amine AS (or any tryptic digest of casein, e.g., tryptone) and 5 g of yeast extract in 925 mL of water and autoclave.
- 2 M MgSO$_4$ stock: dissolve 49.3 g of MgSO$_4$·7H$_2$O in water to a final volume of 100 mL. Autoclave.
- 5052 50× stock: in a beaker place 250 g of glycerol. Then, add 730 mL water, and while stirring, add 25 g of glucose and 100 g of α-lactose. Lactose dissolves slowly; stirring over low heat will fasten the process. Autoclave once dissolved.
- NPS 20× stock: to 900 mL of water, add (in the following order) 66 g of (NH$_4$)$_2$SO$_4$, 136 g of KH$_2$PO$_4$, and 142 g of Na$_2$HPO$_4$. Stir until dissolved, then autoclave.

2.2 Equipment

- Microplate incubator (e.g., Titramax, Heidolph, Germany).
- Multichannel pipettes (e.g., Eppendorf, Germany).
- Benchtop centrifuge with capacity for microplates (e.g., 5810-R, Eppendorf Germany).
- Thermoshaker (e.g., Thermomixer Comfort, Eppendorf, Germany).

3   Methods

3.1 Transformation of Libraries

1. Thaw a number of aliquots of chemically competent BL21 (DE3) strain (or appropriate derivative, *see* Table 1) on ice.
2. Supplement with β-mercaptoethanol if required by the manufacturer's instructions.
3. Carefully pipette the adequate amount of DNA (following the manufacturer's instructions) on the surface of the cell suspension, tap the Eppendorf tube and incubate on ice for 30 min.
4. Heat shock for 30 s on a water bath or thermoshaker at 42 °C.
5. Incubate on ice for 5 min.
6. Add 500 μL of preheated SOC medium and transfer to a 10 mL tube.
7. Incubate in an orbital shaker for 1 h at 37 °C, or as directed by the manufacturer depending on the antibiotic used for transformant selection.
8. Plate a suitable amount of cell suspension on LB-agar plates supplemented with the required antibiotic for transformant selection.
9. Incubate overnight at 37 °C.

3.2 Preinoculum

1. Fill the microplates to half the capacity with LB medium supplemented with the required antibiotic.

2. Using sterile toothpicks, pick single colonies and deposit each one in a separate well.

3. Close the microplate with a sterile lid or a breathable adhesive film and incubate for 24 h at 37 °C in a microplate shaker (or until saturation).

3.3 Recombinant Expression

1. Fill each well of the production microplates to half the capacity with autoinduction medium supplemented with the required antibiotic (*see* **Notes 1–3**).

2. Using multichannel pipettes, transfer 1/100 of the total volume from the preinoculum plate to the production plate.

3. Close the microplate with a sterile lid and incubate for 24–36 h at 20 °C in a microplate shaker at 200 rpm (*see* **Note 4**).

4. Supplement the remainder of the preinoculum plate with 85% v/v glycerol to a final concentration of 20% v/v. Mix thoroughly and store at −80 °C as a master plate.

5. After 24–36 h centrifuge the production microplate for 10 min at 3000 ×g and 4 °C. Discard the supernatant by vigorously inverting the plate and freeze the pellet at −20 °C until further use.

4   Notes

1. The cell density at which autoinduction occurs can be effectively regulated through the concentration of glucose. Additionally, the cell density obtained at the end of that phase will influence the duration of the autoinduction period, as higher densities will consume lactose faster [59].

2. Oxygen concentration exerts an effect on the diauxic growth pattern depending on the promoter used, to the point of reverting the order of utilization of glycerol and lactose, thus delaying induction even further [59]. It is advisable to check that the recombinant protein was correctly expressed in sufficient amount by SDS-PAGE analysis of randomly selected wells.

3. In general, it is advisable to apply filling volumes of half the well capacity with regard to optimize the oxygen transfer. However, larger filling volumes may be applied in order to ensure that sufficient protein is obtained for further analysis. The resulting oxygen limitation can be mitigated by increasing incubation time [60]. Nevertheless, the concentration of 0.2% lactose chosen for autoinducing media seems likely to be high enough to induce full expression of target protein at almost any rate of aeration likely to be encountered [16].

4. It is important to note that autoinduction and saturation often occur at considerably higher density at 20 °C than at 37 °C (perhaps due to the higher solubility of oxygen at the lower temperature). Higher saturation densities combined with slower growth at 20 °C means that cultures may be quite dense after overnight incubation but not yet be induced, so care must be taken not to collect low-temperature cultures before they have saturated. The incubation time can be shortened by incubating at 37 °C for a few hours, until cultures become lightly turbid, and then transferring to 20 °C for autoinduction [16].

References

1. Graslund S, Nordlund P, Weigelt J et al (2008) Protein production and purification. Nat Methods 5:135–146

2. Demain AL, Vaishnav P (2009) Production of recombinant proteins by microbes and higher organisms. Biotechnol Adv 27:297–306

3. Makrides SC (1996) Strategies for achieving high-level expression of genes in *Escherichia coli*. Microbiol Rev 60:512–538

4. Sørensen HP, Mortensen KK (2005) Advanced genetic strategies for recombinant protein expression in *Escherichia coli*. J Biotechnol 115:113–128

5. Costa S, Almeida A, Castro A et al (2014) Fusion tags for protein solubility, purification and immunogenicity in *Escherichia coli*: the novel Fh8 system. Front Microbiol 5:63

6. Studier FW, Rosenberg AH, Dunn JJ et al (1990) Use of T7 RNA polymerase to direct expression of cloned genes. Methods Enzymol 185:60–89

7. Dubendorff JW, Studier FW (1991) Controlling basal expression in an inducible T7 expression system by blocking the target T7 promoter with lac repressor. J Mol Biol 219:45–59

8. Studier FW, Moffatt BA (1986) Use of bacteriophage T7 RNA polymerase to direct selective high-level expression of cloned genes. J Mol Biol 189:113–130

9. Zhang X, Studier FW (1997) Mechanism of inhibition of bacteriophage T7 RNA polymerase by T7 lysozyme. J Mol Biol 269:10–27

10. Wagner S, Klepsch MM, Schlegel S et al (2008) Tuning *Escherichia coli* for membrane protein overexpression. Proc Natl Acad Sci U S A 105:14371–14376

11. Miroux B, Walker JE (1996) Over-production of proteins in *Escherichia coli*: mutant hosts that allow synthesis of some membrane proteins and globular proteins at high levels. J Mol Biol 260:289–298

12. Novy R, Drott D, Yaeger K et al (2001) Overcoming the codon bias of E. *coli* for enhanced protein expression. Innov 12:1–3

13. Prinz WA, Aslund F, Holmgren A et al (1997) The role of the thioredoxin and glutaredoxin pathways in reducing protein disulfide bonds in the *Escherichia coli* cytoplasm. J Biol Chem 272:15661–15667

14. Ferrer M, Chernikova TN, Timmis KN, Golyshin PN (2004) Expression of a temperature-sensitive esterase in a novel chaperone-based *Escherichia coli* strain. Appl Environ Microbiol 70:4499–4504

15. Baneyx F, Mujacic M (2004) Recombinant protein folding and misfolding in *Escherichia coli*. Nat Biotechnol 22:1399–1408

16. Studier FW (2005) Protein production by auto-induction in high density shaking cultures. Protein Expr Purif 41:207–234

17. Deutscher J (2008) The mechanisms of carbon catabolite repression in bacteria. Curr Opin Microbiol 11:87–93

18. Correa A, Oppezzo P (2011) Tuning different expression parameters to achieve soluble recombinant proteins in E. *coli*: advantages of high-throughput screening. Biotechnol J 6:715–730

19. Nilsson J, Stahl S, Lundeberg J et al (1997) Affinity fusion strategies for detection, purification, and immobilization of recombinant proteins. Protein Expr Purif 11:1–16

20. Wood DW (2014) New trends and affinity tag designs for recombinant protein purification. Curr Opin Struct Biol 26:54–61

21. Correa A, Oppezzo P (2015) Overcoming the solubility problem in E. *coli*: available approaches for recombinant protein production. Methods Mol Biol 1258:27–44

22. Bolanos-Garcia VM, Davies OR (2006) Structural analysis and classification of native proteins from E. coli commonly co-purified by immobilised metal affinity chromatography. Biochim Biophys Acta 1760:1304–1313

23. Schmidt TG, Skerra A (2007) The Strep-tag system for one-step purification and high-affinity detection or capturing of proteins. Nat Protoc 2:1528–1535

24. Skerra A, Schmidt TG (2000) Use of the Strep-Tag and streptavidin for detection and purification of recombinant proteins. Methods Enzymol 326:271–304

25. Kapust RB, Waugh DS (1999) Escherichia coli maltose-binding protein is uncommonly effective at promoting the solubility of polypeptides to which it is fused. Protein Sci 8:1668–1674

26. Kellermann OK, Ferenci T (1982) Maltose-binding protein from Escherichia coli. Methods Enzymol 90:459–463

27. Nikaido H (1994) Maltose transport system of Escherichia coli: an ABC-type transporter. FEBS Lett 346:55–58

28. Bach H, Mazor Y, Shaky S et al (2001) Escherichia coli maltose-binding protein as a molecular chaperone for recombinant intracellular cytoplasmic single-chain antibodies. J Mol Biol 312:79–93

29. Smith DB, Johnson KS (1988) Single-step purification of polypeptides expressed in Escherichia coli as fusions with glutathione S-transferase. Gene 67:31–40

30. Kaplan W, Husler P, Klump H, Erhardt J, Sluis-Cremer N, Dirr H (1997) Conformational stability of pGEX-expressed Schistosoma japonicum glutathione S-transferase: a detoxification enzyme and fusion-protein affinity tag. Protein Sci 6:399–406

31. Malhotra A (2009) Tagging for protein expression. Methods Enzymol 463:239–258

32. Hammarstrom M, Hellgren N, van Den Berg S et al (2002) Rapid screening for improved solubility of small human proteins produced as fusion proteins in Escherichia coli. Protein Sci 11:313–321

33. Dyson MR, Shadbolt SP, Vincent KJ et al (2004) Production of soluble mammalian proteins in Escherichia coli: identification of protein features that correlate with successful expression. BMC Biotechnol 4:32

34. Boyer TD (1989) The glutathione S-transferases: an update. Hepatology 9:486–496

35. LaVallie ER, DiBlasio EA, Kovacic S et al (1993) A thioredoxin gene fusion expression system that circumvents inclusion body formation in the E. coli cytoplasm. Nat Biotechnol 11:187–193

36. LaVallie ER, Lu Z, Diblasio-Smith EA et al (2000) Thioredoxin as a fusion partner for production of soluble recombinant proteins in Escherichia coli. Methods Enzymol 326:322–340

37. Kim S, Lee SB (2008) Soluble expression of archaeal proteins in Escherichia coli by using fusion-partners. Protein Expr Purif 62:116–119

38. Terpe K (2003) Overview of tag protein fusions: from molecular and biochemical fundamentals to commercial systems. Appl Microbiol Biotechnol 60:523–533

39. Dummler A, Lawrence AM, de Marco A (2005) Simplified screening for the detection of soluble fusion constructs expressed in E. coli using a modular set of vectors. Microb Cell Fact 4:34

40. Gusarov I, Nudler E (2001) Control of intrinsic transcription termination by N and NusA: the basic mechanisms. Cell 107:437–449

41. Marblestone JG, Edavettal SC, Lim Y et al (2006) Comparison of SUMO fusion technology with traditional gene fusion systems: enhanced expression and solubility with SUMO. Protein Sci 15:182–189

42. De Marco V, Stier G, Blandin S et al (2004) The solubility and stability of recombinant proteins are increased by their fusion to NusA. Biochem Biophys Res Commun 322:766–771

43. Davis GD, Elisee C, Newham DM et al (1999) New fusion protein systems designed to give soluble expression in Escherichia coli. Biotechnol Bioeng 65:382–388

44. Butt TR, Edavettal SC, Hall JP (2005) SUMO fusion technology for difficult-to-express proteins. Protein Expr Purif 43:1–9

45. Li SJ, Hochstrasser M (1999) A new protease required for cell-cycle progression in yeast. Nature 398:246–251

46. Khorasanizadeh S, Peters ID, Roder H (1996) Evidence for a three-state model of protein folding from kinetic analysis of ubiquitin variants with altered core residues. Nat Struct Biol 3:193–205

47. Malakhov MP, Mattern MR, Malakhova OA et al (2004) SUMO fusions and SUMO-specific protease for efficient expression and purification of proteins. J Struct Funct Genomics 5:75–86

48. Nallamsetty S, Waugh DS (2007) Mutations that alter the equilibrium between open and closed conformations of Escherichia coli maltose-binding protein impede its ability to

enhance the solubility of passenger proteins. Biochem Biophys Res Commun 364:639–644

49. Douette P, Navet R, Gerkens P et al (2005) *Escherichia coli* fusion carrier proteins act as solubilizing agents for recombinant uncoupling protein 1 through interactions with GroEL. Biochem Biophys Res Commun 333:686–693

50. Fox JD, Kapust RB, Waugh DS (2001) Single amino acid substitutions on the surface of *Escherichia coli* maltose-binding protein can have a profound impact on the solubility of fusion proteins. Protein Sci 10:622–630

51. Zhang YB, Howitt J, McCorkle S et al (2004) Protein aggregation during overexpression limited by peptide extensions with large net negative charge. Protein Expr Purif 36:207–216

52. Su Y, Zou Z, Feng S et al (2007) The acidity of protein fusion partners predominantly determines the efficacy to improve the solubility of the target proteins expressed in *Escherichia coli*. J Biotechnol 129:373–382

53. Esposito D, Chatterjee DK (2006) Enhancement of soluble protein expression through the use of fusion tags. Curr Opin Biotechnol 17:353–358

54. Arnau J, Lauritzen C, Petersen GE, Pedersen J (2006) Current strategies for the use of affinity tags and tag removal for the purification of

recombinant proteins. Protein Expr Purif 48:1–13

55. Pina AS, Lowe CR, Roque AC (2014) Challenges and opportunities in the purification of recombinant tagged proteins. Biotechnol Adv 32:366–381

56. Koehn J, Hunt I (2009) High-throughput protein production (HTPP): a review of enabling technologies to expedite protein production. Methods Mol Biol 498:1–18

57. Young CL, Britton ZT, Robinson AS (2012) Recombinant protein expression and purification: a comprehensive review of affinity tags and microbial applications. Biotechnol J 7:620–634

58. Waugh DS (2011) An overview of enzymatic reagents for the removal of affinity tags. Protein Expr Purif 80:283–293

59. Blommel PG, Becker KJ, Duvnjak P et al (2007) Enhanced bacterial protein expression during auto-induction obtained by alteration of lac repressor dosage and medium composition. Biotechnol Prog 23:585–598

60. Kensy F, Engelbrecht C, Büchs J (2009) Scale-up from microtiter plate to laboratory fermenter: evaluation by online monitoring techniques of growth and protein expression in *Escherichia coli* and *Hansenula polymorpha* fermentations. Microb Cell Fact 8:68

Chapter 8

Library Growth and Protein Expression: Optimal and Reproducible Microtiter Plate Expression of Recombinant Enzymes in *E. coli* Using MTP Shakers

Sandy Schmidt, Mark Dörr, and Uwe T. Bornscheuer

Abstract

Escherichia coli (*E. coli*) as heterologous host enables the recombinant expression of the desired protein in high amounts. Nevertheless, the expression in such a host, especially by utilizing a strong induction system, can result in insoluble and/or inactive protein fractions (inclusion bodies). Furthermore, the expression of different enzyme variants often leads to a diverse growth behavior of the *E. coli* clones resulting in the identification of false-positives when screening a mutant library. Thus, we developed a protocol for an optimal and reproducible protein expression in microtiter plates showcased for the expression of the cyclohexanone monooxygenase (CHMO) from *Acinetobacter* sp. NCIMB 9871. By emerging this protocol, several parameters concerning the expression medium, the cultivation temperatures, shaking conditions as well as time and induction periods for CHMO were investigated. We employed a microtiter plate shaker with humidity and temperature control (Cytomat™) (integrated in a robotic platform) to obtain an even growth and expression over the plates. Our optimized protocol provides a comprehensive overview of the key factors influencing a reproducible protein expression and this should serve as basis for the adaptation to other enzyme classes.

Key words Reproducible protein expression, Microtiter plates, *Escherichia coli*, High-throughput, Screening, Library growth, Baeyer-Villiger monooxygenase, CHMO

1 Introduction

The reproducible expression of recombinant proteins, e.g., Baeyer-Villiger monooxygenases (BVMOs) and all other biotechnologically relevant enzymes has become a standard method, but still comprises challenging issue. This is not only highly important for the increased interest of such enzymes in the growing field of applications, but particular for the evolvement of the enzymes' properties using directed-evolution methods [1]. On the one hand, millions of gene sequences are constantly discovered, resulting also in numerous new protein families as recently shown [2]. These yet uncharacterized proteins potentially provide unexplored

Uwe T. Bornscheuer and Matthias Höhne (eds.), *Protein Engineering: Methods and Protocols*, Methods in Molecular Biology, vol. 1685, DOI 10.1007/978-1-4939-7366-8_8, © Springer Science+Business Media LLC 2018

enzymes harboring new activities, which can serve as valuable bio-catalysts for "green" production processes for the synthesis of bulk chemicals, pharmaceutical building blocks, new materials or nanos-tructured materials [3–6]. A mandatory prerequisite for the screening of such rich protein archives toward new enzymatic activities is a working expression system leading to high amounts of soluble protein for the functional characterization of the respective new activity. On the other hand, such enzymes can be altered and improved, respectively, by protein engineering concerning for example substrate specificity and stereoselectivity. Moreover, such approaches can also address the stability or product inhibition issue of already known and well-characterized enzymes to finally adapt these biocatalysts for industrial purposes [5, 7]. Such directed evolution experiments generate even more enzyme variants, which are often less solubly expressable than the wild-type counter-part. Thus, a high demand for reliable protein expression methods for mutant libraries arises too. To lower the screening effort and, most importantly, the expenditure of time to investigate such (often huge) libraries, the protein expression in 96-well microtiter plates (MTPs) is usually the method of choice. The protein expression in MTPs is often challenging relating to the investigated class of enzyme and the warranty of an equal expression level of each enzyme variant. Otherwise the unequal amount of enzyme can lead to an apparent higher activity, which is not due to the better performance of the enzyme variant, but a result of a varied (better) expression. Consequently this leads to the identification of false-positives, which are then only discovered and discarded after labo-rious follow-up experiments (expression on larger scale, purifica-tion, biochemical assays). When performing and comparing variations in different growth and expression experiments, we found that using a microtiter plate shaker with humidity and tem-perature control (Cytomat™) gave most reproducible results with smallest variations within one or multiple microtiter plates. Although such a device is an investment, we recommend its usage, especially for larger screening campaigns. It significantly helps to get uniform and reproducible expression pattern, as it provides a well-controlled environment for 32 standard 96-well plates regarding humidity, temperature, and shaking behavior. The protocol we describe in this chapter is designed for using this automated Cytomat™ shakers. They are included in our robotic platform [7], but can in principle also be operated separately.

We have chosen the cyclohexanone monooxygenase (CHMO) from *Acinetobacter* sp. NCIMB 9871 as "not-easy to handle" model enzyme concerning the expression level and stability. How-ever, this enzyme is well characterized in terms of substrate speci-ficity, analysis of substrate and product inhibition, comparison of different expression hosts, stability, analysis of oxygen supply, kinetic data and the analysis of different microscale processing techniques [8–17]. These studies pointed out the importance of the expression host as well as the chosen expression system. Both must be efficient but are known to be strongly influenced by different factors such as the expression temperature, time of induction, and total cultivation time. Although valuable, these studies do not always provide a clear, general and easy reproducible method for the optimal library growth and expression of the CHMO due to the variety of the investigated expression hosts, the different expression systems and the overall conditions used. Furthermore, most of the reported results were focused on the improvement of biotransformation for the conversion of a respective substrate based on *E. coli* whole-cells harboring overexpressed enzyme.

2 Materials

In general, most of the materials needed are commonly required reagents and solutions for enzyme library screening. The media and the solution for induction should be prepared under sterile conditions whereas it is not necessary to prepare the cell lysis solution sterile.

2.1 Chemicals, Solutions and Materials

1. Agar plates containing the clones from a transformed mutant library in *E. coli* (*see* **Notes 1** and **2**).

2. Lysogeny broth (LB) medium: 0.5% (w/v) yeast extract (5 g/L), 1% (w/v) tryptone (10 g/L), 1% (w/v) NaCl (10 g/L), sterilized by autoclaving.

3. Terrific broth (TB) medium: 2.4% (w/v) yeast extract (24 g/L), 1.2% (w/v) peptone/tryptone (12 g/L), 0.8% glycerol (8 g/L). Prepare separately 10× TB-salts: 0.72 M $K_2HPO_4 \times 3\ H_2O$ (164.4 g/L), 0.17 M KH_2PO_4 (23.2 g/L), both the medium and the salts are sterilized by autoclaving. For use add 100 mL of autoclaved 10× TB-salts to 900 mL of autoclaved TB medium.

4. Auto-induction medium (ZYP-5052) [18]: mixture of 20× ZYP salts, 20× ZYP sugars and ZYP medium. For 0.4 L ZYP salts: 54 g KH_2PO_4 (1 M, MW 136.09 g/mol), 143 g Na_2HPO_4 (1 M, MW 358 g/mol), 26.4 g $(NH_4)_2SO_4$ (0.5 M). For 0.4 L 20× ZYP sugars: 40 g glycerol (1.08 M, 10%), 4.4 g D-glucose-monohydrate (55.5 mM 1%), 16 g lactose X M (4%). For 50 mL 100× $MgSO_4$: 2.46 g $MgSO_4$ (200 mM). For 712 mL ZYP medium: 8 g tryptone, 4 g yeast extract. Every solution must be sterilized by autoclaving. Finally add 40 mL 20× ZYP salts, 40 mL 20× ZYP sugars and 8 mL 100× $MgSO_4$ to the 712 mL prepared medium.

5. 1000x Antibiotics stock solutions: 50 mg/mL kanamycin.

6. 1000x Inducer stock solution: Dissolve 2.62 g of isopropyl-beta-D-thiogalactoside (IPTG) (*see* **Note 1**) in 8 mL of Aq. dest. Adjust volume to 10 mL with Aq. dest to get a final IPTG concentration of 1.1 M. Filter sterilize with a 0.22 μm syringe filter.

7. Assay buffer: 50 mM sodium phosphate buffer, pH 8 (*see* **Note 3**).

8. Assay solution: 0.56 mM NADPH, 1.11 mM cyclohexanone in assay buffer. Add cyclohexanone as a 100× stock solution (10.9 mg/mL in DMF).

9. Cell lysis buffer: 1 mg/mL lysozyme, 1 μL/mL DNase I (from a 10 mg/mL stock solution) in assay buffer (*see* **Note 4**).

10. 60% Glycerol solution in distilled water (sterile).

11. 96-Well clear microtiter plates (flat bottom, with lid), sterile or treated for 5 min by UV.

12. AeraSeal™ Sealing Films for sealing 96-well MTPs.

13. Toothpicks (sterilized by autoclaving).

2.2 Equipment

1. Absorbance microplate reader with computer control and the corresponding software.

2. 96-Channel pipetting robot (if possible with a cooled station).

3. Handheld multichannel pipette (if possible with electronic control, capable of accurate multiple deliveries in μL range).

4. Autoclavable solution basin (sterile).

5. Microtiter plate shaker with humidity and temperature control (ideal: Cytomat™ 24 C Automated Incubator, Thermo Fisher Scientific, Waltham, Massachusetts, USA, *see* **Note 5**).

6. Centrifuge with rotor designed to spin 96-well plates.

7. 96-pin plate replicator, a 70% ethanol bath and a Bunsen burner or the like.

8. If available: a colony-picking robot instead of a 96-pin plate replicator.

3 Methods

When optimizing the enzyme expression in MTPs, important variables have to be considered including the growth conditions, lysis procedures, and concentrations and volumes of each used component (e.g., medium, inducer, and lysis solution). Thereby it should be taken into account for the assay used to determine the enzyme activities (kinetics or end-point, colorimetric assay, etc.), because this can influence the parameters for the enzyme expression protocol (*see* **Note 3**).

Before starting the screening of a mutant library, the suitable enzyme expression conditions should be determined in prescreens using 1–6 MTPs containing just the wild-type enzyme and blanks (Subheading 3.1).

The preparation of libraries for screening concerning the protein expression and cell lysis/protein extraction is described in the Subheadings 3.2 and 3.3 (see also Fig. 2).

3.1 Determining Suitable Expression Conditions

To determine suitable conditions for the best expression of the respective enzymes (in our case the CHMO), 1–6 MTPs containing just the wild-type enzyme (positive control) and negative controls should be tested with different media and inducer concentrations. As "good" negative controls serves the empty vector transformed in the respective *E. coli* strain and wells containing just the medium (blanks). Figure 1 recommends the distribution of the controls in a 96-well MTP (*see* **Note 6**). Also in the prescreens, enzyme activities should be measured for each plate and checked for reproducibility and cross contaminations. When starting with the establishment of the CHMO library screening using the standard expression protocol, nearly no enzyme activity was measurable. Although the cells showed a really good growth (during the bacterial growth as well as after induction), nearly no band corresponding to the CHMO could be detected in the SDS-PAGE analysis, which was made after cell lysis with lysozyme. In accordance with the low enzyme amount, no activity was measurable in the obtained cell lysate. Thus, it was necessary to improve the expression protocol for

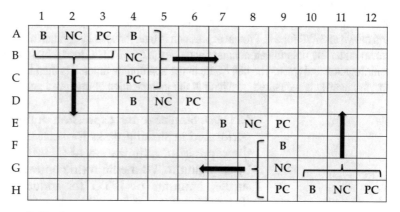

B: blank
NC: negative control
PC: positive control

Fig. 1 Representative scheme of a microtiter plate showing a suggestion for the distribution of the blanks (B, medium), negative controls (NC, empty vector) and positive controls (PC, wild-type enzyme). When screening a library of at least 5 MTPs, the "hiking" of the controls is indicated by the given *arrows* to cover each position of the plate

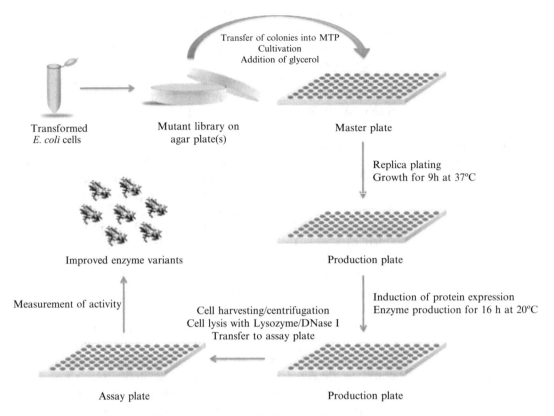

Fig. 2 Illustration of a typical library screening round. After transformation, the clones were plated and grown on LB agar plates. The colonies were transferred using sterile toothpicks (or a colony-picking robot) to a preculture MTP and cultured to saturation. Glycerol is added resulting in the master plate serving as stored back-up of the library (at $-80\ °C$). A new MTP is inoculated from the master plate(s) using a 96-pin replicator and grown at $37\ °C$ for 9 h. Protein expression is then induced by addition of the induction solution and further grown under the determined optimal conditions (e.g., at $20\ °C$ for 16 h). Afterward, cells were harvested by centrifugation and lysed. The cell lysate in the appropriate amount is then transferred into an assay plate and enzyme activity is measured, resulting in the identification of improved enzyme variants

CHMO. Similar to the expression optimization, which is normally performed in shaking flasks for such not well expressing enzymes, the expression optimization of CHMO in MTPs was investigated by changing to TB medium (in contrast to LB medium) and using smaller amounts of IPTG for induction of protein expression (0.1 mM instead of 1.0 mM). Moreover, the time for growth of the *E. coli* cells and the protein expression was prolonged to ensure a slow, but soluble enzyme expression. Additionally, the washing step after harvest of the cells was omitted to avoid that cells get lost.

For the determination of appropriate expression conditions follow these steps:

1. Prepare 6 MTPs (already sterile or treat them for 5 min under UV light) with LB, TB, and auto-induction medium (containing the respective antibiotic, *see* **Note 7**). In each case two

plates were filled with 180 μL LB, 180 μL TB, and 200 μL autoinduction medium. Follow the procedure for preparing a library for protein expression (Subheading 3.2) but inoculate your wells by picking wild-type enzyme from freshly transformed LB agar plates instead of the mutant library.

2. Seal the plates with AeraSeal™ Sealing Films and let the cultures grow at 37 °C for 9 h. When using auto-induction medium, let cultures grow for 20–24 h (*see* **Note 8**).

3. Prepare two different inducer solutions in the respective medium under sterile conditions, e.g., 11 mM or 1.1 mM IPTG in LB/TB by diluting the 1000× IPTG stock solution with LB or TB medium 1:100 and 1:1000.

4. After growth of the cultures for 9 h, add 20 μL of the respective inducer solution to the plates containing the LB/TB medium, respectively. One plate containing LB/TB, respectively, is than induced using a final IPTG concentration of 1 mM, whereas the other plate with LB/TB medium is induced with 0.1 mM IPTG.

5. After further growth/protein expression for 16 h cultures were harvested as described in Subheading 3.2.

6. Cell pellets are usually washed with buffer to get rid of remaining medium. This step can be omitted when recognizing a great loss of cells (check if remaining medium not disturbs the enzyme activity measurement).

7. Follow the procedure given in Subheading 3.3 for cell lysis.

8. Transfer all solutions necessary for the activity assay into a fresh plate, preferentially by using a pipetting robot (final volume in one well should be not more than 200 μL). For CHMO assay, mix 180 μL of assay solution with 20 μL of cell lysate and measure activities in the microplate reader. (*see* **Note 9**).

9. If the cell lysate contains too much enzyme for the respective activity measurement, e.g., when applying NADPH assay in case of CHMO, always consider a predilution of the lysate (*see* **Note 10**). Determine a suitable volume of lysate to use for screening activity.

10. Perform the initial activity assay of the entire wild-type enzyme containing plates using the adapted lysate volume/dilution (*see* **Note 11**).

11. Perform a SDS-PAGE analysis using two (or more) samples from each plate.

12. The best identified condition concerning the medium and the IPTG concentration leading to the highest amount of soluble enzyme should be used for investigation of further improvements by a prolonged growth/expression time and/or of the

cell lysis procedure using BugBuster® as described in Subheading 3.2 (if further improvements are necessary).

3.2 Preparation of Libraries for Screening—Protein Expression

1. By using sterile toothpicks (or if available a colony picker), pick colonies from your transformed mutant library into 96-well master culture plates. When using an *E. coli* expression system, the master plate is typically prepared with LB medium (e.g., 200 μL per well) supplemented by the respective antibiotic. Follow the distribution of the controls as depicted in Fig. 1 (*see* **Note 6**).

2. The master plate cultures were grown to saturation (until the stationary phase is reached, determined by OD_{600} measurements) in a humidified shaker. With a 96-pin plate replicator, these cultures are then used to inoculate another set of plates used for expression of the mutant library (*see* **Note 12**).

3. The master plates are stored in 20% glycerol at −80 °C to ensure a recovery of those mutants with improved properties (*see* **Note 13**).

4. The new set of cultures containing the library are grown and induced for protein expression using the best-identified conditions in the prescreen (e.g., enzyme production at 20 °C for 16–20 h).

5. After expression, pellet the cells by centrifugation for 10 min at $2200 \times g$ and 4 °C (*see* **Note 14**).

6. After centrifugation, the cell pellet is normally washed with 200 μL assay buffer (the buffer, which is also used for preparing the lysis solution, e.g., sodium phosphate buffer). Therefore, the cell pellets are resuspended by pipetting the solution ten times up and down using either a multichannel pipette or by the pipetting robot and centrifuged again for 10 min at $2200 \times g$ and 4 °C. If remaining medium does not disturb the activity assay, it is recommended to waive this step.

3.3 Preparation of Libraries for Screening—Cell Lysis/ Protein Extraction

The cell lysis can be accomplished in a number of possible ways. A common detergent is the BugBuster® Protein Extraction Reagent (Novagen), which releases the protein of interest by disrupting the bacterial cell wall. The manufacturer's protocol can be applied to microliter-scale suitable for disruption of cells in MTPs. However, it should be considered that a high amount of BugBuster® is necessary to screen large mutant libraries. In our hands, we could reduce the suggested amount of the BugBuster® reagent from 10% (v/v) to 1% (v/v) without losing lysis efficiency to save costs (For 90 € around 100 plates can be screened when using 100 μL lysis buffer with 1% BugBuster®). Secondly, such detergents can be harmful to the assay system (e.g., formation of bubbles can disturb absorbance measurements, *see* **Note 15**). Freeze-thaw cycles are also often used for disruption of cells. Nevertheless, for screening large mutant libraries we recommend using cell disruption by lysozyme/DNase I in terms of handling, costs, and effectiveness. The following protocol describes how to use lysozyme to disrupt the cell pellets from 96-well plate (also suitable for deep-well plates) cultures:

1. The pelleted cells (Subheading 3.2) can be frozen at −20 °C (*see* **Note 16**).

2. The cell pellets are resuspended in 100–200 μL of lysis buffer by a 96-channel pipetting robot.

3. Cell pellets are lysed for 3 h at 30 °C with gentle agitation.

4. After lysis, MTPs are centrifuged for 10–15 min at 2200 × g to pellet the cell debris and clarify the lysate. If possible, the usage of a cooled centrifuge (4 °C) is recommended.

5. The lysates should be kept cooled (on ice) and stored at 4 °C, but should be assayed as fast as possible to avoid enzyme inactivation.

4   Notes

1. In our study, we used a CHMO gene inserted in the pET28a (+) plasmid [5]. The plasmid contains a T7 promoter; gene expression can be induced with IPTG. Cells harboring the plasmid are selected by using kanamycin (50 μg/mL). In case you use other plasmids, adopt the antibiotic or inducer.

2. Typically, *E. coli* BL21 (DE3) as efficient expression strain is used. If it is necessary to screen a library for 95% coverage of every single amino acid substitution in saturation mutagenesis, it is recommended to transform the library by electrotransformation to achieve a high number of clones. Furthermore, also alternative *E. coli* strains should always be considered for the expression of the protein of interest, e.g., *E. coli* BL21 Star™ (DE3) pLysS (high mRNA stability results in increased protein yield, and therefore it is ideal for use with high copy number based T7 promoter plasmids showing low background expression in uninduced cells).

3. If you adapt this protocol for enzymes other then CHMO, choose a suitable assay buffer. Pay attention that in the assay buffer must be compatible with the conditions of the cell lysis step. DNase I requires divalent metal cations for activity. Normally, the metal ions present in the cell lysate are apparently sufficient to activate the DNaseI. Additional salts (e.g., $MgCl_2$) can be added to the lysis solution, but depending on the

applied assay after cell lysis this may cause complications in the screening.

4. We recommend using dialyzed and purified lysozyme and DNase I, respectively, to avoid contaminations by proteases. The lysis buffer should be freshly prepared in the assay buffer, which is later used for the enzymes' activity assay and stored at 4 °C (or on ice) until use.

5. It is very important that a sufficient humidity is ensured to prevent evaporation and thus dehydration of the cultures. Furthermore, the water in cultures in wells close to the edge of each plate evaporates faster than cultures in the center of the plates. This causes variations in the enzyme concentration across each plate resulting in a lowered reproducibility and increased identification of false-positives during the screen. Therefore, we highly recommend using a Cytomat™ due to the regulation capabilities in terms of temperature, humidity and stable CO_2 conditions. When using a microtiter plate shaker without humidity control, insert (if possible) the plates in a plastic box (or the like) with enough wet tissues and additionally put beakers with water inside of the shaker. Otherwise evaporation will be a huge problem.

6. It is highly recommended that these negative controls should be distributed over the 96-well MTP in a way that they statistically occur in each well of the plate when screening a whole library. For instance, when screening ≥ 5 MTPs, the blanks are hiking from well A1 in the first plate to well B2 in the second plate and so on. This is also true for the negative control (empty vector) as well as the positive control (wild type). Such controls are important to follow cross-contaminations, differences in the growth, which derives from the position on the plate and to control the activity assay.

7. When using auto-induction medium, adjust the final antibiotic concentration as following: kanamycin 100 μg/mL (high phosphate induces Kanamycin resistance, 100 μg/mL is sufficient when using media described in Subheading 2.1), ampicillin 50 μg/mL and chloramphenicol 35 μg/mL.

8. Auto-induction is a result of lactose in the medium. Glucose prevents induction by lactose. Adjusting glucose/lactose levels in the medium can regulate autoinduction. Suitable *E. coli* strains are: BL21 (DE3) (T7 polymerase present in chromosome), also compatible with B834 (DE3) and C41 (DE3). Not recommended are cell types expressing lysozyme (e.g., *pLysS*).

9. Each pipetting step (whether by hand or by using a pipetting robot) should be performed very carefully to avoid bubble formation as this can disturb the activity assay. Moreover, it is recommended to perform a shaking step after all reaction

components were mixed together (30–60 s, many photometers suitable for MTPs are able to shake). This can additionally be useful to get rid of formed bubbles.

10. When using the NAD(P)H assay to determine monooxygenase activity, the consumption of NAD(P)H (decrease in absorption) might be too fast leading to no analyzable slopes. In some cases, it is necessary to decrease the amount of cell lysate to obtain evaluable slopes. This can be done by predilution of the cell lysate and then subjection to the NAD(P)H assay using the same amounts and concentrations of buffer, substrate and NAD(P)H.

11. Make sure that the wild-type activity is consistent across the whole plate. The standard deviation in activity should be not more than 10–15% across the entire plate when using a Cytomat^{TM}. Check also the controls (blank and empty vector) if no activities are detectable.

12. Depending on the results of the expression optimization, you may have to use deep-well plates (1 mL or 2 mL total volume) for an increased culture volume (up to 1 mL).

13. The glycerol should be added by hand using a multichannel pipette. For an easier pipetting you can cut off a small part from the tip (around 3–4 mm) for a wider opening of the tip. In a sterile solution basin the sterile glycerol solution is filled (under a sterile bench). Ensure short distances from the wells on the MTP you would like to fill and the solution basin (the glycerol drops out of the tip).

14. Make sure that your centrifuge is suitable for the spinning of MTPs. Follow the manufacturer's specifications for the centrifuge and the used plates to avoid cracking the plates. In case of doubt, centrifuge in steps of 5 min at $\leq 2200 \times g$ and check the pellet formation in between.

15. The performance can be enhanced using Benzonase® nuclease and rLysozyme™ (Merck Millipore, Darmstadt, Germany) if necessary. Using the BugBuster® protocol, resuspension is problematic due to the repeated pipetting steps. This detergent will create air bubbles, thereby hindering the following assay procedure (absorbance-based as well as fluorescence-based assays). Therefore, repeating pipetting steps are not recommended for 96-well plate cultures. Resuspension can be achieved by shaking of the MTPs and/or by stirring the cell extract with the metal tips from the 96-pin replicator (do not forget the sterilization by 70% ethanol and heat in between each plate to avoid cross-contaminations).

16. Freezing the cell pellet first enhances the cell disruption efficiency.

References

1. Bornscheuer UT, Huisman GW, Kazlauskas RJ et al (2012) Engineering the third wave of biocatalysis. Nature 485:185–194

2. Venter JC, Remington K, Heidelberg JF et al (2004) Environmental genome shotgun sequencing of the Sargasso Sea. Science 304:66–74

3. Woodley JM (2008) New opportunities for biocatalysis: making pharmaceutical processes greener. Trends Biotechnol 26:321–327

4. Tao JH, Xu JH (2009) Biocatalysis in development of green pharmaceutical processes. Curr Opin Chem Biol 13:43–50

5. Schmidt S, Scherkus C, Muschiol J et al (2015) An enzyme cascade synthesis of ε-caprolactone and its oligomers. Angew Chem Int Ed 54:2784–2787

6. Sattler JH, Fuchs M, Mutti FG et al (2014) Introducing an *in situ* capping strategy in systems biocatalysis to access 6-aminohexanoic acid. Angew Chem Int Ed 53:14153–14157

7. Dörr M, Fibinger MPC, Last D et al (2016) Fully automatized high-throughput enzyme library screening using a robotic platform. Biotechnol Bioeng 113:1421–1432

8. Barclay SS, Woodley JM, Lilly MD et al (2001) Production of cyclohexanone monooxygenase from *Acinetobacter calcoaceticus* for large scale Baeyer-Villiger monooxygenase reactions. Biotechnol Lett 23:385–388

9. Doig SD, O'Sullivan LM, Patel S et al (2001) Large scale production of cyclohexanone monooxygenase from *Escherichia coli* TOP10 pQR239. Enzym Microb Technol 28:265–274

10. Staudt S, Bornscheuer UT, Menyes U et al (2013) Direct biocatalytic one-pot-transformation of cyclohexanol with molecular oxygen into ε-caprolactone. Enzym Microb Technol 53:288–292

11. Walton AZ, Stewart JD (2002) An efficient enzymatic Baeyer-Villiger oxidation by engineered *Escherichia coli* cells under non-growing conditions. Biotechnol Prog 18:262–268

12. Walton AZ, Stewart JD (2004) Understanding and improving NADPH-dependent reactions by nongrowing *Escherichia coli* cells. Biotechnol Prog 20:403–411

13. Lee WH, Park JB, Park K et al (2007) Enhanced production of epsilon-caprolactone by overexpression of NADPH-regenerating glucose 6-phosphate dehydrogenase in recombinant *Escherichia coli* harboring cyclohexanone monooxygenase gene. Appl Microbiol Biotechnol 76:329–338

14. Baldwin CVF, Woodley JM (2006) On oxygen limitation in a whole cell biocatalytic Baeyer–Villiger oxidation process. Biotechnol Bioeng 95:362–369

15. Ferreira-Torres C, Micheletti M, Lye GJ (2005) Microscale process evaluation of recombinant biocatalyst libraries: application to Baeyer–Villiger monooxygenase catalysed lactone synthesis. Bioprocess Biosyst Eng 28:83–93

16. Doig SD, Pickering SCR, Lye GJ et al (2002) The use of microscale processing technologies for quantification of biocatalytic Baeyer-Villiger oxidation kinetics. Biotechnol Bioeng 80:42–49

17. Doig SD, Avenell PJ, Bird PA et al (2002) Reactor operation and scale-up of whole cell Baeyer-Villiger catalyzed lactone synthesis. Biotechnol Prog 18:1039–1046

18. Studier F (2005) Protein production by auto-induction in high density shaking cultures. Protein Expr Purif 41:207–234

Chapter 9

Normalized Screening of Protein Engineering Libraries by Split-GFP Crude Cell Extract Quantification

Javier Santos-Aberturas, Mark Dörr, and Uwe T. Bornscheuer

Abstract

The different expression level and solubility showed by each protein variant represents an important challenge during screening campaigns: Usually, the total activity measurement constitutes the only criterion for identifying improved variants. This hampers the chances of finding interesting mutants, especially if the aim is to improve activity: On the one hand, interesting but poorly soluble variants will remain undetectable. On the other hand, a mutation might not increase activity, but improve expression level or solubility. The split-GFP technology offers an affordable and technically simple manner for overcoming that constraints, making protein library screening more efficient through the normalization of the detected enzymatic activities in relation to the quantified protein contents responsible for them.

Key words High-throughput screening, Directed evolution, Protein engineering, Data normalization, Split-GFP, Protein solubility, Mutant library

1 Introduction

The screening of large protein engineering libraries in the search for improved mutants usually ignores the big variations in the solubility of the different protein variants across the explored mutational landscape. As proteins normally have evolved as marginally soluble macromolecules, a vast majority of the individual amino-acid substitutions affect negatively their solubility (in other words, their effective expression level) [1, 2]. That fact substantially hampers protein library screening campaigns, because many interesting but poorly soluble protein variants remain undetectable, as normally only total protein activity is employed to evaluate them. Thus, a methodology enabling the exploration of mutant libraries in a normalized way, taking into account the quantity of protein responsible for the observed activity in each individual case, would provide a highly informative, clearer, and more realistic view of the properties of the generated mutants. Such approach would allow the

Uwe T. Bornscheuer and Matthias Höhne (eds.), *Protein Engineering: Methods and Protocols*, Methods in Molecular Biology, vol. 1685, DOI 10.1007/978-1-4939-7366-8_9, © Springer Science+Business Media LLC 2018

researchers to rescue interesting mutants invisible during ordinary screening campaigns.

Given that the purification of each individual protein variant at large library scales constitutes an expensive and highly time-consuming process (to which a quantification step should be added), this approach is not an option in the case of large screenings requiring high-throughput standards. Similarly, protein quantification based on antibody blot only provides very limited screening capacity. Thus, the ideal normalization tool should enable the quantification of the expression level of each mutant without its purification and directly from the cell crude extract. A number of chimeric systems have been developed to quantify the expression level of the protein of interest through the detection of the fluorescent or catalytic properties of a second protein fused to it [3]. However, in many cases the big size of that label leads to substantial and unpredictable disturbances in both the solubility and the activity of the protein of interest, constraining the usefulness of those chimeric systems to particular experimental cases.

The split-GFP technology [4, 5] offers an excellent alternative to solve all those issues in an affordable and technically simple manner. This system is based on the reconstitution of the Green Fluorescent Protein (GFP) fluorescence signal when a GFP truncated form (GFP 1–10) self-assembles with a small GFP fragment (GFP 11, also known as S11), consisting in a 16 amino acids tag (RDHMVLHEYVNAAGIT) fused to the protein of interest through a flexible linker. Compared to large fusion tags, this small fragment has a much lower probability to affect the folding or the solubility of the protein of interest, and can be fused both to the N-terminal or C-terminal ends of the protein (or inserted into flexible loops in any region along the protein). In the presence of a sufficient excess of GFP 1–10, the generated fluorescence is proportional to the amount of tagged protein, thus providing a straightforward tool for its quantification from cell crude extracts (Fig. 1). The coefficient between the enzymatic activity measured for each individual variants during the library screening and the amount of protein responsible for it ($R_{a/f}$ ratio) can then be calculated, and employed for a realistic comparison of the variants respect to the original scaffold protein.

Even when originally conceived to be employed in the context of *E. coli* heterologous protein expression, the split-GFP screening normalization system can potentially be adapted to any other expression system and to any conceivable screening scale, from single-position saturation mutagenesis to large random mutagenesis libraries for directed evolution.

In this chapter, we provide the general information required for the application of the split-GFP normalization principle to the screening of a protein variants library [6]. It must be taken into account that the particular conditions employed in the screening of

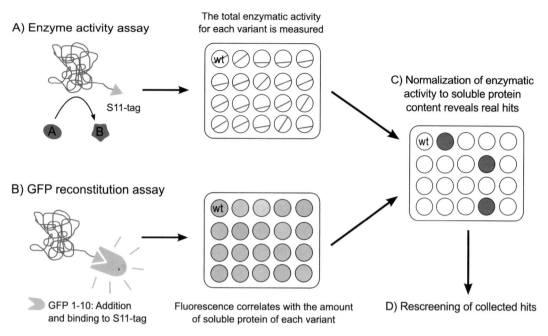

A) Enzyme activity assay

The total enzymatic activity
for each variant is measured

S11-tag

A   B

C) Normalization of enzymatic
activity to soluble protein
content reveals real hits

B) GFP reconstitution assay

GFP 1-10: Addition
and binding to S11-tag

Fluorescence correlates with the amount
of soluble protein of each variant

D) Rescreening of collected hits

Fig. 1 General overview of the split GFP normalized screening workflow. The normalized enzyme screening requires the separated measurement of the enzymatic activity for each variant (**a**) and the amount of soluble split GFP-tagged protein in each well via the addition of purified GFP 1–10 (**b**). The calculated activity ratio between the observed activity and the fluorescence ($R_{a/f}$) permits a more realistic comparison of each variant in relation to the starting scaffold (wt) and allows the detection of previously overlooked hits (**c**), which should be finally collected in a new plate for confirmation through a repetition of the normalized screening (**d**)

a given enzyme (culture conditions of the library clones, requirements for the induction of the heterologous protein, enzymatic assay for the detection of the activity) as well as the method employed for the generation of the library (error-prone PCR, QuikChange™ or gene shuffling, for example) must be decided by the researcher according to the requirements of every individual experimental plan. Luckily, a generous amount of literature is available to be interrogated about all that issues [7, 8].

Even when this protocol is oriented to the optimization of the screening of protein engineering libraries based on industrially relevant enzymes, the same principles are applicable to any kind of protein activity, as far as its activity measurement is compatible with in vitro microtiter plate screenings.

2   Materials

Unless otherwise indicated, prepare all the solutions employing ultrapure water as solvent (prepared by purifying deionized water to attain 18 MΩ of resistance at 25 °C) and analytical grade reagents. Ordinary laboratory glassware and plastic material (1.5,

10, and 50 mL tubes, micropipette tips, syringes, syringe filters, and petri dishes) will be required during different steps of the protocol and should be available.

2.1 Cloning and Expression

1. High-fidelity PCR purification kit (e.g., a Pfu polymerase-based kit).

2. PCR product purification kit.

3. Plasmid DNA purification kit.

4. Expression vector for the cloning of the gene of interest (e.g., pET11a and pET22a) (*see* **Note 1**).

5. Restriction enzymes and buffers.

6. T4 DNA ligase and buffer.

7. Chemically competent DH5α and BL21 (DE3) *E. coli* cells from a commercial supplier, or prepared following a standard protocol (i.e., rubidium chloride method) [9].

8. Lysogenic Broth (LB) medium: 0.5% (w/v) yeast extract (5 g/L), 1% (w/v) tryptone (10 g/L), 1% (w/v) NaCl (10 g/L), sterilized by autoclaving.

9. Lysogeny Broth (LB) ampicillin agar plates: LB medium with 1.5% (w/v) agar. Before pouring the plates, add the antibiotic stock solution (1 mL/L medium) (*see* **Note 1**).

10. 1000× Antibiotic stock solution: 50 mg/mL ampicillin. Sterilize by filtration (*see* **Note 1**).

11. Inducer solution: 1 M IPTG. Sterilize by filtration (*see* **Note 1**).

12. Microtiter plates (96 round wells, flat bottom).

13. Lysis buffer: 1 mg/mL lysozyme, 1 μg/mL DNAse in saline buffer.

14. Saline buffer: 50 mM Tris–HCl, pH 7.5, 300 mM NaCl.

2.2 Enzymatic Activity Screening

The choice of the reagents for activity screening depends entirely on the studied enzymatic activity in each particular screening. One of the big advantages of the split-GFP normalization is its compatibility with most activity assays, as protein quantification is performed independently of activity measurement.

2.3 Expression and Purification of the GFP 1–10 Reporter Protein

1. Plasmid encoding the GFP 1–10 reporter protein: pET GFP 1–10 [6] (*see* **Note 2**).

2. Chemically competent BL21 (DE3) *E. coli* cells.

3. LB kanamycin medium: 50 mg/L kanamycin in LB medium.

4. BugBuster™ (Novagen) or an alternative protein extraction reagent.

5. TNG Buffer: 100 mM Tris–HCl, 150 mM NaCl, 10% (v/v) glycerol. Dissolve 12.1 g of Tris in 800 mL of water, adjust the pH to 7.4 with HCl, add 8.76 g of NaCl and after dissolution add 100 mL of glycerol. Transfer to a 1 L cylinder and adjust the volume up to 1 L with water.

6. 9 M Urea: 545.4 g/L urea.

7. 1 M DTT: 154 mg/mL dithiothreitol.

2.4 Equipment

1. Microcentrifuge.

2. Thermal cycler.

3. Orbital shaking incubator for flasks.

4. Centrifuge (including 50 mL tubes and 96-well plate adapters).

5. Ultracentrifuge.

6. Probe sonicator.

7. Microtiter plate shaking incubator (preferably including air humidity control).

8. Microtiter plate reader (should be able to measure fluorescence and absorbance).

9. Ideally, a pipetting robot for 96-well plates would be perfect for the handling of the solutions, but for small or medium screening efforts, multichannel pipettes (preferably electronic) can be employed.

10. The colonies to be screened can be picked manually into the 96-well plates, but for large screening campaigns the employment of an automated colony picker would be highly recommendable.

3 Methodology

3.1 Cloning of the GFP 11-Tagged Gene of Interest

The GFP 11 fused fragment can be placed both to the N-terminal or C-terminal of the protein. Of course it is possible to order a synthetic construct including the GFP 11 tagging. In this section, we describe the general steps for the PCR-based tagging of the original gene of interest and its cloning into any conceivable *E. coli*-expression vector.

3.1.1 Primer Design

1. The total length of the GFP 11 tag added to the gene of interest consists of the tag itself (16 residues RDHMVLHEYV-NAAGIT) plus a flexible linker required to connect the tag with the protein of interest. Any flexible linker could be useful, but we have regularly employed the one indicated by Cabantous and Waldo [5], consisting of the 14 amino acids LIGSDGGSGGGSTS. Thus, the primer requires (1) 48 nt encoding the GFP 11 fragment, (2) 42 nucleotides encoding

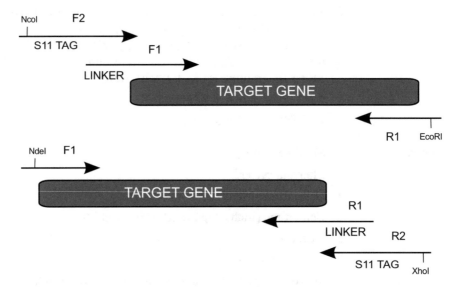

Fig. 2 PCR strategy for connecting the GFP11 tag to the target gene at the N terminus (**a**) or C terminus (**b**). Two overlapping primers are used in two consecutive PCRs against a fixed opposite primer

the linker, (3) nucleotides required for a translation start (or end, in the case of C-terminal tagging) codon, and (4) the addition of the restriction site required for the cloning of the PCR product (ca. 12 nucleotides), and around 20 nucleotides annealing inside of the template. This sums up to a total number of ca. 120 nucleotides in a single primer.

2. We recommend ordering this sequence as a gene string (small synthetic DNA), though we have used a two-PCR strategy to generate the desired construct. For doing so, the nucleotide sequence to be attached is split into two overlapping forward primers (or two overlapping reverse primers, in the case of the C-terminal tagging). Figure 2 illustrates the general tagging PCR strategy. For recommendations on the tag positioning, *see* **Note 3**.

3.1.2 PCR Amplification and Cloning into the Expression Vector

1. First PCR: In the case of N-terminal tagging of the gene of interest, the first fragment of the desired tag, corresponding mainly to the flexible linker, will be added to the gene of interest employing the first forward primer (F1) and the reverse primer (R). Ideally, the gene of interest would have been previously cloned (i.e., into an expression vector and fused to a purification tag). Amplification from any other DNA template (as genomic DNA) is also possible, but could require additional optimization of the PCR conditions. In the case of C-terminal tagging of the gene of interest, that first step will employ for the PCR the forward primer (F) and the first reverse primer (R1).

2. Perform a standard agarose gel electrophoresis in order to separate the desired PCR product from possible unspecific amplifications, together with a proper DNA ladder. After DNA-staining (employing for example ethidium bromide), visualize the gel and check it for the presence of the desired amplicon size. Extract the amplified band with a scalpel.

3. Purify the desired PCR product from the agarose gel employing a PCR-purification kit, preferably eluting the DNA with 20–25 μL of hot water (ca. 65 °C).

4. Quantify the purified PCR product (for example with a Nano-Drop™ spectrophotometer).

5. Perform the second PCR, using 30–50 ng of the previously purified PCR-product as template. In that case, the second reverse primers (F2) should be employed together with the R primer, completing the GFP 11 tag and adding at the same time the restriction site required for the cloning of the GFP 11-tagged gene into the chosen expression vector. In the case of the C-terminal tagging of the gene of interest, this PCR reaction would employ the primer F and the reverse primer 2 (R2).

6. Perform a standard agarose gel electrophoresis in order to separate and purify the final PCR product, as described in **steps 4** and **5**.

7. Quantify the purified DNA of the second PCR.

8. Digest the GFP 11-tagged gene of interest (insert) and the expression vector with the chosen restriction enzymes (if possible, that enzymes should be suitable for double digestion). In principle, we would recommend to digest 5–6 μg of the vector overnight in a final volume of 50 μL, employing 10 units of each restriction enzyme. The double digested, linearized vector must undergo agarose electrophoresis and be purified from the gel, eluting from the purification column with 30 μL of hot water (65 °C). Finally, the purified DNA should be quantified. In the case of the insert, we recommend the digestion of 500–1000 ng in a final volume of 15–20 μL for 2 h in the presence of 5 units of each restriction enzyme. If possible, perform a heat inactivation after the digestion. Otherwise, we recommend to freeze the digestion mixture at −20 °C and melt it again just before its ligation into the expression vector (*see* **Note 4**).

9. Ligate the GFP 11-tagged gene of interest into the chosen expression vector. We recommend the usage of a T4 DNA ligase provided together with a polyethyleneglycol containing buffer, as this additive improves substantially the reaction efficiency. Adjust the reaction mixture to a final volume of 10 μL containing 20–50 ng of digested vector, a 4–6 fold molar excess of insert (without representing more than 4 μL in the

final mixture) and 1 unit of T4 DNA ligase. Incubate the reaction for 2–3 h at 16–20 °C.

10. Employ 10 μL of the ligation mixture for the transformation of 100 μL of *E. coli* DH5α chemically competent cells, following a standard heat-shock based method: (a) melt a chemically competent cells vial in ice, (b) add the 10 μL of ligation mixture to the cells and mix gently by pipetting, (c) incubate in ice for 20 min, (d) perform a 45 s heat shock by introducing the vial of cells into a 42 °C water bath, (e) incubate on ice for 2 min, (f) add 700 μL of room temperature LB medium (without antibiotics) to the cells and place it in a shaker at 37 °C during 40 min. Finally, (g) spread the cells on LB ampicillin plates (*see* **Note 1**) and (h) incubate the agar plates at 37 °C for 12 h.

11. Check some (10–20) of the antibiotic-resistant clones by colony-PCR or by restriction analysis after plasmid purification. Select 1–3 positive clones and sequence their inserts to verify the absence of undesired mutations and the correct GFP 11-tagging of the gene of interest.

3.2 Preparation of the GFP 1–10 Reporter Protein Solution

This protocol is only slightly modified from the original one proposed by Cabantous and Waldo [5].

1. Transform 1 μL of pET GFP 1–10 expression vector into 50 μL of *E. coli* BL21 (DE3) chemically competent cells following the method previously described (Subheading 3.1.2, **step 10**).

2. Select a few (1–5) antibiotic-resistant clones and check them by colony PCR. Select one of the positive clones for protein expression.

3. Employ the selected clone to inoculate 10 mL preculture of LB medium supplemented with the kanamycin and incubate overnight at 37 °C in an orbital shaker.

4. Inoculate 500 mL of LB kanamycin medium with 5 mL of the preculture, and grow it at 37 °C in an orbital shaker (250 rpm) until the culture reaches $OD_{600} = 0.6$.

5. Induce the expression of GFP 1–10 by adding 0.5 mL of IPTG stock solution and continue the incubation of the culture at 37 °C in an orbital shaker (250 rpm) for 3 h. Under these conditions, a substantial amount of the GFP 1–10 will be directed to inclusion bodies, which will be employed for the purification of the reporter protein.

6. Centrifuge the culture in a refrigerated centrifuge (4 °C) at $4000 \times g$, for 10 min and discard the supernatant.

7. Resuspend the cells in 20 mL of cold TNG buffer, transfer them to a preweighted 50 mL Falcon tube, and sonicate them on ice for 10 min (50% working/resting cycle, 50% power).

8. Centrifuge the suspension to recover the inclusion bodies (in the pellet) at 30,000 × *g* for 30 min, and discard the supernatant.

9. Add 10 mL of Bugbuster 1× (or employ an alternative protein extraction reagent) and sonicate for 2 min in order to resuspend the inclusion bodies (if additional resuspension is required, repeat the sonication step). Centrifuge at 30,000 × *g* for 30 min and discard the supernatant. Repeat this step three times.

10. Add 10 mL of cold TNG Buffer and sonicate for 2 min for resuspension and detergent removal. Centrifuge at 30,000 × *g* for 30 min and discard the supernatant. Repeat this step two times.

11. Calculate the weight of the washed inclusion bodies pellet.

12. Calculate the volume required to reach a final concentration of 75 mg inclusion bodies/mL of TNG buffer, add that volume of TNG buffer and resuspend the pellet by sonication.

13. Split the resuspended inclusion bodies into 1 mL aliquots in 1.5 mL tubes, and centrifuge them at 16,000 × *g* for 10 min in a micro centrifuge, and discard the supernatant. The pellets can be frozen at −80 °C until use up to 6 months.

14. Dissolve the pellet contained in one tube in 1 mL of 9 M urea containing 5 mM DTT. Resuspend by pipetting and incubate at 37 °C until the complete dissolution of the inclusion bodies.

15. Centrifuge the tube for 1 min at 16,000 × *g* in a microcentrifugue in order to precipitate any GFP 1–10 misfolded aggregates.

16. Transfer 1 mL of the GFP 1–10 urea-dissolved GFP 1–10 to a 50-mL Falcon tube and add 25 mL of TNG buffer. Mix gently by inversion and pass the solution through a 0.2 μm filter to a new tube. This will be the reporter solution employed for the split-GFP protein quantification assays required for the normalization of the protein library screenings. It is ready to use and can be stored up to 2 weeks at −20 °C, or up to 2 months at −80 °C.

3.3   Performance of the Normalized Library Screening

3.3.1   Generation of the Library

1. Create the library of protein variants by your method of choice (employing the GFP 11-tagged gene of interest as starting scaffold) and introduce the created library into an *E. coli* strain suitable for heterologous expression of proteins, like *E. coli* BL21 (DE3). If you intend to use an expression system different from *E. coli*, please *see* **Note 5**.

3.3.2   Preparation of Crude Cell Extracts in MTP

1. Pipette 200 μL of LB ampicillin medium into the wells of each 96-welll microtiter plate.

2. Pick one colony from the library into each well, keeping four wells for the GFP 11-tagged wild type (that will serve as a reference during the screening), four for the expression strain carrying the empty expression plasmid (that will serve for the calculation of the background enzymatic activity and fluoresce required for statistic subtractions) and keep another four wells without inoculation, just with medium (preferably in the inner region of the plate, in order to control possible cross contaminations due to the plate shaking).

3. Grow the inoculated 96-well plates shaking overnight at 37 °C and 700 rpm.

4. Replicate the overnight grown 96-well microtiter plates (that will serve as library backup) into new 96-well plates containing expression medium (normally LB or TB, supplemented with the required antibiotics), using a 96-well replicator or a robotic colony picker. These plates will be the expression plates for the screening.

5. Grow the expression plates for 4–6 h until OD_{600} reaches a value of 0.5 (controlling the average OD_{600} of the wells can be useful to establish the required time) under shaking and trigger the protein expression adding IPTG. Shaking frequency, time of induction and inducer concentration has to be optimized for each protein of interest.

6. Incubate the induced expression plates shaking them overnight at the optimal temperature required for the expression of the protein of interest.

7. Centrifuge the expression plates for 20 min at $4000 \times g$ to pellet the cells and eliminate the supernatant (by decantation or with the help of a pipetting robot).

8. Wash the cell pellets once by resuspending them in saline buffer. Centrifuge the 96-well plates once again ($4000 \times g$, 20 min) and discard the supernatant.

9. For cell lysis, resuspend the cell pellets in 200 μL of lysis buffer and incubate the 96-well plates shaking at 700 rpm for 2 h.

10. Centrifuge the 96-well plates for 20 min at $4000 \times g$, and transfer the supernatants to a new 96-well plate, in which that cell crude extracts will be preserved at 4 °C until usage.

3.3.3 Split-GFP Protein Quantification and Activity Assays

1. In a 96-well microtiter plate, pipette for each well (including all the control wells) 20 μL of crude extract and 180 μL of the GFP 1–10 superfolder reporter protein solution obtained in the Subheading 3.2, **step 16**.

2. Incubate the protein quantification assays overnight at 4 °C (*see* **Note 6**) and measure fluorescence generated by the reconstituted split-GFP (λ_{ex} = 488 nm/λ_{em} = 530 nm). The

fluorescence value will later be used for normalizing the activities. However, it is also possible to calibrate the assay and to determine specific activities directly from the crude extracts (*see* **Note 7**).

3. The enzymatic activity assay depends on the particular studied protein of interest. Perform your assay method of choice in a new microtiter plate using the cell crude extracts containing the studied GFP 11-tagged enzyme variants prepared in Subheading 3.3.2, **step 10**.

3.3.4   Data Analysis

1. The data analysis is based simply on the calculation of the ratio between the observed enzymatic activity and the fluorescent signal ($R_{a/f}$) and its comparison with the values obtained for the four wild-type control wells present in the plate. Once the $R_{a/f}$ ratio has been calculated for every clone, different conditions can be established in the statistical analysis in order to identify positive hits with different levels of stringency. A variant will be considered as a hit if it shows a certain level improvement of $R_{a/f}$ compared to the wild type. This threshold has to be set wisely (*see* **Note 8**).

2. As manual calculations are very time consuming, we recommend to apply automated data analysis. A useful R-code for split-GFP screening of protein engineering libraries data analysis is available, and can be inspiring for the design of additional tools [6].

3.3.5   Confirmation of Hit Clones

1. To confirm the putative hit clones identified in the first round of quantitative screening, a confirmation 96-well plate should be prepared starting from the original clones preserved in the library backup plates (*see* Subheading 3.3.2, **step 4**). A useful experimental design for the confirmation plate will include once again four replicates of the reference/positive control (the GFP11-tagged wild-type protein of interest), four negative controls for background subtractions and cross-contamination controls. In addition to that, this plate should include four replicates of each putative hit identified during the data analysis of the original normalized screening. The confirmation plate must be processed and analysed exactly in the same way than the plates of the original screening, starting from Subheading 3.3.2, **step 5**.

4   Notes

1. Any suitable vector can be employed for the expression of your gene of choice. Inducer and antibiotic must be adapted accordingly.

2. The sequence of the GFP 1–10 reporter protein is: MSKGEELFTGVVPILVELDGDVNGHKFSVRGEGEGDA-TIGKLTLKFICTTGKLPVPWPTLVTTLTYGVQCFSRYPD-HMKRHDFFKSAMPEGYVQERTISFKDDGKYKTRAVVK-FEGDTLVNRIELKGTDFKEDGNILGHKLEYNFNSHNV-YITADKQKNGIKANFTVRHNVEDGSVQLADHYQQNT-PIGDGPVLLPDNHYLSTQTVLSKDPNEK (214 amino acids). A gene encoding this amino acid sequence can be cloned in other suitable vectors. In this protocol, we would assume that the expression system employed is inducible by the addition of IPTG. Adapt the inducer and antibiotic to your system.

3. The GFP 11 tag can theoretically be placed at any position in the protein (even inserted into loops), given that the activity of the protein is not disturbed while being accessible for the GFP 1–10 truncated reporter. Usually, placing the tag in one of the terminal ends of the protein of interest is technically easier and more feasible. As most of the times the proteins employed as scaffolds for protein engineering are fused to terminal affinity tags, the most pragmatic decision would be to simply place the GFP 11 tag before the affinity tag (in N-terminal affinity tagged proteins) or after it (in C-terminal affinity tagged proteins). In any case, every available structural data about the protein of interest would be valuable in order to make the right decision. In principle, the small size of the GFP 11 tag usually does not affect significantly the expression level, solubility or folding of the protein of interest, but before starting a resource-consuming screening, it is necessary to verify that the GFP 11-tagged protein has similar properties like the untagged starting protein in terms of activity and expression. This should be done preferably after the purification of both proteins by affinity chromatography, but evaluation from crude extracts could be possible if the intention is only to detect the activity under the usual expression conditions for the protein of interest. Similarly, the generation of fluorescence after the addition of the GFP 1–10 reporter must be verified, in order to ensure the accessibility of the GFP 11 for the split GFP reconstitution. The ideal sample for that should be a cell crude extract obtained under the conditions that will be employed in the screening (culture medium, volume, protein expression induction), in order to test the efficiency of the method at that scale.

4. The purification of the digested DNA directly from the digestion mixture using a DNA-purification kit (like the one employed for previous purifications from agarose gels along this protocol) could be another option, as well as the phenol-chloroform extraction and ethanol precipitation, but those procedures normally lead to a considerable decrease of the

final insert yield. In our experience, the direct usage of a certain amount of the digestion mixture (not exceding 40% of the final ligation reaction volume) usually ensures a successful DNA ligation reaction.

5. *E. coli* protein expression strains are by far the most widely used for the screening of protein engineering libraries (especially in the case of enzymatic activity screenings). However, the expression of certain proteins must be performed in different organisms (i.e., *Pichia pastoris* or other yeast systems). The only recommendable modifications in these cases would be the codon optimization of the GFP 11 tag according to the requirements of the specific expression host and the utilization of an adequate cell disruption protocol. In case that the protein of interest has to be secreted to the culture medium, the GFP 11 tag should not constitute a problem, as it is very flexible, and has been shown to be fully compatible with translocation systems in the past [10].

6. The time for reaching the maximum fluorescence varies. Usually, incubation over night ensures that reconstitution of the full GFP has completed, but in some cases this process might take a longer time. For the first assays we recommend to perform additional fluorescence measurements after 24 and 48 h incubation, in order to establish the incubation time required for the generation of the maximal fluorescent signal in a particular screening.

7. During the description of the normalized screening procedure, we have shown how hidden hits can be rescued from protein variant libraries by the relative comparison of the $R_{a/f}$ values of the mutants and the wild-type reference clones. However, the split-GFP principle allows the precise quantification of soluble protein amounts, and can be employed for the calculation of the specific activity for every single protein variant directly from cell crude extracts. For that, an affinity purification tag must be available in addition to the GFP 11 tag, and employed for the isolation of the GFP 11-tagged protein. Purity should be assessed by standard SDS-PAGE. The concentration of a dilution series of the purified protein (from the original solution to a 1024-fold diluted sample, by a ten times twofold dilution series, for example) must then be calculated by a standard protein quantification method (i.e., Bradford- or Biuret-based protocols), employing BSA for the generation of a standard curve. Each diluted sample should then be employed in a split-GFP reconstitution assay, as described in Subheading 3.3. The relationship between the calculated protein concentration and the fluorescent signal generated by each dilution in the split-GFP assay is linear within a wide range of protein concentrations [4]. Thus, the fluorescence value for each protein variant

can be employed for the calculation of protein content during the screening, making it possible to calculate the specific activity values for each protein variant without its purification. An excellent guideline about how to perform in vitro molar calculations for the protein content employing the split-GFP technology is provided by Cabantous and Waldo [5].

8. In general, it is more reliable to consider as hits the variants which show an $R_{a/f}$ value higher than any of the wild-type wells, not only bigger than the average of them. However, in order to increase the chance of finding actually improved mutants, being simply better than any of the wild-type replicates is not enough. Establishing a certain threshold (a $R_{a/f}$ 20–30% better than any of the wild-type replicates, for example) helps to reduce substantially the number of false positives. The fluorescence value by itself is also an interesting feature that can indicate the presence of mutations favoring the solubility of the protein of interest, and can be considered for an independent analysis based on similar threshold improvement criteria than the $R_{a/f}$.

References

1. Taverna DM, Goldstein RA (2002) Why are proteins marginally stable. Proteins 46:105–109

2. Romero PA, Arnold FH (2009) Exploring protein landscapes by directed evolution. Nat Rev Moll Cell Biol 10:866–876

3. Waldo GS (2003) Genetic screens and directed evolution for protein solubility. Curr Opin Chem Biol 7:33–38

4. Cabantous S, Terwilliger TC, Waldo GS (2005) Protein tagging and detection with engineered self-assembling fragments of green fluorescent protein. Nat Biotechnol 23:102–107

5. Cabantous S, Waldo GS (2006) In vivo and in vitro protein solubility assays using split GFP. Nat Methods 3:845–854

6. Santos-Aberturas J, Dörr M, Waldo GS et al (2015) In-depth high-throughput screening of protein engineering libraries by split-GFP direct crude cell extract data normalization. Chem Biol 22:1406–1414

7. Gillam EMJ, Copp JN, Ackerley DF (eds) (2014) Directed evolution library creation: methods and protocols, 2nd edn. New York, Humana Press

8. Ruff AJ, Dennig A, Schwaneberg U (2013) To get what we aim for–progress in diversity generation methods. FEBS J 280:2961–2978

9. Green R, Rogers EJ (2013) Chemical transformation of E. coli. Methods Enzymol 529:329–336

10. Hyun SI, Maruri-Avidal L, Moss B (2015) Topology of endoplasmic reticulum-associated cellular and viral proteins determined with split-GFP. Traffic 16:787–795

Chapter 10

Functional Analysis of Membrane Proteins Produced by Cell-Free Translation

Srujan Kumar Dondapati, Doreen A. Wüstenhagen, and Stefan Kubick

Abstract

Cell-free production is a valuable and alternative method for the synthesis of membrane proteins. This system offers openness allowing the researchers to modify the reaction conditions without any boundaries. Additionally, the cell-free reactions are scalable from 20 μL up to several mL, faster and suitable for the high-throughput protein production. Here, we present two cell-free systems derived from *Escherichia coli* (*E. coli*) and *Spodoptera frugiperda* (*Sf*21) lysates. In the case of the *E. coli* cell-free system, nanodiscs are used for the solubilization and purification of membrane proteins. In the case of the *Sf*21 system, endogenous microsomes with an active translocon complex are present within the lysates which facilitate the incorporation of the bacterial potassium channel KcsA within the microsomal membranes. Following cell-free synthesis, these microsomes are directly used for the functional analysis of membrane proteins.

Key words Cell-free protein synthesis, *Sf*21, *E. coli*, Nanodiscs, Lipid bilayers, Membrane proteins, Proteoliposomes

---

1 Introduction

Membrane proteins (MPs) represent one third of the total proteins encoded by the human genome. These include receptors, ion channels, transporters, and porins. They play an important role in a wide range of biological processes like cell-to-cell communication, extracellular and intracellular ligand recognition, signal transduction, ion-channel conductance, and transport of a range of substrates across the membranes which are vital for survival of any organism. Any functional defect in the MPs could affect the cellular activities which could often lead to a wide range of diseases like Alzheimer's disease, cystic fibrosis, epilepsy, cardia arrhythmia, and migraine [1–3]. Due to their medical importance, MPs have become more than 50% of the total drug targets from

The original version of this protocol was revised. An erratum to this protocol can be found at DOI 10.1007/978-1-4939-7366-8_21

Uwe T. Bornscheuer and Matthias Höhne (eds.), *Protein Engineering: Methods and Protocols*, Methods in Molecular Biology, vol. 1685, DOI 10.1007/978-1-4939-7366-8_10, © The Author(s) 2018

pharmaceutical companies. However, despite their significance in cellular physiology, complete structural information is only known for a small percentage of MPs. This is due to the lack of methods to synthesize high quality MPs essential for structural and functional analysis. Functional synthesis of MPs in vivo is challenging due to low yields, solubilization and purification problems, and overexpression often leads to toxicity. Having a flexible approach and faster synthesis method are crucial for synthesizing a wide range of high quality MPs which might help the researchers and drug companies to develop functional assays and to design new therapeutics. For more detailed understanding of the protein function, one needs to have an open isolated system where one can vary the parameters regulating the protein expression systematically. Cell-free system offers all the conveniences required for proper synthesis of MPs. This method offers a high degree of controllability and provides a completely open system allowing direct manipulation of the reaction conditions to optimize protein folding, disulfide bond formation, incorporation of noncanonical amino acids and the synthesis of toxic proteins [4–9]. In comparison to conventional cell-based systems, cell-free systems offer rapid protein synthesis, purification and functional analysis. One of the most widely used cell-free systems is based on *E. coli* extracts. This system is widely used for synthesis of MPs in the presence of membrane solubilization supplements in the form of nanodiscs, detergents, proteoliposomes etc. added externally into the cell-free reaction [5, 10]. Nanodiscs are synthetic discoidal nanoparticles consisting of a phospholipid bilayer surrounded by two copies of membrane scaffold proteins (MSPs). MSPs are modified apolipoproteins consisting of a hydrophobic part toward the lipid bilayer and a hydrophilic part outside thus providing stability to the nanodiscs and make them soluble without any detergents [8]. These nanoparticles can be added directly into the cell-free reaction system. Another existing cell-free system derived from insect (*Spodoptera frugiperda Sf*21) extracts is also used for synthesizing MPs. This eukaryotic cell-free system offers additional advantages in the form of native, ER-derived endogenous microsomes. Such microsomes contain the entire translocon machinery responsible for proper folding of MPs [8–13]. Recently, the potassium channel KcsA was synthesized successfully in this system [13]. The synthesized MP showed tetrameric configuration and exhibited single-channel activity characteristic to the protein.

In this chapter, we will present a general method for measuring the functionality of MPs derived from *E. coli* and insect-based cell-free systems. The expression and analysis will be shown exemplary with the proteins bacteriorhodopsin (BR) and mannitol permease (MtlA) using the *E. coli* cell-free system (Subheading 3.2), whereas the insect cell-free system will be used to produce the potassium channel protein KcsA (Subheading 3.3). We recommend using these proteins as positive controls when establishing the described protocols for your proteins of interest. The methodology we present here is also suitable for functional analysis of MPs synthesized by additional cell-free systems not discussed in this chapter. The main objective of this chapter is to propose two simple methods for the functional analysis of MPs derived from cell-free systems. These protocols can also be applied for screening protein variants.

2 Materials

2.1 Preparation of Nanodiscs

1. Membrane scaffold protein (MSP) (MSP1D1) (Sigma-Aldrich).

2. Dimyristoyl phosphatidylcholine (DMPC) (Avanti Polar Lipids).

3. 1,2-dioleoyl-3-trimethylammonium-propane (DOTAP).

4. 1,2-dioleoyl-*sn*-glycero-3-phospho-(1′-*rac*-glycerol) (DOPG).

5. Biobeads SM-2 (Bio-Rad).

6. Purified water (Milli-Q system).

7. Cholate buffer: 100 mM sodium cholate, 20 mM Tris, 100 mM NaCl, pH 7.4.

2.2 Prokaryotic Cell-Free Synthesis

1. *E. coli* lysate and reaction buffer (EasyXpress *E. coli* kit, BR1402001) (Biotechrabbit GmbH, Germany).

2. ^{14}C-labeled leucine (PerkinElmer) (100 dpm/pmol).

3. Genes encoding proteins bacteriorhodopsin (BR) and mannitol permease (MtlA) cloned in the pIX3.0 plasmid (100 nM stock solutions).

4. Ni-NTA magnetic beads (Qiagen).

2.3 Synthesis of MPs in the Insect Cell-Free System

1. Transcription reaction mixture containing 80 mM HEPES–KOH buffer, pH 7.6, 15 mM $MgCl_2$, 3.75 mM NTPs, 0.5 mM m^{7}G(5′)ppp(5′)G-CAP analog, and 1 U/μL T7 RNA polymerase.

2. DyeEx spin columns (Qiagen).

3. Insect (*Spodoptera frugiperda Sf*21) lysates used for the cell-free reaction have to be prepared as described [5–7] (*see* **Note 1**).

4. Translation mixture containing 25% (v/v) Sf21 lysate, 30 mM HEPES–KOH, pH 7.6, 2.5 mM Mg(OAc)2, 75 mM KOAc, 0.25 mM spermidine, 200 μM amino acids and energy regeneration components 20 mM creatine phosphate, 1.75 mM ATP, and 0.45 mM GTP.

5. Genes encoding the protein KcsA cloned in the pIX3.0 plasmid.

2.4 Preparation of Proteoliposomes and Lipid Bilayers	1. 2.5 mM Puromycin (Sigma-Aldrich) dissolved in 500 mM KCl.

2.4 Preparation of Proteoliposomes and Lipid Bilayers

1. 2.5 mM Puromycin (Sigma-Aldrich) dissolved in 500 mM KCl.

2. 1,2-diphytanoyl-*sn*-glycero-3-phosphocholine   (DPhPC) (Avanti Polar Lipids).

3. Octane.

4. Multi electrode cavity arrays (MECA) chips (Nanion GmbH, Germany).

5. 0.1 M phosphate buffered saline (PBS) pH 7.0.

2.5 Analysis of Synthesized Proteins

1. Trichloroacetic acid.

2. Filter paper (MN GF-3, Macherey-Nagel).

3. Scintillation tubes (Zinsser Analytic).

4. Scintillation cocktail (Quicksafe A, Zinsser Analytic).

5. NuPAGE® LDS Sample Buffer (Invitrogen).

6. Precast SDS-PAGE gels (NuPAGE 10% Bis–Tris Gel with MES SDS buffer) (Invitrogen).

2.6 Instrumentation and Software

1. Thermomixer comfort.

2. NanoDrop 2000c.

3. Vacuum filtration system.

4. Orbital shaker.

5. Unigeldryer.

6. Size measurements are done by using dynamic light scattering measurements, e.g., using the Zetasizer Nano ZS instrument (Malvern, UK).

7. The functionality of MPs is analyzed with the Port-a-Patch system using borosilicate glass chips with an aperture diameter of approximately 1 μm and chip-based, parallel bilayer recording setup Orbit 16 System (Nanion Technologies GmbH, Munich, Germany) with multi-electrode-cavity-array (MECA) chips (Ionera Technologies). This also requires a single channel amplifier (EPC-10, HEKA Electronic Dr. Schulze GmbH, Lambrecht, Germany) and the data acquisition software Patchmaster (HEKA).

8. Electrophysiology recordings are analyzed by Clampfit 10.7 software (Molecular devices, Sunnyvale, California, USA).

9. For quality control of the synthesized proteins, typical radionucleotide laboratory equipment is required, including for example a LS6500 Multi-Purpose scintillation counter (Beckman Coulter) and a phosphorimager system (Typhoon TRIO + Imager, GE Healthcare).

3   Methods

3.1   Preparation of Nanodiscs (NDs)

1. First prepare 5 mg/mL of MSP protein solution (200 μM) containing 20 mM Tris, pH 7.4, with 0.1 M NaCl and 0.5 mM EDTA.

2. Dissolve DMPC, DOTAP and DOPG lipids in cholate buffer at a concentration of 15 mM (*see* **Note 2**).

3. Set up the reactions: mix the lipid solutions (*see* **Notes 3** and **4**) with MSP as indicated in Table 1.

4. Incubate the reactions at 25 °C for 30 min. Then, biobeads SM-2 are added almost to 80 vol % of the Lipid-MSP mixtures and left to react at 25 °C for 45 min and then centrifuged at 1000 × *g* for 1 min. Remove the upper supernatants.

5. For each reaction, add a second set of 80 vol % biobeads-SM2 and incubate for 15 min. After centrifugation at 1000 × *g* for 1 min at RT, recollect the top solution. The finally recovered solution is further centrifuged at 5000 × *g* for 5 min and just the top solution (leaving the lower 5 μL volume) is recollected and analyzed by zetasizer.

6. Place 20 μL of the recovered solution in a zeta sizer cuvette and measure the size intensity of the particles present in the solution. Figure 1 shows an exemplary measurement: the presence of a strong peak with maximum intensity at around 11.5 nm indicates the presence of NDs.

3.2   Membrane Protein Synthesis in Prokaryotic Cell-Free Systems

1. BR and MtlA proteins are synthesized in cell-free systems using *E. coli* lysates. A typical 50 μL standard reaction comprises 35% (v/v) *E. coli* lysate containing T7 RNA-polymerase, 40% reaction buffer containing the complete amino acid mix (1.2 mM

Table 1
Pipetting scheme for reaction setup of NDs

Reaction	DMPC (15 mM) (μL)	DOTAP (15 mM) (μL)	DOPG (15 mM) (μL)	MSP (200 μM) (μL)	Total (μL)
1	40	10		50	100
2	25	25		50	100
3		50		50	100
4	50			50	100
5			50	50	100
6	40		10	50	100
7	25		25	50	100

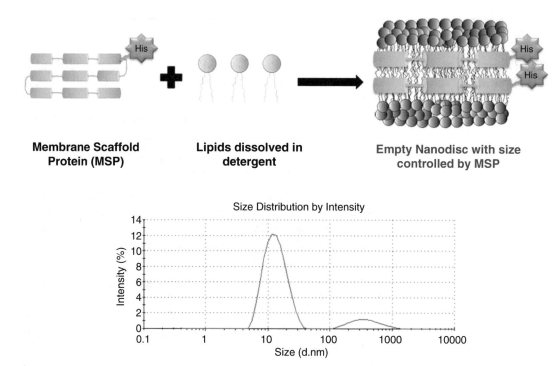

Fig. 1 Scheme depicting the preparation of NDs by detergent based reconstitution. Zetasizer measurements showing the presence of a single significant peak at around 11.5 nm indicating the presence of nanodiscs

each), 25 × XE-solution (EasyXpress *E. coli* kit), 1 µL of the respective plasmid (0.75 µg/mL) and ^{14}C-labeled leucine, final concentration: 50 µM; 4 DPM/pmol) (*see* **Note 5**), and 10% (v/v) of the prepared ND. Set up the following reactions and controls (without NDs) (Table 2).

2. Perform the protein synthesis in a thermomixer with shaking at 500 rpm and 37 °C for 90 min.

3. Once the reaction is completed, all the qualitative and quantitative measurements are done by SDS-PAGE combined with autoradiography and TCA precipitation using ^{14}C-leucine. Initially 2 × 5 µL of the reaction mixture (suspension) is collected for TCA precipitation and 1 × 5 µL for SDS-PAGE analysis. Next, centrifuge the remaining reaction mixture at 16,000 × *g* for 10 min at 4 °C.

4. Collect the supernatant into a separate Eppendorf tube. 2 × 5 µL of the supernatant is collected for TCA precipitation and 1 × 5 µL for SDS-PAGE analysis.

5. For TCA precipitation, mix 5 µL aliquots (both suspension and supernatant) with 3 mL trichloroacetic acid (TCA) and incubate the mixture in a water bath at 80 °C for 15 min. Afterward, keep the reaction vessel on ice for 30 min.

Table 2
Setup of the *E. coli* based cell-free synthesis of BR and MtlA proteins in the presence of NDs

Protein	*E. coli* lysate (μL)	Reaction buffer (μL)	^{14}C-Leucine (μL)	Plasmid (5 nM) (μL)	NDs (μL)	Water (μL)	Total (μL)
BR	17.5	20	1.25	2.5	5	3.75	50
MtlA	17.5	20	1.25	2.5	5	3.75	50
BR	17.5	20	1.25	2.5	0	8.75	50
MtlA	17.5	20	1.25	2.5	0	8.75	50

6. Filter the solution using the vacuum filtration system to retain the radiolabeled proteins on the surface of the filter paper. Wash proteins twice with TCA and twice with acetone, dry the filter papers for some minutes under the hood.

7. Transfer the dried protein-enriched filter papers to scintillation tubes (Zinsser Analytic), overlay with 3 mL of scintillation cocktail and let it shake on the orbital shaker for at least 1 h. Measure the incorporation of ^{14}C-leucine by liquid scintillation counting using the scintillation counter.

8. For SDS-PAGE analysis, precipitate 5 μL aliquots (both suspension and supernatant) of the cell-free reaction mixtures by cold acetone precipitation: Add 45 μL of the water and 150 μL of ice cold acetone to the 5 μL aliquot and incubate on ice for 15 min. Next, centrifuge the mixture at $16,000 \times g$ for 10 min at 4 °C. Discard the supernatant containing the acetone.

9. Resuspend the pellets in 20 μL of NuPAGE® LDS Sample Buffer and load the samples on precast SDS-PAGE gels. Run the gel at 200 V for 35 min. After drying the gels for 70 min at 70 °C using the Unigeldryer, visualize radioactively labeled proteins with the phosphorimager.

10. BR in the presence of NDs is folded in a correct form and shows a purple color due to the conversion of all-*trans* retinal to 11-*cis* retinal (Fig. 2). Confirm this by analyzing the supernatant samples of BR (Subheading 3.2, **step 4**) by UV-Visible spectroscopy for measuring the BR-specific peak.

An absorbance peak at around 550 nm corresponds to the purple color of the BR protein. Supernatants from the control reaction (without NDs) should neither produce a purple color nor show an absorbance peak at 550 nm. This indicates that NDs help in correct folding of the BR. Figure 3 shows the synthesis of functional BR in the presence of NDs with saturated DMPC lipids doped with cationic (DOTAP) and anionic lipids (DOPG) (Reactions 1 and 6, Table 1). The intensity of the purple color, which indicates the presence of functional BR, varies with lipid

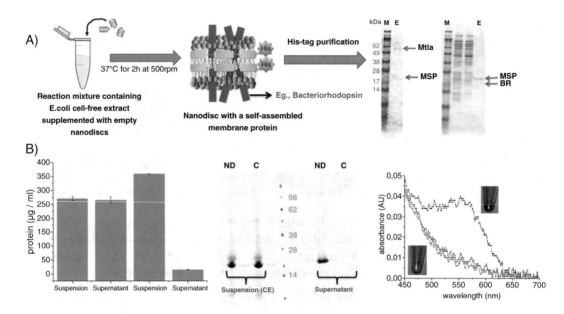

Fig. 2 (**a**) *E. coli* based cell-free synthesis of MPs in the presence of NDs. Proteins solubilized in NDs can be purified by using the His-tag available on the MSP. The SDS-PAGE gel (*right*) shows the presence of two bands corresponding to the MSP and the synthesized protein (MtlA and BR). (**b**) Quantification of de novo synthesized BR based on TCA-precipitation. The cell-free system shows the recovery of the synthesized BR MP only in the presence of NDs in the supernatant fraction. SDS-PAGE combined with autoradiography shows the presence of bands corresponding to the synthesized BR protein in the supernatant fraction only in the presence of NDs. UV-Visible measurements showing the presence of a peak around 550 nm corresponding to the presence of BR. *Inset* shows the purple color formation corresponding to the BR incorporated in the NDs

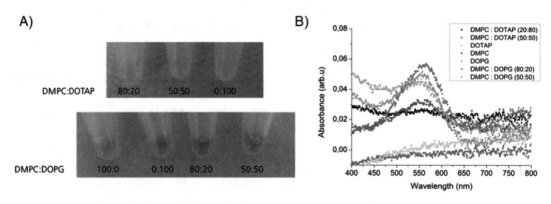

Fig. 3 (**a**) Reaction tubes showing the BR synthesized in the presence of nanodiscs containing different lipid composition by using *E. coli* based cell-free system. (**b**) UV-visible absorbance peak of BR synthesized in the presence of nanodiscs with different lipid compositions at 550 nm

composition with higher intensity in the presence of DOPG lipids. This clearly shows that doping of DOPG lipids along with DMPC helps in a better folding of the BR compared to DOTAP lipids. These results suggest that ND with different lipids can have influence on the functionality of the protein [14].

3.3 Membrane Protein Synthesis in Eukaryotic Cell-Free Systems

Cell-free protein synthesis is performed in the linked reaction mode by performing transcription and translation individually (*see* **Notes 5** and **6**). Details can be found in Refs. [4, 7]:

1. In the first step, mRNA is generated by transcription from the plasmid DNA (plasmid encoding the KcsA protein, final concentration: 60 μg/mL) or E-PCR product (final concentration: 8 μg/mL) and is incubated for 2 h at 37 °C. The reaction is performed with 80 mM HEPES–KOH buffer, pH 7.6, 15 mM $MgCl_2$, 3.75 mM NTPs, 0.5 mM $m^7G(ppp)G$-CAP analog and 1 U/μL T7 RNA polymerase.

2. Once the reaction is completed, purify the mRNA using DyeEx spin columns according to the manufacturer's instructions.

3. Quantify the purified mRNA using the NanoDrop 2000c. After estimating the concentration, mRNA is used for protein synthesis.

4. In the second step, translation is performed by adding mRNA at a final concentration of 260 μg/mL. To monitor protein quality and quantity, add ^{14}C-labeled leucine to the translation reaction mixture to yield a final concentration of 60 μM. The translation mixture is incubated for 90 min at 27 °C.

5. Quantification of the synthesized protein is done by hot trichloroacetic acid (TCA) precipitation followed by radioactivity measurements as described in Subheading 3.2, **steps 4–7**. To analyze homogeneity and molecular weight of in vitro translated proteins, take 5 μL aliquots of radiolabeled cell-free translation reaction mixtures for SDS-PAGE analysis using precast gels as explained in Subheading 3.2, **step 8** and in Refs. [4–7].

6. KcsA forms a stable tetramer as seen in Fig. 4a. A prolonged time of translation leads to an increase in the intensity of the tetramer band in the vesicular fraction (VF). These observations correlate with the protein yields for different time periods (Fig. 4b).

3.4 Preparation of Proteoliposomes

Once the protein is synthesized, the vesicular fraction (VF) harboring the synthesized protein of interest is used for functional analysis. One can add either the translation reaction mixture harboring the microsomes and synthesized proteins directly onto the lipid bilayer or prepare proteoliposomes for a faster fusion, as described in the following steps:

1. 50 μL of the cell-free translation mixture (TM) is centrifuged at $16,000 \times g$ for 10 min at 4 °C. Supernatant (SN) is separated from the pellet (VF). VF contains the microsomes incorporating the MP of interest.

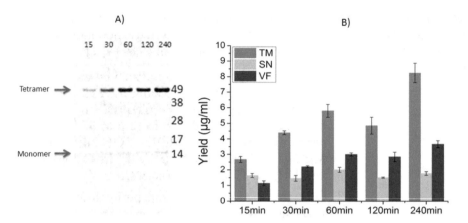

Fig. 4 Cell-free synthesis of KcsA. (**a**) Time course analysis of KcsA assembly during cell-free translation. Products of the cell-free reaction mixture were separated in a 10% SDS polyacrylamide gel. De novo synthesized proteins were labeled with ^{14}C-leucine and visualized by autoradiography showing the KcsA tetramer band at approximately 49 kDa. (**b**) Time course analysis of protein yields in different fractions. Quantification is performed by hot TCA precipitation of ^{14}C-leucine labeled KcsA. Translation mixture (TM) is separated by centrifugation into supernatant (SN) and vesicular fractions (VF)

2. Resuspend the VF in 0.1 M PBS pH 7.0 by pipetting up and down. Repeat the centrifugation step. After the second centrifugation, separate the SN from the VF. The VF contains the washed microsomes.

3. Dissolve the washed VF in 50 μL of the cholate buffer. Resuspend vigorously.

4. Mix 50 μL of the detergent resuspended microsomal fraction with 50 μL of the 15 mM lipid solution of choice (DOPG in our case) and keep the solution for rotation at 300 rpm at 4 °C for 1 h (*see* **Note 3**).

5. Add Biobeads-SM2 up to 80 vol% of the lipid mixture and incubate further overnight (ON) at 4 °C.

6. After ON incubation at 4 °C, spin down all the Biobeads-SM2 by using a short centrifugation step for few seconds and collect the supernatant from the top. The proteoliposomes can be stored at 4 °C for few days when measured continuously (*see* **Note 7**).

7. For measuring the activity of the native translocon sec61 pore naturally present in the microsomes, add 30 μL of 2.5 mM puromycin dissolved in 500 mM KCl to the vesicular fraction and incubate on ice for 45 min (VF from **step 1** in this section) to get a final concentration of 250 μM in 500 mM KCl. Puromycin combined with high salt concentration unplugs the ribosome from the microsomes and opens the sec61 pore [18].

8. Centrifuge the microsomal fraction once again at $16,000 \times g$ for 10 min at 4 °C and remove the SN. Prepare the proteoliposomes by repeating the **steps 2–6** in this section.

3.5 Formation of Lipid Bilayers

1. Lipids of interest are dissolved in octane at a concentration of 2 mg/mL. All the stocks of lipids are stored at -20 °C.

2. Lipid bilayers are formed on MECA array chips mounted on the Orbit 16 System [16].

3. Lipid bilayers are formed as described in [16, 17]: Briefly, 200 μL of electrolyte solution is added to the measurement chamber containing the MECA chip. Once the buffer is added, all the electrodes will be in open (seal resistance of few MΩ).

4. For the automated bilayer formation on the 16 cavities in parallel, a small amount of approx. 0.1 μL of DPhPC at 2 mg/mL in octane is pipetted on to the chip surface and painted with the help of a magnetic bar lying on the MECA chip [16]. Following pipetting of lipids, the bar is moved across the apertures in a circuitous fashion by performing one slow (45–180°/s) rotation of a counter magnet positioned below the chip with the help of the electromotor.

5. Lipid bilayer formation will be indicated by the change in resistance from MΩ to GΩ (for confirmation of lipid bilayer formation, *see* **Notes 8** and **9**).

3.6 Functional Assessment of MPs

1. Once the lipid bilayer is formed, add 4 μL of the proteoliposomes prepared from the puromycin treated native microsomes (no protein synthesized) directly into the buffer chamber containing the lipid bilayers and wait for fusion.

2. After incubating the lipid bilayer with the proteoliposomes, measure activity from the voltage-clamped lipid bilayers of the sec61 pore by a single channel amplifier single channel amplifier (EPC-10) and the data acquisition software Patchmaster connected to the multiplexer electronics port of the Orbit16 system [14–16]. Recordings are done at a sampling rate of 50 kHz with a 10 kHz Bessel filter (see Subheading 2.6, **step** 7). Unitary currents are recorded from the voltage-clamped lipid bilayers.

3. Analyze data: Parameters like voltage ramps for monitoring currents are measured by using Patchmaster software and analyzed by the Clampfit software version 10.7.

3.7 Case Study: Exemplary Data for sec61 and KcsA Proteins

Figure 5 shows the functional analysis of native endogenous translocon sec61 protein activity from the microsomal membranes. The diagram shows the typical voltage-gating behavior of native sec61 translocon pore typically present in the microsomes obtained after fusion to the planar lipid bilayer. At lower voltages of -20 and

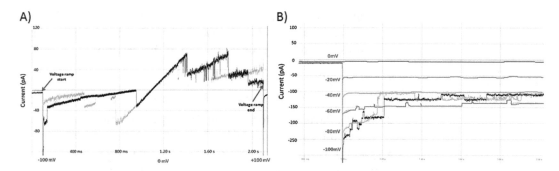

Fig. 5 Measurement of native endogenous translocon activity from the proteoliposomes formed from PG and microsomal lipids derived from the insect cell-free system. (**a**) Two Voltage ramp traces (shown in *black* and *grey*) showing the translocon currents (−100 mV to +100 mV). (**b**) Current voltage relationship of a single sec61 channel unit from the microsomal proteoliposomes at different negative voltages (−100 mV shown in *black* and other voltages shown in *different colors*) (observe the sub conductance states at larger potentials)

−40 mV, the channel is open without any subconductance states. When the potential is increased to −60, −80, and −100 mV, one can notice the subconductance states. The probability of the channel to be either closed or in one of its open states was constantly reduced at larger potentials. From these activity studies, we observed that the rate of fusion of the protein to the lipid bilayer is increased with modified microsomes (proteoliposomes) when compared to the unmodified. This also shows us the efficiency of the fusion of microsomal proteoliposomes to the lipid bilayer. From Fig. 5a, we can see that with increase in the holding potential, the probability of the channel to be open continuously is reduced [18].

The next step was to prove that this observation is applicable to the newly expressed proteins into the microsomes. The results of the activity measurements of KcsA synthesized in the eukaryotic cell-free system are presented in Fig. 6. After the synthesis reaction, **steps 1** and **2** in the Subheading 3.4 were repeated and the microsomes were extracted and used for the functionality measurements. Electrophysiology measurements were recorded and analyzed as shown in Subheading 3.6, **steps 2** and **3**. KcsA present in the microsomes showed the functional activity after fusion with the lipid bilayer. Parameters like single-channel conductance events (in the case of KcsA), all point histograms (*see* **Note 10**) were measured and analyzed. The protein exhibits a typical single-channel gating characteristic of KcsA at a transmembrane voltage of +60 mV, with a varying mean open time of few ms followed by the closure of the channel (inactivation) [13] (*see* **Note 11**).

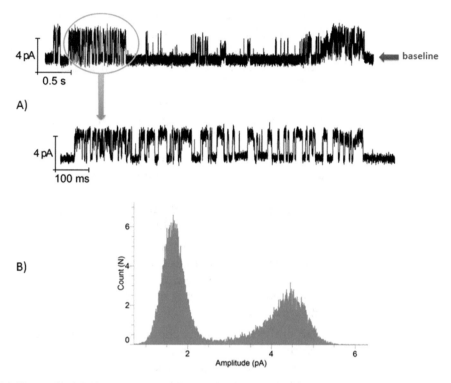

Fig. 6 (**a**) Electrophysiological characterization of KcsA demonstrating the single-channel activity. All the electrophysiology measurements were done in symmetrical 10 mM Tris, 200 mM KCl solutions with asymmetric pHs 4.0 and 7.0 on either side of the lipid bilayer. The closed state of the channel is marked with a red arrow indicating a baseline. Single-channel KcsA trace obtained at +60 mV. Zoomed-in view of the 1 s current trace obtained at +60 mV (pointed by *blue arrow*). (**b**) All-point histograms of the current trace plotted as amplitudes (pA) of the KcsA currents against counts of current amplitudes at +60 mV

4   Notes

1. Preparation of *Sf21* lysate: Fall army worm (*Spodoptera frugiperda 21, Sf21*) cells which were grown exponentially in well-controlled fermenters at 27 °C in an animal component free insect cell medium (*Sf21*: Insect-XPRESS medium, Lonza) were harvested at a density of approximately 4×10^6 cells/ mL. *Sf21* cells were collected by centrifugation at $200 \times g$ for 10 min, washed and resuspended with a HEPES-based [4-(2-hydroxyethyl)-1-piperazineethanesulfonic acid] buffer consisting of 40 mM HEPES–KOH (pH 7.5), 100 mM KOAc, 4 mM DTT to reach a final cell density of approximately 2×10^8 cells/mL. *Sf21* cells were disrupted mechanically by passing the cell suspension through a 20-gauge needle or by high-pressure homogenization followed by centrifugation at $10,000 \times g$ for 10 min to remove the nuclei and cell debris. The supernatant was applied to a Sephadex G-25 column (GE Healthcare, Freiburg, Germany), equilibrated with the above mentioned resuspension buffer, and the fractions with the highest RNA

content, as measured by absorbance at 260 nm, were pooled. Cell lysates were treated with micrococcal nuclease (S7) in order to degrade residual mRNA. Finally, aliquots of the cell lysate were immediately shock-frozen in liquid nitrogen and stored at -80 °C to preserve maximum activity.

2. Always aliquot the samples (lysates, cell-free components, NDs, proteoliposomes etc) to avoid repeated freeze thaw cycles.

3. The choice of using the lipids is flexible and can be changed according to the protein of interest. Some proteins, especially voltage-gated ion channels, are sensitive to the lipid composition [19]. The above-explained strategy can be modified with different lipids to improve the activity of the protein.

4. NDs can be prepared with different lipids and the proteins functionality can be improved by carefully adjusting the lipid composition. BR is shown as an example as it emits purple color when folded correctly but the NDs protocol can be applied to any type of MP.

5. Wait until all samples are completely thawed and fully dissolved before use. Do not keep the cell-free components outside on ice for longer periods. Once the cell-free components are mixed, it is recommended to shock-freeze them immediately in liquid N_2 and store at -80 °C.

6. In order to increase the protein yields, plasmids can be designed with additional regulatory elements such as CrPV IRES etc. [6].

7. For longer storages we recommend to store them by shock freezing in liquid N_2 at -80 °C. Always thaw the proteoliposomes or microsomes on ice when taken from -80 °C.

8. Once we notice the formation of Gigaseals indicating the formation of lipid bilayers, it is better to confirm by applying the ZAP pulse button (900 mV for 200 ms) to the underlying microelectrode cavities. Application of ZAP pulse clearly breaks the lipid bilayer (Resistance changes from GΩ to few MΩ). If there is no effect on the seal current, it indicates the clogging of MEC by the solvent or formation of multiple layers of lipids.

9. If there is an indication of clogging, it is recommended to turn the magnet multiple times until we notice the lipid bilayer presence (ZAP function test).

10. All point histograms presents the total number of opening and closing events within a defined time scale.

11. This lipid bilayer reconstitution protocol for studying the functionality of the synthesized proteins is applicable only with proteoliposomes or microsomes. In the case of soluble proteins (e. g., porins), one can directly add the purified protein over the lipid bilayer.

Acknowledgment

This work is supported by the European Regional Development Fund (EFRE) and the German Ministry of Education and Research (BMBF, No. 031B0078A).

References

1. Bear CE, Li CH, Kartner N et al (1992) Purification and functional reconstitution of the cystic fibrosis transmembrane conductance regulator (CFTR). Cell 68:809–818

2. Caterina MJ, Schumacher MA, Tominaga M et al (1997) The capsaicin receptor: a heat-activated ion channel in the pain pathway. Nature 389:816–824

3. Sanguinetti MC, Tristani-Firouzi M (2006) hERG potassium channels and cardiac arrhythmia. Nature 440:463–469

4. Sachse R, Wüstenhagen D, Šamalíková M et al (2012) Synthesis of membrane proteins in eukaryotic cell-free systems. Eng Life Sci 13:39–48

5. Brödel AK, Raymond JA, Duman JG et al (2013) Functional evaluation of candidate ice structuring proteins using cell-free expression systems. J Biotechnol 163:301–310

6. Brödel AK, Sonnabend A, Roberts LO et al (2013) IRES-mediated translation of membrane proteins and glycoproteins in eukaryotic cell-free systems. PLoS One 8:e82234

7. Zeng A, Stech M, Broedel AK et al (2013) Cell-free systems: functional modules for synthetic and chemical biology. Adv Biochem Eng Biotechnol 137:67–102

8. Sachse R, Dondapati SK, Fenz SF et al (2014) Membrane protein synthesis in cell-free systems: from bio-mimetic systems to bio-membranes. FEBS Lett 588:2774–2781

9. Quast RB, Kortt O, Henkel J et al (2015) Automated production of functional membrane proteins using eukaryotic cell-free translation systems. J Biotechnol 203:45–53

10. Bechlars S, Wüstenhagen DA, Drägert K et al (2013) Cell-free synthesis of functional thermostable direct hemolysins of *Vibrio parahaemolyticus*. Toxicon 76:132–142

11. Quast RB, Claussnitzer I, Merk H et al (2014) Synthesis and site-directed fluorescence labeling of azido proteins using eukaryotic cell-free orthogonal translation systems. Anal Biochem 451:4–9

12. Quast RB, Mrusek D, Hoffmeister C et al (2015) Cotranslational incorporation of nonstandard amino acids using cell-free protein synthesis. FEBS Lett 589:1703–1712

13. Dondapati SK, Kreir M, Quast RB et al (2014) Membrane assembly of the functional KcsA potassium channel in a vesicle-based eukaryotic cell-free translation system. Biosens Bioelectron 59:174–183

14. Cappuccio JA, Blanchette CD, Sulchek TA et al (2008) Cell-free co-expression of functional membrane proteins and apolipoprotein, forming soluble nanolipoprotein particles. Mol Cell Proteomics 7:2246–2253

15. Baaken G, Ankri N, Schuler AK et al (2011) Nanopore-based single-molecule mass spectrometry on a lipid membrane microarray. ACS Nano 5:8080–8088

16. Baaken G, Halimeh I, Bacri L et al (2015) High-resolution size-discrimination of single nonionic synthetic polymers with a highly charged biological nanopore. ACS Nano 9:6443–6449

17. del Rio Martinez JM, Zaitseva E, Petersen S et al (2015) Automated formation of lipid membrane microarrays for ionic single-

molecule sensing with protein nanopores. Small 11:119–125

18. Erdmann F, Jung M, Maurer P et al (2010) The mammalian and yeast translocon complexes comprise a characteristic Sec61 channel. Biochem Biophys Res Commun 396:714–720

19. Lee SY, Lee A, Chen J et al (2005) Structure of the KvAP voltage-dependent K+ channel and its dependence on the lipid membrane. Proc Natl Acad Sci U S A 102:15441–15446

Part III

Screening and Selection Assays

Chapter 11

Practical Considerations Regarding the Choice of the Best High-Throughput Assay

Carolin Mügge and Robert Kourist

Abstract

All protein engineering studies include the stage of identifying and characterizing variants within a mutant library by employing a suitable assay or selection method. A large variety of different assay approaches for different enzymes have been developed in the last few decades, and the throughput performance of these assays vary considerably. Thus, the concept of a protein engineering study must be adapted to the available assay methods. This introductory review chapter describes different assay concepts on selected examples, including selection and screening approaches, detection of pH and cosubstrate changes, coupled enzyme assays, methods using surrogate substrates and selective derivatization. The given examples should guide and inspire the reader when choosing and developing own high-throughput screening approaches.

Key words Protein engineering, High-throughput screening, Selection, pH assay, Selective derivatization, Surrogate substrates

1   Introduction

Since the first successful directed evolution studies in the 1990s, experience has shown that the success of protein engineering for an enzyme of interest mainly depends on the extent of knowledge of its structure–function relationships and the ability to characterize a sufficiently large number of enzyme variants [1]. A good understanding of the catalytic property that lies in the focus of the engineering project makes it possible to achieve the desired improvements with very few amino acid exchanges. In contrast, a powerful assay makes it possible to extend the number of investigated variants and thus to increase the likelihood to find an improved variant, especially in cases where details of the catalytic mechanism are poorly or not understood. A typical example is the bacterial arylmalonate decarboxylase, which is able to decarboxylate arylmethyl malonic acids in an enantiospecific manner [2]. A valid hypothesis on the mechanistic basis of enantioselectivity made it possible to invert the enzyme's enantioselectivity from 99%ee R to

Uwe T. Bornscheuer and Matthias Höhne (eds.), *Protein Engineering: Methods and Protocols*, Methods in Molecular Biology, vol. 1685, DOI 10.1007/978-1-4939-7366-8_11, © Springer Science+Business Media LLC 2018

99%ee$S$ with two simultaneous amino acid exchanges [3]. However, structure–function relationships regarding the activity of the enzyme were less understood, and it took the screening of several thousand variants to create S-selective decarboxylase variants having an activity comparable to the wild type [4, 5]. In his history of the elucidation of the protein biosynthesis machinery, Hans-Jörg Rheinberger pointed out that many unexpected discoveries do not so much depend from mere chance, but rather from the ability of the scientists to positively interpret results they had not anticipated [6]. Often, a tedious proceeding with numerous and systematic control reactions plays a crucial role in order to be able to "see" the potential in an unexpected finding. An efficient high-throughput screening method is an important prerequisite to develop truly "open" experimental systems, i.e., systems without fixed expectations on the results, for the identification of enzyme variants that cannot be anticipated or accurately predicted with the available knowledge.

The decisive criteria for such high-throughput screening (HTS) systems are the number of variants to be screened, the signal-to-noise ratio, the reproducibility and the practical effort needed to realize the screening [7]. All these technical parameters summarize our ability to break down a complex chemical reaction to a change of one single property that can be easily measured. In the following, a few practical considerations regarding the expression of enzymes in microtiter plates are discussed. Then, different measurement principles are presented.

2 High-Throughput Cultivation and Screening

Figure 1 shows the typical procedure of a high-throughput screening. The first crucial step is the transformation of the DNA library into a suitable expression strain. As some expression strains such as the derivatives of *E. coli* BL21 show a poor DNA uptake after mutagenesis, an intermediary transformation of a highly competent strain is often recommended [8]. The cells are then used for the inoculation of a master plate, which is often stored as glycerol stock for backup purposes. After inoculation of the production plate, the expression of the enzyme is induced (often by using timesaving autoinduction media) and the cells are harvested after a few hours. After lysis, parts of the cell-free extracts are transferred to the assay plate. In this step, knowing the concentration of the active enzyme is important for further dilution. Undiluted cell-free extracts are prone to cause background reactions and would thus reduce the signal-to-noise ratio. Purification in microtiter plates of enzymes bearing a polyhistidine tag is in principle possible, but tedious.

A first—often overlooked—requirement for a successful high-throughput screening is an efficient expression of the target enzyme

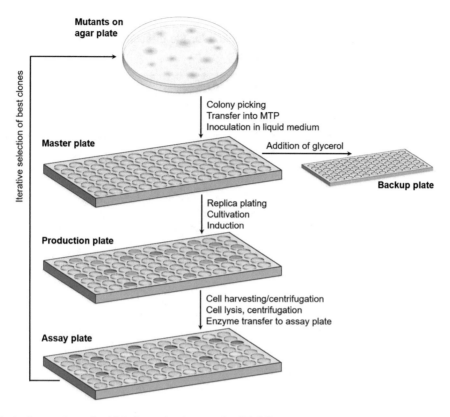

Fig. 1 Typical procedure of a high-throughput screening [7, 22]

in microtiter plates. The majority of technically used enzymes can be expressed in *E. coli*. Due to its ease of manipulation and the rich availability of tools, this is the ideal host for microtiter expression. However, the soluble expression of a considerable number of bio-catalysts is very challenging in *E. coli*. As the amount of solubly expressed enzyme is a crucial factor for a HTS assay, the optimization of cultivation conditions is often tedious, but decisive for the success of the experiment. In cases where the expression in bacteria is difficult, fungal expression systems are often an effective alternative [9]. A typical case is the expression of fungal lipases such as the widely applied lipases from the yeast *Candida antarctica*. While the technical production in fungal expression systems is straightforward, the production of lipases in microtiter plates is often challenging. Table 1 compares three approaches for the microtiter expression of the lipase A from the yeast *C. antarctica* (CAL-A). The first two tried to solve expression problems in *E. coli*, while the third chose the yeast *Pichia pastoris* as expression system.

The presence of several disulfide bonds made the soluble expression in the reductive cytosol from *E. coli* challenging. A first approach used the strain *E. coli* Origami, in which a deficient glutathione reductase system avoids the reduction of cystins to

Table 1
Comparison of different problem-solving approaches for the difficult expression of lipase A from *Canidida antarctica* (CAL-A) in microtiter plates

	Cytoplasmic expression in gram-negative bacteria	Periplasmic expression in gram-negative bacteria	Expression in yeast cells
Expression system	*E. coli* Origami	*E. coli* CR41	*P. pastoris*
Optimization tools	Fusion with thioredoxin tag; coexpression of chaperones; expression at 15 °C under control of a cold-shock promoter	Expression in the periplasma using the pelB-tag	Use autonomously replicating episomal plasmid; export of the lipase using N-terminal fusion with the *S. cerevisiae* α-mating factor.
Expression efficiency	Low	Low	High
Transformation efficiency	High	High	Very low; requires intermediary transformation of libraries to *E. coli*
Directed evolution of CAL-A	Not successful	Successful	Successful
Reference	Pfeffer et al. [10]	Brundiek et al. [11]	Sandström et al. [9]

cysteins [10]. In a second approach, fusing the lipase to a signal tag allowed the export of the peptide in the periplasm [11]. The oxidative environment of the periplasm of gram-negative bacteria facilitates the soluble expression of enzymes containing disulfide bonds. A third approach avoided the disadvantages of the bacterial expression system by using the yeast *Pichia pastoris* as expression host. Key to success was the use of an autonomously replicating plasmid [12] instead of an insertion into the yeast genome, which greatly facilitated the handling of the system [9]. As the periplasmic expression in *E. coli* and the expression in yeast were both successfully applied for the directed evolution of the enzyme, it is difficult to judge which system is more suitable. As a rule of thumb, the genetic modification of *E. coli* is usually easier, while the better expression of lipases in yeasts makes the actual high-throughput screening easier. In cases where the expression of enzymes in bacteria or yeast cells fails, cell-free expression systems are a perhaps expensive, but efficient solution [13].

3   Role of Active Biocatalyst Concentration

In many cases, particularly when the activity of an enzyme toward a substrate is of interest, it is imperative that the enzyme concentration in all wells of a microtiter plate is the same, or that the concentration of the enzyme is known in order to relate the measured activity to the actual concentration of the catalyst. Different growth in the high-throughput cultivation format and different soluble expression introduces variation in the amounts of active enzyme. The optical density or the total protein concentration is often used as an indicator for the enzyme concentration but of limited value since the expression rate of the cells may differ. His-tag purification and subsequent spectroscopic quantification of the protein is possible in microtiter plates, but often not practical. The use of reference reactions, e.g., the conversion of a second substrate [14] or the comparison of two enantiomers [15] avoids the necessity to quantify the catalyst amount in each single well as the different enzyme concentrations in the wells can be normalized. This makes the measurement of selectivity often a more gratifying task than that of actual enzyme activity.

Fusion of the enzyme to the green-fluorescent protein (GFP) is a powerful method to quantify enzyme concentration and is possible even in the presence of other fluorophores [16]. Fusion with GFP might lead to folding problems, but variants such as super folder GFP (sfGFP) cause less problems [17]. The decrease of the expressed enzyme's specific activity is usually reasonable. A problem here is the tendency of N-terminally fused GFP constructs to lead to fragments of free GFP from premature termination of translation. This bears the risk that the fluorescence signal may not correlate with the amount of correctly folded enzyme. Recently, split-GFP has been introduced to overcome this limitation [18]. Here, the enzyme of interest is fused to a fragment of GFP that reassembles with core GFP after expression and only then induces fluorescence, thus ensuring that only properly fused and folded enzyme constructs contribute to the measured fluorescence intensity.

4   Screening or Selection?

Selection and screening are alternative strategies for the characterization of a library of enzyme variants [19]. For a comparison of both approaches, Fig. 1 shows two recently developed assays for the identification of variants of N-acyl amino acid racemase (NAAAR) variants with higher racemizing activity. When coupled to enantioselective N-acyl amino acid acylases, NAAAR can be applied for the dynamic kinetic resolution of optically pure amino acids. However, the low activity of NAAAR represents a bottleneck for industrial applications. The determination of racemizing activity usually requires chiral analytics, which is not easy to perform in high-throughput-scale.

In growth assay selection assays, an enzymatic reaction is coupled to a growth advantage of a microorganism. By deletion of pathways leading to L-methione (Fig. 2a), an auxotrophic *E. coli* strain was generated [7]. While auxotrophes usually grow fast on complex media, they grow much slower on miminal media lacking an L-methionine source. By supplying *N*-acyl-D-methionine, only clones with the ability to racemize this compound are selected. On the one hand, the generation of an auxotrophic strain is not trivial, as it requires the deletion of endogenous pathways (often bacteria have more than one pathway leading to an important metabolite) and the presence of a suitable transporter for the complementation of the nutrient. On the other hand, once it is established, a selection strain is a powerful tool that allows the characterization of vast libraries with up to 10^9 clones with very simple means. The main limitation of selection approaches is that the target reaction must be coupled to the metabolism of the cell. In the example of NAAAR, the strain can be used for the directed evolution of the activity toward conversion of *N*-acyl-methionine, but not for any other amino acid. The development of selection assays for non-natural substrates is particularly challenging. Another disadvantage is that beyond a certain threshold concentration, a metabolite ceases to be a limiting factor, which makes it sometimes difficult to increase an enzymatic activity above a certain level. Selection assays are therefore highly suitable to improve a poor biocatalyst, but an activity increase of an already active and well-expressed enzyme is quite difficult to achieve. While selection assays often rely on simple experimental means such as agar plate assays, the use of a fluorescence-activated cell sorter (FACS) allows a vast throughput combined with a more accurate characterization of the individual clones [20]. Notably, FACS can be used both for screening and selection.

In order to be able to engineer NAAAR for the conversion of a wider spectrum of non-proteinogenic *N*-acyl amino acids, an enzyme-coupled screening assay was developed in which the racemization induces downstream formation of hydrogen peroxide via a cascade, which can then easily be detected by adding horseradish peroxidase and a chromogenic substrate (Fig. 2b) [21]. This assay can be applied to any amino acid that is converted by the amino acid acylase and amino acid oxidase, which increases the flexibility of the assay considerably. Moreover, the assay can also be used for the determination of kinetic parameters [22]. On the downside, the cultivation of the cells in microtiter plates and the addition of all enzymes and co-reagents for this screening method has a much higher requirement for consumables, handling, and automation than the selection assay described above, which can basically be done on agar plates. This reduces the throughput and hence the

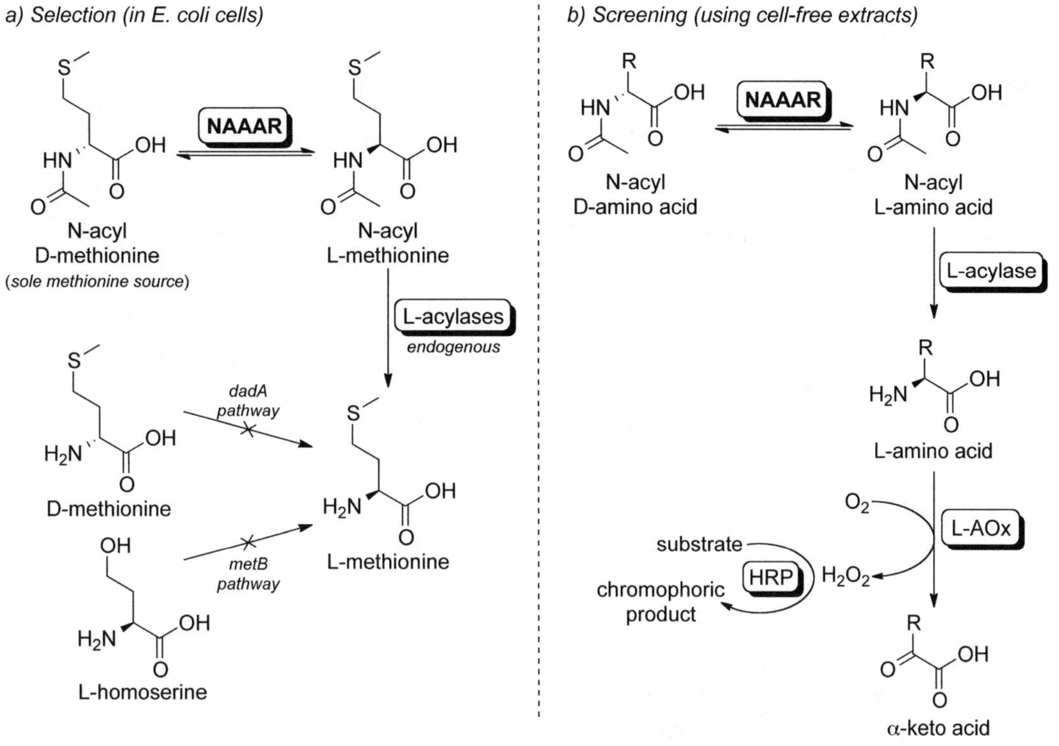

a) Selection (in E. coli cells)

b) Screening (using cell-free extracts)

Fig. 2 Comparison of (**a**) a selection [17] and (**b**) a screening assay [18] for the characterization of *N*-acyl amino acid racemases [21, 49]. *NAAAR N*-Acetyl amino acid racemase, *L-AOx* L-amino acid oxidase, *HRP* horseradish peroxidase

combinatorial diversity that can be screened. Tables 2 and 3 exemplarily list screening assays for different type of enzyme reactions and screening methods.

Albeit the limitations of selection and screening methods, a combination of both can lead to a reasonable reduction of experimental efforts. Libraries that vary catalytically relevant amino acids often contain a considerable number of inactive clones. Sorting out these clones with a rapid preselection can efficiently reduce the overall screening effort. Therefore, rather than being the actual screening tool, selection assays can be coupled to high-throughput screenings for an efficient preselection of active clones.

Table 2
Examples for typical colorimetric assays for known enzyme classes

Enzyme	EC class	Method	Reference
Dehydrogenases	EC 1 (Oxidoreductases)	NADPH-detection	[36]
P450 monooxygenases	EC 1 (Oxidoreductases)	CO-spectra; p-nitrophenyl ethers	[44]
Amino acid oxidases	EC 1 (Oxidoreductases)	Detection of hydrogen peroxide	[39]
Alcohol oxidases	EC 1 (Oxidoreductases)	Detection of hydrogen peroxide	[23]
Amine transaminases	EC 2 (Transferases)	UV-detection of acetophenone	[54]
Lipases, esterases	EC 3 (Hydrolases)	p-nitrophenyl esters (for acyl substrates)	[11]
Lipases, esterases	EC 3 (Hydrolases)	pH shift, Acetate assay	[15, 34]
Amidases	EC 3 (Hydrolases)	p-Nitroanilide derivatives	[43]
Proteases	EC 3 (Hydrolases)	p-Nitroanilide derivatives	[55]
Hydroxynitrile lyases	EC 4 (Lyases)	Detection of the cyanide ion	[40]
Arylmalonate decarboxylases	EC 4 (Lyases)	Detection of pH shift	[56]
Amino acid decarboxylase	EC 4 (Lyases)	Derivatization of *prim*-amine function	[51]
Racemase	EC 5 (Isomerases)	Use of a L-methionine auxotroph selection strain	[49]

5   Instrumental Analytics

Particularly for emerging and novel enzyme classes, often no satisfactory HTS method is available. Moreover, several reactions are difficult to monitor due to the absence of suitable markers or characteristics that can be measured directly. This holds true for example for the composition of a mixture of sesquiterpenes that all have similar physical and chemical properties. So far, the available true high-throughput screens for terpenes are specific for selected products and not generally applicable [23]. Gas chromatography is here a tedious but reliable tool for medium-throughput screenings [24].

Another notoriously challenging parameter for screening is the optical purity of compounds and hence the enantioselectivity of enzymes. While the enantioselectivity in kinetic resolutions can conveniently be determined by measuring the conversion of the pure enantiomers in separate wells [15], an asymmetric synthesis or

Table 3
Different methods available for the screening of enzymatic activity in the racemization of optically pure compounds

Method	Enzyme	Comment	Weekly throughput	Reference
Polarimetry	Glutamate racemase from *B. subtilis*	Requires purification of racemase	n.d.	[31]
Polarimetry	Mandelate racemase	Medium throughput	<100	[32, 33]
In situ NMR	Arylpropionate racemase AMDase G74C	Limited by availability of NMR measurement time and intensive handling	<10	[26]
Chiral HPLC	Arylpropionate racemase AMDase G74C	Feasible with cell-free extracts; various substrates	1000	[28]
Coupled enzyme assay	Alanin racemase	Requires purification of racemase	<100	[47]
Coupled enzyme assay	Lactate racemase		<500	[48]
Coupled enzyme assay	N-Acetylamino acid racemases	Feasible with cell-free extracts; various substrates		[21]
Selection	N-Acetylamino acid racemases	Limited to methionine		[49]

a racemization reaction requires the measurement of the enantiomeric excess of a large number of samples. Use of selectively isotope-labeled racemates and prochiral compounds, usually referred to as "pseudoracemates" or "pseudoprochiral molecules," allows to detect optical purity by mass spectrometry (Fig. 3) [25].

Similarly, an enzymatic racemization can be followed by in situ NMR by monitoring the exchange of the proton at the stereocenter with a deuteron (Fig. 3c) [26]. However, an efficient HTS method requires screening formats that ideally enable the cultivation and analysis of many samples at the same time, e.g., by scaling each individual experiment down to microtiter plate level. The downscaling and automation of MS-based assays usually requires much effort, and often an appropriate device is not available. Assays using detector enzymes (vide infra) are very useful for such a problem, but are restricted by the availability of detectable enzymes with an appropriate substrate spectrum and selectivity. Gas and liquid chromatography (GC [24, 27] and HPLC [28]) have been successfully applied in some cases for medium-throughput screens. While being

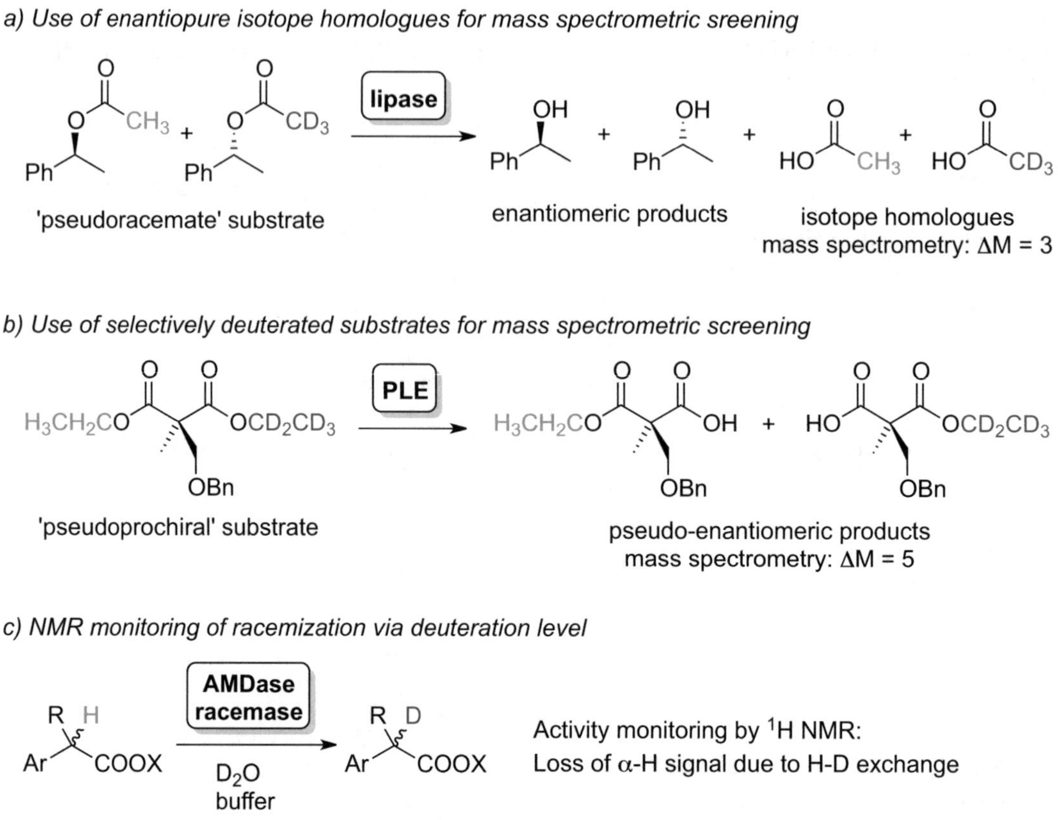

a) Use of enantiopure isotope homologues for mass spectrometric sreening

'pseudoracemate' substrate   enantiomeric products   isotope homologues
mass spectrometry: $\Delta M = 3$

b) Use of selectively deuterated substrates for mass spectrometric screening

'pseudoprochiral' substrate   pseudo-enantiomeric products
mass spectrometry: $\Delta M = 5$

c) NMR monitoring of racemization via deuteration level

Activity monitoring by ^{1}H NMR:
Loss of α-H signal due to H-D exchange

Fig. 3 Examples of typical approaches to use instrumental analytics for enantioselective conversions. (**a**) Enantioselective ester hydrolysis with enantiopure isotope homologues for product determination with mass spectrometry [53]. (**b**) Enantioselecitive ester hydrolysis monitored by selectively deuterated substrates [53]. (**c**) Racemization activity monitored as H-D exchange by in situ NMR [26]. *PLE* pig liver esterase, *AMDase* arylmalonate decarboxylase

quite expensive, these robust and reliable methods can be optimized for the measurement of several hundred samples per week. As these devices are widely available, GC and HPLC methods are usually the benchmark for the practicability of a high-throughput assay. A successful HTS should have a higher throughput or a higher cost–benefit ratio than instrumental analytics. A typical case is the optimization of the enoate reductase YqjM [8]. While the enzyme is NADPH-dependent, decoupling (i.e., the oxidation of NADPH without product formation) under aerobic conditions makes it difficult to use the consumption of NADPH as indicator for the conversion. GC-analysis of the formed product is here the preferred option. Instrumental analysis therefore presents a perhaps tedious, but often practical approach for the screening of medium-sized libraries. Pooling techniques [8] or preselection approaches [29] can be successfully used to decrease the sample number while still being able to monitor several thousand clones.

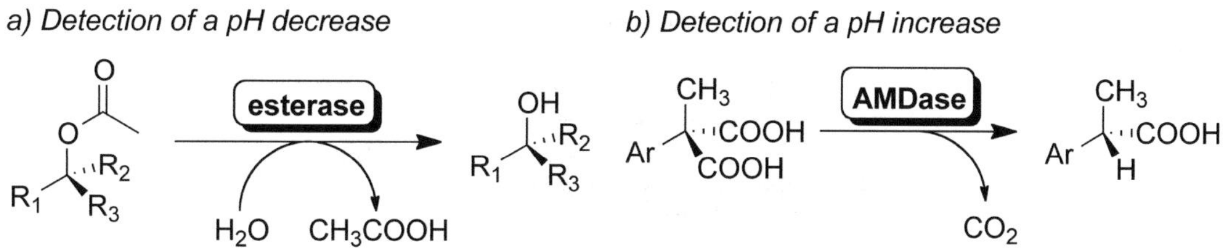

a) Detection of a pH decrease   *b) Detection of a pH increase*

Fig. 4 Monitoring of reactions by following (**a**) acidification and (**b**) basification. The change of the pH value is made visible by suitable pH indicators. *AMDase* arylmalonate decarboxylase

As many enzymatic reactions lead to a change of the pH value, monitoring the pH value with a pH indicator is an easy and widely applicable measurement principle. It is possible to detect an increase or a decrease of the pH. Figure 4 shows an acidification of the reaction medium by an enzymatic hydrolysis and a basification by cleavage of carboxyl groups through a decarboxylase. Despite the simplicity of the measurement principle, the practical performance of such an assay is often challenging. Measuring the pH usually requires working at low buffer concentrations in order not to suppress the pH changes induced by the monitored reaction. It is difficult to deduce quantitative conversions from the color change of the indicator because the reaction needs to overcompensate the buffer capacity. In addition to the pH value, also other properties such as the conductivity [30] or the rotation of plane-polarized light (with optically pure substrates) [31–33] can be efficiently used for screening purposes.

Measuring the change of a physical or chemical property of the reaction usually has to cope with an often very low sensitivity. This is a particular challenging problem in the case of low enzyme activities. Consequently, lipases and esterases are usually screened using chromogenic surrogate substrates [9] or coupled detector enzymes [15] (vide infra). If no alternative option is available, repetition of pH screens for the samples and the control reaction can increase the reproducibility, but drastically reduce the throughput [4]. Nevertheless, with a sufficiently high quantity of the biocatalyst available, pH changes are an excellent tool for the prescreening of libraries or enzyme collections [34, 35]. The great advantage is that pH screens are completely independent from the substrate and do not require the application of "surrogate" substrates.

6   Detection of a Coreagent or Side Product

The detection of coproducts that are formed in the enzymatic reaction bears the advantage that all enzymes of a class can be utilized and the true substrate (as opposed to surrogate substrates,

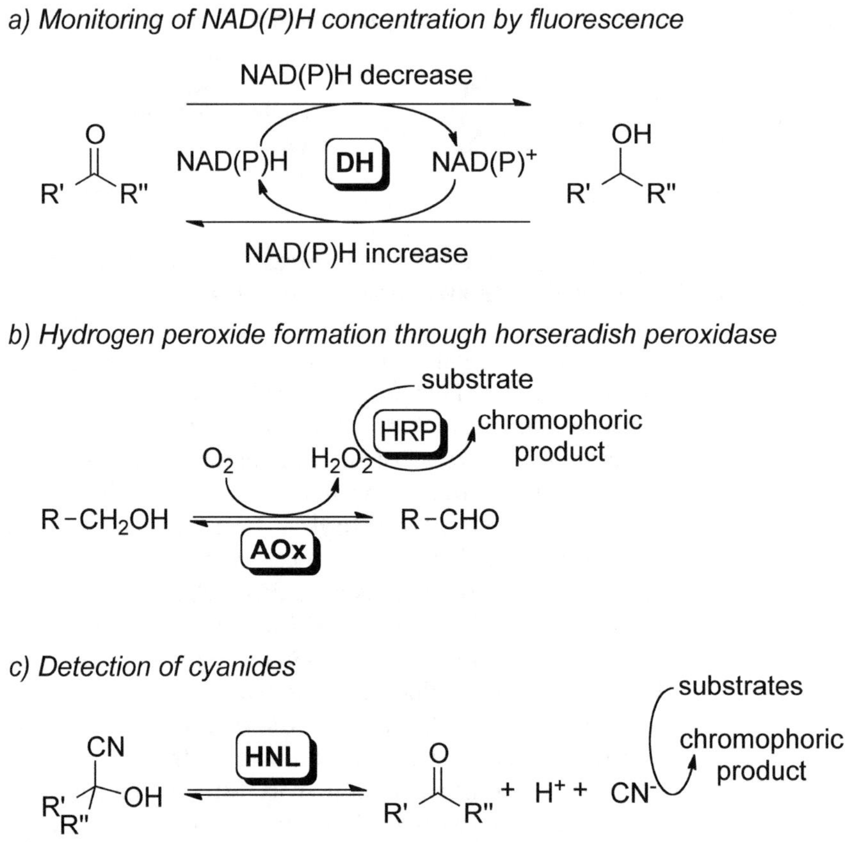

a) *Monitoring of NAD(P)H concentration by fluorescence*

b) *Hydrogen peroxide formation through horseradish peroxidase*

c) *Detection of cyanides*

Fig. 5 Detection of side products that are inherently formed from all substrates of the enzyme class. Unlike assays using surrogate substrates, the shown assays are applicable for all substrates of an enzyme class. (**a**) Monitoring the concentration of the coreagents NAD(P)H and NAD(P)$^+$ by absorption or fluorescence [36]; (**b**) detection of hydrogen peroxide by hydrogen peroxidase [39]; (**c**) detection of cyanide through a chromophoric assay [40]. *DH* dehydrogenase, *AOx* alcohol oxidase, *HRP* horseradish peroxidase, *HNL* hydroxynitrile lyase

vide infra) can be used for high-throughput screening. The perhaps most prominent example is the detection of the reduced forms of the nicotinamide cofactors NADPH and NADH by absorption or fluorescence (Fig. 5a). While this is a robust and widely applied approach, a few practical considerations should be kept in mind. The cell-free extracts of the expression host contain small amounts of cofactors and several NAD(P)H-dependent enzymes that might lead to a strong noise. It is therefore important to dilute the cell-free extracts after cultivation. An important requirement for the detection of nicotinamide cofactors is thus a high expression yield that allows several dilution steps. Using undiluted cell-free extracts requires a careful handling and numerous control reactions [36]. Moreover, in case of a low enzymatic activity it is easier to detect NAD(P)H formation than a decrease of the NAD(P)H concentration. Reversing the reaction for screening purposes is therefore often an efficient option. The same holds true for the equilibrium of the reaction. Here it is advisable to conduct the reaction in the direction of a favourable equilibrium. While monitoring the reaction of alcohol dehydrogenases is straightforward, several NAD(P)H-dependent enzymes (such as enoate reductases, vide supra) show decoupling. In such a case, the quantification of side products might not be the method of choice (vide supra).

Another widely used side product is hydrogen peroxide [37], that can be conveniently detected by using horseradish peroxidase (HRP) [38]. This is perhaps one of the most widely used enzymatic assays [39] and can be coupled to any enzyme that generates hydrogen peroxide as side product (Fig. 5b) [23]. In this case, the formed hydrogen peroxide is used by HRP to convert an additional substrate such as luminol, 10-acetyl-3,7-dihydroxyphenoxazine (AmplexRed or Ampliflu), or 3,5,3′,5′-tetramethylbenzidine (TMB) into a chromophoric product which can then easily be monitored by spectrophotometry or fluorescence spectroscopy [38].

Also less frequent side products like cyanide can be efficiently detected in colorimetric HTS assays (Fig. 5c) [40]. Again, the method relies on the selective conversion of an assay substrate by the formed side product, here cyanide, to give a chromophoric product that can easily be detected using spectrophotometric methods.

7   Chromophoric Surrogate Substrates

Assays using chromophoric surrogate substrates stand out for their high sensitivity and favourable signal-to-noise ratio [41]. However, the model substrate is usually too expensive to be used for production purposes. Usually, surrogate substates are applied for protein engineering only. Then the improved variants are characterized with the target substrate of the desired process. A typical example for the use of surrogate substrates is the screening of esterases and lipases, where the p-nitrophenyl group allows a simple screening of subtrates with different acyl groups. Hydrolysis of p-nitrophenyl esters releases p-nitrophenol, which is brightly yellow. Its anion, the p-nitrophenolate, shows strong absorption around 400 nm, which can easily be monitored spectrophotometrically (Fig. 6). Numerous p-nitrophenyl esters are commercially available and the synthesis of p-nitrophenylesters and p-nitrophenyl anilides is simple, which makes this a flexible and widely applicable assay. Consequently, numerous studies have applied this assay for the investigation of lipase activity [13], enantioselectivity [9, 42], cis-trans-selectivity (Fig. 6a) [11], and chemoselectivity [43]. Interestingly, several alcohols with a strong chromophore are available for the screening of esters with different acids, but in reverse, there are no useful acids

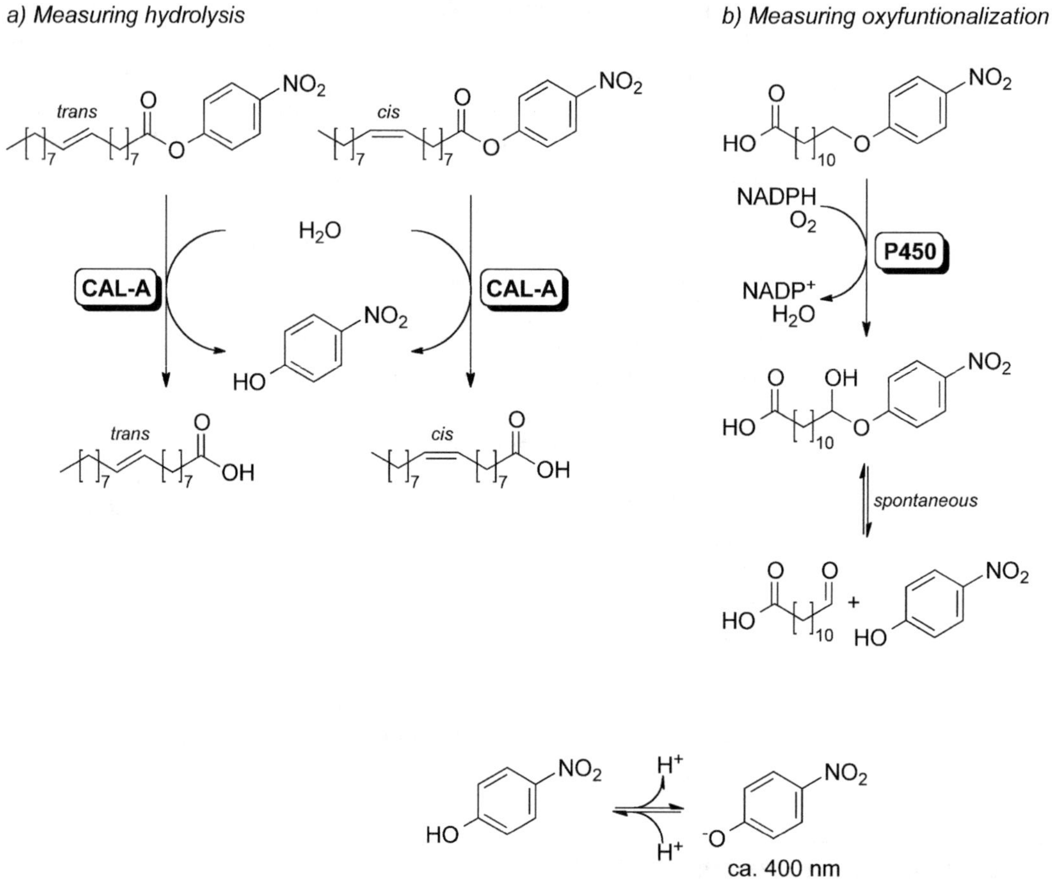

Fig. 6 Application of the chromophore of *p*-nitrophenol for high-throughput screening. (**a**) The hydrolysis of *p*-nitrophenyl fatty acid esters leads to the release of *p*-nitrophenol. By measuring the reaction rate of lipase A from *Candida antarctica* toward different isomers in separate wells, the enzyme's *trans*-selectivity could be increased [11]; (**b**) the hydroxylation of terminal *p*-nitrophenyl ethers of fatty acids by P450 monooxygenases leads to the release of *p*-nitrophenol [44]

that bear a suitable chromophore to be used in the screening of different alcohols. Studies on the directed evolution of hydrolases for the conversion of alcohols have therefore mainly relied on pH-shift [34] or coupled enzyme assays [15] (vide infra). Another disadvantage of the *p*-nitrophenyl group is its low thermostability, and thus similar surrogate substrates with higher stability have been developed [41].

Reaction principles leading to the formation or release of chromophores can often be transferred to other enzyme classes. An interesting example is the measurement of P450 monooxygenases that hydroxylate fatty acids in (ω-1)-position. Using ω-bound *p*-nitrophenyl ethers, the hydroxylation of these surrogate substrates leads to a hemiacetal, which undergoes spontaneous decomposition and thus releases *p*-nitrophenol (Fig. 6b) [44].

Optimizing an enzyme for a surrogate substrate bears the risk that the results cannot be transferred to the real target compound of interest. In an early example, directed evolution of the *p*-nitrobenzyl esterase from *B. subtilis* for the conversion of a derivative of the antibiotic compound loracarbef using a surrogate *p*-nitrophenyl ester instead of the "true substrate," a *p*-nitrobenzyl ester, yielded variants with up to 24-fold increased activity [45]. In the conversion of the true substrate, however, the best improvement obtained from this approach was 4-fold. This example underlines the risk of using surrogate substrates. A second limitation is the availability of model compounds for a given reaction. Here, commercially and inexpensive screening substrates are preferred.

8 Coupled Enzyme Assays and Selective Derivatization

While the combination of an investigated reaction with a second enzymatic step adds complexity, the high selectivity of enzymes can be utilized to selectively monitor a specific enzymatic reaction. The main advantages of coupled enzyme assays is the often outstanding selectivity of enzymes, and the fact that they are very efficient for the detection of frequent coproducts such as hydrogen peroxide (Fig. 5b) [23]. Combining two enzymatic reactions requires a careful optimization of the reaction conditions, and often makes it necessary to conduct the assay as end-point method. The main strength of coupled enzyme assays lies in the high selectivity of enzymes and the fact that enzymes from different organisms can be very efficiently combined to multienzyme one-pot reactions [46]. In the case of amino acid racemization (vide supra), a determination of enzymatic activity requires the measurement of the optical purity of a compound. While this is very difficult in microtiter scale, nature offers several enzymes that can easily distinguish between the substrate and product of a racemization. It is therefore not surprising that coupled enzyme assays [21, 47, 48] and selection methods [49, 50] are often applied for the characterization of racemases (Fig. 2). When enantioselective reactions can be monitored from enantiopure starting materials, as shown for the esterase activity screen in Fig. 7a, also kinetic resolutions can be measured without the need of actual chiral analytics [15]. In this case, the enzyme's substrate activity is determined by detection of formed acetate via an acetate assay and the activities in two separate wells toward the (*R*)-and (*S*)-enantiomer are compared.

A similar approach is the selective derivatization of analytes [51]. For instance, the decarboxylation of amino acids to their corresponding amines can be elegantly determined by derivatization. *o*-Phthalaldehyde (OPA) is a standard reagent for the conversion of amines and amino groups into fluorogenic derivatives. While OPA converts amines and amino acids with similar efficiency,

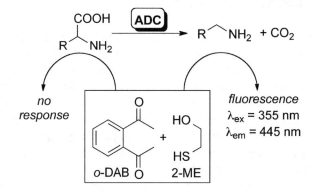

a) Conversion of pure enantiomeric substrates to determine esterase selectivity

enantiopure substrate

b) Chemoselective conversion of derivatization reagents to fluorescent products

Fig. 7 (**a**) Kinetic resolution monitoring of ester hydrolysis by conversion of pure enantiomers [15]. (**b**) HTS assay for amino acid decarboxylases using a chemoselective derivatization with *o*-diacetylbenzene (*o*-DAB)

ketones such as *o*-diacetylbenzene (DAB) can discriminate efficiently between both. Notably, it is in principle also possible to detect the pH shift of the decarboxylation (Fig. 4). For amino acid decarboxylases however, the fluorescence assay is the preferred option due to its much higher sensitivity. Due to the aforementioned drawbacks, pH screenings are usually only used if no alternative is available.

Obviously, every step for the incubation of samples with the derivative agents and their subsequent quenching or removal is an additional source of error. The use of automated pipetting devices is therefore highly recommended in order to increase the assay's accuracy. A more serious setback however, is that the reaction conditions of derivatizations are often not compatible with those of the enzymatic reactions. Quenching the reaction, though, either makes it necessary to withdraw samples at certain time points, resulting in multiple derivatization incidents per reaction monitored, or to conduct the assays as the so-called endpoint assays.

Endpoint assays give much less information than a continuous assay where the development of a reaction can be monitored [7].

9 Quality Assessment of Screens

Once a suitable expression system and an assay reaction have been established, the quality of the assay is crucial for the identification of desired variants. Here it is very helpful to compare different enzymes or enzyme variants in order to see if their different properties are reflected in the assay. It is advisable to use purified enzymes when establishing a new assay in order to define the best sample time, enzyme concentration, substrate concentration, addition of cosolvents, and other parameters. Use of cell-free extracts from cultivations in shake-flasks then gives the first hint on the robustness of the assay regarding possible interference with the metabolism of the expression hosts. Here, the dilution factor is a critical parameter. In the choice of end-point and kinetic assays, the latter is usually the preferred option as it gives more information and makes problems easier to detect. The final test is the cultivation in microtiter plates combined with high-throughput screening. For all these stages, the Z-factor is a viable parameter to assess the quality of an assay [52]. It is calculated using the mean value μ of a positive (μ_p) and negative control (μ_n) and the corresponding standard deviations σ_n and σ_p. A Z-factor of 1 is ideal, and a value higher than 0.5 is considered meaningful. Assays with Z-factors below 0.5 may be used successfully for medium-throughput screens, but are difficult for automation since they require numerous controls and repetition.

$$Z\text{-factor} = 1 - \frac{3\left(\sigma_p + \sigma_n\right)}{\left|\mu_p - \mu_n\right|}$$

Before conducting a comprehensive screening of a mutant collection with several thousand clones, it is advisable to make a prescreening of one plate containing the wild type and one single mutant plate, the so-called "wild-type" and "mutant" landscapes of a library. The amount of inactive clones depends largely on the mutagenesis strategy. Error-prone PCR (epPCR) causes the creation of stop-codons, and iterative saturation mutagenesis or simultaneous saturation mutagenesis of the active site usually generates a considerable number of inactive clones. The prescreening of selected microtiter plates should therefore yield a considerable number (5–20%) of inactive clones, confirming that that the mutagenesis strategy creates diversity. A library containing exclusively active clones usually stems from an unsuccessful mutagenesis or targets a neutral site of the enzyme. Once the Z-factor and the prescreens are satisfactory, the actual screening can start.

References

1. Bornscheuer U, Huisman G, Kazlauskas R et al (2012) Engineering the third wave of biocatalysis. Nature 485:185–194

2. Miyamoto K, Ohta H (1990) Enzyme-mediated asymmetric decarboxylation of disubstituted malonic acids. J Am Chem Soc 112:4077–4078

3. Terao Y, Ijima Y, Miyamoto K et al (2007) Inversion of enantioselectivity of arylmalonate decarboxylase via site-directed mutation based on the proposed reaction mechanism. J Mol Catal B Enzym 45:15–20

4. Yoshida S, Enoki J, Kourist R et al (2015) Engineered hydrophobic pocket of (S)-selective arylmalonate decarboxylase variant by simultaneous saturation mutagenesis to improve catalytic performance. Biosci Biotechnol Biochem 79:1965–1971

5. Gassmeyer S, Wetzig J, Mügge C et al (2016) Arylmalonate decarboxylase-catalyzed asymmetric synthesis of both enantiomers of optically pure flurbiprofen. ChemCatChem 8:916–921

6. Rheinberger HJ (2001) Experimentalsysteme und epistemische Dinge. Wallstein, Göttingen

7. Acker MG, Auld DS (2014) Considerations for the design and reporting of enzyme assays in high-throughput screening applications. Persp Sci 1:56–73

8. Bougioukou DJ, Kille S, Taglieber A et al (2009) Directed evolution of an enantioselective enoate-reductase: testing the utility of iterative saturation mutagenesis. Adv Synth Catal 351:3287–3305

9. Sandström AG, Wikmark Y, Engström K et al (2012) Combinatorial reshaping of the Candida antarctica lipase A substrate pocket for enantioselectivity using an extremely condensed library. Proc Natl Acad Sci U S A 109:78–83

10. Pfeffer J, Rusnak M, Hansen CE et al (2007) Functional expression of lipase A from Candida antarctica in Escherichia coli—a prerequisite for high-throughput screening and directed evolution. J Mol Catal B Enzym 45:62–67

11. Brundiek H, Evitt AS, Kourist R, Bornscheuer UT (2012) Creation of a highly trans fatty acid selective lipase by protein engineering. Angew Chem Int Ed 51:412–414

12. Sandström AG, Engström K, Jyhlén J et al (2009) Directed evolution of Candida antarctica lipase A using an episomaly replicating yeast plasmid. Protein Eng Des Sel 22:413–420

13. Koga Y, Kato K, Nakano H et al (2003) Inverting enantioselectivity of Burkholderia cepacia KWI-56 lipase by combinatorial mutation and high-throughput screening using single-molecule PCR and in vitro expression. J Mol Biol 331:585–592

14. Janes LE, Kazlauskas RJ (1997) Quick E: a fast spectrophotometric method to measure the enantioselectivity of hydrolases. J Org Chem 62:4560–4561

15. Bartsch S, Kourist R, Bornscheuer UT (2008) Complete inversion of enantioselectivity towards acetylated tertiary alcohols by a double mutant of a Bacillus subtilis esterase. Angew Chem Int Ed 47:1508–1511

16. Bartsch M, Gassmeyer SK, Igarashi K et al (2015) Expression of enantioselective enzymes in cyanobacteria. Microb Cell Factories 14:53

17. Pédelacq J-D, Cabantous S, Tran T et al (2006) Engineering and characterization of a superfolder green fluorescent protein. Nat Biotechnol 24:79–88

18. Santos-Aberturas J, Dörr M, Waldo GS et al (2015) In-depth high-throughput screening of protein engineering libraries by split-GFP direct crude cell extract data normalization. Chem Biol 22:1406–1414

19. Acevedo-Rocha CG, Agudo R, Reetz MT (2014) Directed evolution of stereoselective enzymes based on genetic selection as opposed to screening systems. J Biotechnol 191:3–10

20. Fernández-Álvaro E, Snajdrova R, Jochens H et al (2011) A combination of in vivo selection and cell sorting for the identification of enantioselective biocatalysts. Ang Chem Int Ed 50:8584–8587

21. Sánchez-Carrón G, Fleming T, Holt-Tiffin KE et al (2015) Continuous colorimetric assay that enables high-throughput screening of N-acetylamino acid racemases. Anal Chem 87:3923–3928

22. Böttcher D, Bornscheuer UT (2006) High-throughput screening of activity and enantioselectivity of esterases. Nat Protoc 1:2340–2343

23. Lauchli R, Rabe KS, Kalbarczyk KZ et al (2013) High-throughput screening for terpene-synthase-cyclization activity and directed evolution of a terpene synthase. Angew Chem Int Ed 52:5649

24. Yoshikuni Y, Ferrin TE, Keasling JD (2006) Designed divergent evolution of enzyme function. Nature 440:1078–1082

25. Masterson DS, Rosado DA, Nabors C (2009) Development of a practical mass spectrometry

based assay for determining enantiomeric excess. A fast and convenient method for the optimization of PLE-catalyzed hydrolysis of prochiral disubstituted malonates. Tetrahedron Asymmetry 20:1476–1486

26. Kourist R, Miyauchi Y, Uemura D, Miyamoto K (2011) Engineering the promiscuous racemase activity of arylmalonate decarboxylase. Chem Eur J 17:557–563

27. Rabe P, Dickschat JS (2013) Rapid chemical characterization of bacterial terpene synthases. Angew Chem Int Ed 52:1810–1812

28. Gassmeyer SK, Yoshikawa H, Enoki J et al (2015) STD-NMR-based protein engineering of the unique arylpropionate racemase AMDase G74C. Chembiochem 16:1943–1949

29. Yoshikuni Y, Martin VJ, Ferrin TE et al (2006) Engineering cotton (+)-delta-cadinene synthase to an altered function: germacrene D-4-ol synthase. Chem Biol 13:91–98

30. Höhne M, Schätzle S, Jochens H et al (2010) Rational assignment of key motifs for function guides in silico enzyme identification. Nat Chem Biol 6:807–813

31. Schönfeld DL, Bornscheuer UT (2004) Polarimetric assay for the medium-throughput determination of α-amino acid racemase activity. Anal Chem 76:1184–1188

32. Gu J, Liu M, Guo F et al (2014) Virtual screening of mandelate racemase mutants with enhanced activity based on binding energy in the transition state. Enzym Microb Technol 55:121–127

33. Stecher H, Hermetter A, Faber K (1998) Mandelate racemase assayed by polarimetry. Biotechnol Tech 12:257–261

34. Kourist R, Nguyen GS, Strübing D et al (2008) Hydrolase-catalyzed stereoselective preparation of protected α,α-dialkyl-α-hydroxycarboxylic acids. Tetrahedron Asymmetry 19:1839–1843

35. Nguyen GS, Kourist R, Paravidino M et al (2010) An enzymatic toolbox for the kinetic resolution of 2-(pyridinyl)but-3-yn-2-ols and tertiary cyanohydrins. Eur J Org Chem 2010:2753–2758

36. Steffler F, Guterl JK, Sieber V (2013) Improvement of thermostable aldehyde dehydrogenase by directed evolution for application in synthetic cascade biomanufacturing. Enzym Microb Technol 53:307–314

37. Rhee SG, Chang TS, Jeong W et al (2010) Methods for detection and measurement of hydrogen peroxide inside and outside of cells. Mol Cells 29:539–549

38. Veitch NC (2004) Horseradish peroxidase: a modern view of a classic enzyme. Phytochemistry 65:249–259

39. Beyzavi K, Hampton S, Kwasowski P et al (1987) Comparison of horseradish peroxidase and alkaline phosphatase-labelled antibodies in enzyme immunoassays. Ann Clin Biochem 24:145–152

40. Andexer J, Guterl JK, Pohl M et al (2006) A high-throughput screening assay for hydroxynitrile lyase activity. Chem Commun:4201–4203

41. Reymond JL (2008) Colorimetric and fluorescence based screening. In: Lutz S, Bornscheuer UT (eds) Protein engineering handbook. Wiley-VCH, Weinheim

42. Reetz MT, Wilensek S, Zha D et al (2001) Directed evolution of an enantioselective enzyme through combinatorial multiple-cassette mutagenesis. Angew Chem Int Ed 40:3589–3591

43. Kourist R, Bartsch S, Fransson L et al (2008) Understanding promiscuous amidase activity of an esterase from Bacillus subtilis. Chembiochem 9:67–69

44. Otey CR, Landwehr M, Endelman JB et al (2006) Structure-guided recombination creates an artificial family of cytochromes P450. PLoS Biol 4:e112

45. Moore JC, Arnold FH (1996) Directed evolution of a para-nitrobenzyl esterase for aqueous-organic solvents. Nat Biotechnol 14:458–467

46. Enoki J, Meisborn J, Müller A et al (2016) A multi-enzymatic cascade reaction for the stereoselective production of γ-oxyfunctionalyzed amino acids. Front Microbiol 7:425

47. Cassimjee KE, Trummer M, Branneby C et al (2008) Silica-immobilized His(6)-tagged enzyme: alanine racemase in hydrophobic solvent. Biotechnol Bioeng 99:712–716

48. Desguin B, Goffin P, Viaene E et al (2014) Lactate racemase is a nickel-dependent enzyme activated by a widespread maturation system. Nat Commun 5:3615

49. Baxter S, Royer S, Grogan G et al (2012) An improved racemase/acylase biotransformation for the preparation of enantiomerically pure amino acids. J Am Chem Soc 134:19310–19313

50. Chen IC, Lin WD, Hsu SK et al (2009) Isolation and characterization of a novel lysine racemase from a soil metagenomic library. Appl Environ Microbiol 75:5161–5166

51. Médici R, de María PD, Otten LG et al (2011) A high-throughput screening assay for amino acid decarboxylase activity. Adv Synth Catal 353:2369–2376

52. Zhang J-H, Chung TD, Oldenburg KR (1999) A simple statistical parameter for use in evaluation and validation of high-throughput screening assays. J Biom Screen 4:67–73

53. Reetz MT, Becker MH, Klein HW et al (1999) A method for high-throughput screening of enantioselective catalysts. Ang Chem Int Ed 38:1758–1761

54. Baud D, Ladkau N, Moody T et al (2015) A rapid, sensitive colorimetric assay for the high-throughput screening of transaminases in liquid or solid-phase. Chem Commun 51:17225–17228

55. Pinto MR, Schanze KS (2004) Amplified fluorescence sensing of protease activity with conjugated polyelectrolytes. Proc Natl Acad Sci U S A 101:7505–7510

56. Miyauchi Y, Kourist R, Uemura D et al (2011) Dramatically improved catalytic activity of an artificial (S)-selective arylmalonate decarboxylase by structure-guided directed evolution. Chem Commun 47:7503–7505

Chapter 12

High-Throughput Screening Assays for Lipolytic Enzymes

Alexander Fulton, Marc R. Hayes, Ulrich Schwaneberg, Jörg Pietruszka, and Karl-Erich Jaeger

Abstract

Screening is defined as the identification of hits within a large library of variants of an enzyme or protein with a predefined property. In theory, each variant present in the respective library needs to be assayed; however, to save time and consumables, many screening regimes involve a primary round to identify clones producing active enzymes. Such primary or prescreenings for lipolytic enzyme activity are often carried out on agar plates containing pH indicators or substrates as triolein or tributyrin. Subsequently, high-throughput screening assays are usually performed in microtiter plate (MTP) format using chromogenic or fluorogenic substrates and, if available, automated liquid handling robotics. Here, we describe different assay systems to determine the activity and enantioselectivity of lipases and esterases as well as the synthesis of several substrates. We also report on the construction of a complete site saturation library derived from lipase A of *Bacillus subtilis* and its testing for detergent tolerance. This approach allows for the identification of amino acids affecting sensitivity or resistance against different detergents.

Key words Screening, Lipase, Esterase, Spectrophotometric assays, Fluorimetric assays, Enantioselectivity, Complete site saturation library, Detergent tolerance

1 Introduction

In biotechnology, screening describes the method used to identify an enzyme or protein with a predefined property within a large library of variants. Theoretically, this is done by testing every single variant present in the library, but in practice, only a small part of a given library will be tested. As the screening throughput can vary greatly, we distinguish medium—(approx. 100–1000 samples) from high—(10^4–10^5 samples) and ultrahigh—($>10^6$ samples) throughput screening methods. In most cases, these assays are performed in a microtiter plate (MTP) format, for which experimental errors are commonly around 10% [1, 2]. Major considerations have to go into the design of a screening assay to ensure (1) a high signal-to-noise ratio, (2) a high significance of the results; and (3) reasonable assay costs. The main cost factors are the MTP

Uwe T. Bornscheuer and Matthias Höhne (eds.), *Protein Engineering: Methods and Protocols*, Methods in Molecular Biology, vol. 1685, DOI 10.1007/978-1-4939-7366-8_12, © Springer Science+Business Media LLC 2018

needed for cultivation, for dilutions, and for the activity assay(s), along with the reagents, e.g., chromogenic substrates for activity measurements, and the necessary equipment for automated liquid handling and measurement. In many cases, screenings will result in a high number of false-positive hits, depending on the sensitivity of the assay. False positives are partially sorted out during rounds of rescreening, with more replicates, or by subsequent more elaborate assays once the number of potential candidates is reduced by pre-screening. Many screening regimes involve a primary screening round to identify the general property or activity searched for. Such primary screenings are often carried out on agar plates containing some sort of indicator, e.g., clones expressing a gene-encoding an esterase can be identified by halo formation on agar plates containing the esterase substrate tributyrin; additional examples include screens for organic solvent tolerance and low temperature stability [3, 4]. Ultrahigh throughput can be achieved if the desired property can be linked to a fluorescence signal inside a cell, or an artificial compartment, which can be assayed using fluorescence-activated cell sorting (FACS). Examples include the directed evolution of sialyltransferases and monitoring the formation of sialosides in intact *Escherichia coli* cells by selectively trapping the fluorescently labeled products [5] and the identification of enantio-selective hydrolases using a cell surface display method [6].

In this chapter, we describe three experimental protocols as examples for high-throughput screening methods to identify lipases and esterases.

1.1 High-Throughput Screening for Lipases and Esterases

Screening assays for different enzymes may involve (1) the direct detection of a reaction product or (2) its analysis after consecutive follow-up reactions or addition of indicators. Hence, a screening assay devised to detect enzyme activities in a metagenomic library would differ strongly from an assay set up to find optimal reaction conditions or to determine enantioselectivities. In the following part, we will discuss high-throughput screening methods developed to identify and characterize biotechnologically relevant hydrolytic enzymes belonging to the family of lipases and esterases. However, the general strategy outlined below is also applicable for a variety of other enzymes.

The identification of new enzymes with novel properties usually requires the screening of large libraries consisting of several thousands to hundred thousands of clones which all carry DNA fragments derived from genomes of existing (micro)organisms or environmental DNA isolated by metagenomics approaches [7]. For screening of such large libraries, high-throughput methods to detect enzymatic activity are needed. Agar plates supplemented with substrates such as tributyrin or olive oil and rhodamine B serve as fast and simple assay methods to screen for the presence of lipases and esterases [8, 9]. Similarly, utilizing a halo formation as an output for enzymatic activity, engineered cellulases can be screened on agar plates supplemented with Azo-CM-Cellulase [1]. Glycosidases can be identified with substrates such as X-Gal or X-Glc turning colonies producing galactosidase or glucosidase activity, respectively, blue for simple identification [10]. This solid phase method can also be implemented using pH indicators if the respective enzymatic activity causes a shift in pH over time (demonstrated for example for lipases and glycosynthases) [11]. Furthermore, agar plates can be used for selection when supplemented with compounds which release a sole carbon source upon enzymatic hydrolysis, as shown for example for identification of lypolytic activity [12, 13]. Subsequently, colonies showing enzymatic activity can be transferred into a deep-well MTP, grown and subjected to further quantitative screening methods. The most widely used assays for screening of enzymatic activity are based on colorimetric and fluorimetric detection methods [14]. These assays can most often be carried out in microtiter or microarray plates allowing for high-throughput analysis. The assays are usually based on surrogate substrates, which release a fluorescent or colored product upon hydrolysis of the substrate. Glycosides, ethers, or esters containing an umbelliferyl or 4-nitrophenyl moiety are the most widely used substrates for glycosidase, monooxygenase, and esterase/lipase activity, respectively [15, 16]. Further chromogenic and fluorogenic moieties exhibiting different absorption and emission properties are resorufin, fluorescein, and 7-hydroxy-9H-(1,3-dichloro-9,9-dimethylacridine) [14, 17, 18]. Even though these substrates are relatively simple to synthesize and well-characterized, some drawbacks exist. The stability, in particular of the esters, can often be low, due to the low pK value of the leaving group (umbelliferyl/nitrophenolate ion), leading to a high rate of autohydrolysis [19]. If bulky moieties are coupled in close vicinity to the bond to be hydrolyzed, the activity of the respective enzyme might be negatively influenced as compared to the natural substrate. In order to overcome such disadvantages various substrate derivatives such as esters with self-immolative chemical functionalities such as mixed carbamates and acyloxymethylethers were synthesized, which upon hydrolysis of the ester spontaneously form carbon dioxide or formaldehyde, respectively, together with chromophore/fluorophore [19–21]. Reymond and coworkers developed an ester-linker, which releases a 1,2-diol moiety after enzymatic hydrolysis [22]. The umbelliferone is subsequently released by β-elimination catalyzed by BSA after cleavage of the diol using sodium periodate. The higher stability of these esters enabled the development of a low-volume assay on solid support by impregnating silica gel plates with the fluorogenic substrates [23]. This allowed the amount of enzyme solution needed for each reaction to be reduced to 1 μL or less. A diverse array of *p*-nitrophenyl- and umbelliferyl-esters are commercially available or can be synthesized making them attractive substrates for fingerprinting of lipase/esterase activities (Table 1).

Such in-depth characterizations can be carried out in MTPs with each well containing a different ester leading to a map of activities (fingerprint) indicating for example selectivity or enantioselectivity, which can be compared with other lipase fingerprints [23, 24]. This method has been further extended to the use of single substrate cocktails reducing the amount of stock solutions [25]. The cocktail is analyzed after the enzymatic reaction via HPLC in a single run. In order to avoid the use of surrogate substrates, which might not reflect activities toward real substrates, many indirect assays have been developed such as those using pH indicators [26–28]. Molecules, as for example p-nitrophenol or bromothymol blue, are added to follow the shift in pH during hydrolysis of the targeted natural substrate resulting in a change of their colorimetric properties. This type of assay can be carried out in MTP or as solid phase assay on agar plates. It is prerequisite, however, for the pK_a of the indicator to be in the same range as the chosen buffers pK_a which can limit the range of screenable conditions [29]. A similar method was developed utilizing the pH-dependent fluorescence quenching of fluorescein [26]. Another approach trying to avoid the use of unnatural bulky groups is the use of an enzymatic coupled reaction. Baumann et al. introduced a method in which the conversion of acetate esters was followed by further converting the produced acetate in an enzymatic cascade [30]. The cascade in turn results in the reduction of NAD$^+$ to NADH, which can be quantified by photometric methods without the need to use substrates containing bulky chromogenic groups. Recently, a novel method was described to determine lipase activity based on fluorescence resonance energy transfer (FRET) [31]. This method used a natural-like triglyceride substrate containing an EDANS fluorophore and a DABCYL quencher at the end of two of the fatty acid chains of the triglyceride. The intact triglyceride does not show fluorescence, but hydrolysis by a lipase removes either the fluorophore or the quencher from the triglyceride thus changing the FRET signal.

When considering enzymes for the use in synthesis of fine chemicals or pharmaceuticals, activity and chemoselectivity are important, but enantioselectivity is also an absolute prerequisite [32]. Screening methods determining the enantioselectivity (usually described by the E-value) are mostly carried out by a preparative scale kinetic resolution and lengthy analysis methods such as GC [33], HPLC [34], MS [35], or NMR [36] measurements, which raise the cost and lower the throughput of the screening method. The MTP assays mentioned above can also be used for screening of enantioselectivity with high throughput when using chiral substrates. Most of the mentioned assays are, however, limited to the

Table 1
A selection of commercially available substrates suitable for activity measurement of lipases

Commercially available substrates	Analytical method
	Absorption; $\lambda_{max} = 410$ nm
	Absorption; $\lambda_{max} = 410$ nm
$n = 5, 7, 11, 13, 15$	Absorption; $\lambda_{max} = 410$ nm
$n = 1, 29$	Fluorescence; $\lambda_{ex} = 385$ nm, $\lambda_{em} = 502$ nm
	Absorption/Fluorescence; $\lambda_{ex} = 571$ nm, $\lambda_{em} = 585$ nm
	Absorption/Fluorescence; $\lambda_{ex} = 571$ nm, $\lambda_{em} = 585$ nm

(continued)

Table 1
(continued)

Commercially available substrates	Analytical method
	Fluorescence; $\lambda_{ex} = 490$ nm, $\lambda_{ex} = 515$ nm

screening of enantioselectivity towards chiral acids, as the alcoholic moiety is usually an achiral chromophore/fluorophore. In these cases, the enantiomers are tested in separate reactions and the E-value is determined by comparison of both initial reaction rates [37, 38]. This E-value represents an apparent approximation as competition between the enantiomers in the active site of the enzyme is neglected. An extension of this method is the Quick E method developed by Janes and Kazlauskas [28, 39]. The group introduced an achiral reference compound (resorufin ester) during the conversion of a specific enantiomer to simulate a competition in the active site. The E-values determined by this method showed closer similarity to the determined true E-values than the apparent E-value calculated by conversion of the single enantiomers. The Quick E methodology has been recently extended even further to the Quick ee by Lima et al. for a closer approximation of the ee-value [40].

The first protocol of this chapter (Subheading 3.1) describes the general synthesis of p-nitrophenylester via the Steglich esterification method, giving easy access to a variety of structurally differing esters for lipase activity measurements. The method activates the carboxylic acid toward a nucleophilic attack in a reaction with EDC·HCl and 4-DMAP. The synthesis is carried out in one step and purified by simple column chromatography, leading to pure product in good yields. The described carboxylic acids vary in chain length, degree of branching, and also structure (alkyl vs. aryl), allowing for the use of these p-nitrophenyl esters for a detailed characterization of esterases (fingerprinting assay).

The second protocol (Subheading 3.2) utilizes the synthesized p-nitrophenylesters for an activity-fingerprinting assay to characterize the lipases in respect to chemoselectivity (chain length, degree of branching, and arylic substituents) and also allowing determination of the apparent enantioselectivity. As described above, p-nitrophenyl esters are common surrogate substrates for enzyme activity assays. The liberated p-nitrophenolate ion is simply detected by measurement of the absorption at 410 nm. Essential condition for the absorption measurements is a buffer pH value above pH 7 to ensure the deprotonation of pNP as the sensitivity of the assay is lowered strongly when carried out below a pH of 7. A slight disadvantage of these esters is their low stability and often low solubility of the compounds. Due to the instability, caused by the low pK_a value of the nitrophenolate ion, the measurement requires control reactions without addition of enzyme in order to determine the rate of autohydrolysis. The low solubility of the esters can be circumvented by the use of cosolvents such as acetonitrile or DMSO. The content of cosolvent should not exceed a concentration of 10% in order to avoid an influence on the enzymatic activity.

The third protocol (Subheading 3.3) describes a variation of the assay method to screen the influence of every possible single amino acid substitution in the *Bacillus subtilis* lipase A towards a change in tolerance against different surfactants based on the loss of activity after 2 h incubation. For this particular study we have created a high-quality mutant library, which was fully sequenced to ensure the complete coverage of all possible single amino acid substitutions at every amino acid positions of the protein.

2 Materials

2.1 Synthesis of Colorimetric Assay Substrates (p-Nitrophenyl Esters)

1. The ester synthesis requires a carboxylic acid. We suggest to synthesize esters using the following acids: Propanoic acid, butyric acid, caprilyc acid, lauric acid, palmitic acid, dimethylpropanoic acid (pivalic acid), and 2-methylpropanoic acid (isobutyric acid).

   (R)- and (S)-2-Metyldecanoic acid (*see* **Note 1**).

   (S)- and (R,S)-2-(4-(2-Methylpropyl)phenyl)propanoic acid (ibuprofen).

   (-)-(R)- and (+)-(S)-2-(6-Methoxynaphthalen-2-yl) propanoic acid (naproxen).

2. 4-Dimethylaminopyridine (4-DMAP).

3. Ethylcarbodiimide hydrochloride (EDC HCl).

4. p-Nitrophenol (pNP).

5. Organic solvents: dichloromethane (DCM), petroleum ether (PE), and ethyl acetate (EA) (*see* **Note 2**).

6. Saturated sodium bicarbonate solution.

7. Silica gel for chromatography (Silica 60, 0.040–0.063 mm, 230–400 mesh).

8. Anhydrous magnesium sulfate.

9. Funnel with Celite™ filter material.

10. Glassware and Equipment: *Schlenk*-flask with magnetic stirrer and rubber septum, *Schlenk*-line providing dry nitrogen atmosphere and vacuum, Erlenmeyer- and round bottom flasks, separation funnel, chromatography column, magnetic stirrer, ice bath (0 °C), thermometer, and heat gun.

2.2 Colourimetric Fingerprinting Assay

1. Substrate stock solution: 20 mM *p*-nitrophenyl ester in acetonitrile (*see* **Note 3**).

2. Assay buffer: 50 mM KH_2PO_4, 50 mM K_2HPO_4, adjust to pH 8 using NaOH.

3. *p*-nitrophenol stock solution: 10 mM *p*-nitrophenol in acetonitrile.

4. Enzyme solution in assay buffer.

5. Microtiter plates (flat bottom, clear, nonsterile, without lid).

2.3 HTS for Detergent Tolerance

1. Gene encoding the lipolytic enzyme and mutant library, cloned in pET22 or pET28 plasmid (*see* **Notes 4** and **5**). Prepare the libraries in 96-well MTP before the start of the screening and store the MTP at −80 °C for convenient use once the screening is under way (*see* **Note 6**).

2. *E. coli* strains BL21(DE3), Tuner(DE3), and JM109(DE3).

3. Antibiotic stock solutions: 50 mg/mL ampicillin or kanamycin in aq. dest. Store at −20 °C.

4. LB medium: 10 g tryptone, 10 g NaCl, 5 g yeast extract, ad 1000 mL aq. dest., adjust pH to 7.5 with NaOH, autoclave at 121 °C, 20 min and store at RT (*see* **Note 7**).

5. 1.25× TB medium: 12 g casein, 24 g yeast extract, 5 g glycerol, ad 800 mL aq. dest., autoclave at 121 °C, 20 min and store at RT.

6. Autoinducing medium: 800 mL TB-medium (1.25×), 100 mL lactose (20 g/L), 90 mL potassium phosphate buffer pH 7 (1 M), and 10 mL glucose (50 g/L), store at 4 °C (*see* **Note 8**).

7. LB agar: 15 g agar, ad 1000 mL LB medium (*see* **Note 9**). After autoclaving allow the LB agar bottles to cool to approximately 45 °C. Add the required amount of 1000× antibiotic stock solution and mix well. Pour onto plates immediately.

8. KPi buffer (50 mM, pH 7): 39 mL KH_2PO_4 solution (0.5 M), 61 mL K_2HPO_4 solution (0.5 M), ad 1000 mL aq. dest., autoclave at 121 °C, 20 min and store at RT.

9. Tris–HCl buffer (1 M, pH 8): 1 M Tris base, adjust to pH 8 with HCl.

10. Sørensen buffer: 42.5 mL Na_2HPO_4 (8.9 g/L), 2.5 mL KH_2PO_4 (6.8 g/L).

11. 2× *p*NPB substrate solution: 17.2 μL *p*NP-butyrate, 5 mL acetonitrile, and 45 mL Sørensen buffer (*see* **Note 10**).

12. 2× *p*NPP substrate solution: 30 mg *p*NP-palmitate, 5 mL 2-propanol, 45 mL Sørensen buffer, 103.5 mg sodium deoxycholate, and 50 mg gum arabic (*see* **Note 11**).

13. Surfactant solution: 0.01–0.03% (w/v) surfactant, 50 mM Tris–HCl pH 8 (*see* **Note 12**).

14. Pipetting tips for multichannel pipettes: epTips 1200 μL (blue) and 300 μL (yellow) (*see* **Note 13**).

15. Disposable reagent reservoirs (*see* **Note 14**).

16. Disposable polystyrene weighing dishes.

17. Microtiter plates: 96-well flat bottom (clear, nonsterile, without lid), 96-well flat buttom (any color, sterile, with lid), 96 V-well (any color, sterile, with lid).

2.4 Equipment

1. Xplorer plus (Eppendorf, Germany) 8-channel automatic pipette 100–1200 μL.

2. Xplorer plus (Eppendorf, Germany) 12-channel automatic pipette 5–100 μL.

3. 96-Pin steel replicator and ethanol bath.

4. Spectramax 250 (Molecular Devices, USA) microtiter plate reader for absorbance, and Infinite M1000pro (Tecan, Switzerland) microplate reader for absorbance and fluorescence measurements.

5. Rotanta 460 R centrifuge equipped with a 5264 Rotor (4 × 1 kg) for four microtiter plates (Andreas Hettich GmbH & Co. KG, Germany).

6. TiMix5 with RCS-550 microtiter plate shaker and incubation chamber with cooling option (LTF Labortechnik GmbH & Co KG., Germany).

7. Freedom EVO 200 (Tecan, Switzerland) for automated liquid handling.

3 Methods

3.1 Synthesis of p-Nitrophenyl Esters

1. Connect the *Schlenk*-flask (containing a magnetic stirrer and closed with a rubber septum) to the *Schlenk*-line in order to prepare the flask for the reaction under inert conditions. The evacuated flask is heated carefully using a heat gun while attached to the vacuum. After cooling down to room temperature, the flask is then flushed with dry nitrogen. This process of evacuating, heating and flushing is repeated twice again (*see* **Note 15**).

Table 2
Required weight of carboxylic acid for 1 mmol product

Carboxylic acid	Weight [mg]/(1 mmol)
Propanoic acid	74.1
Butyric acid	88.1
Caprylic acid	144.2
Lauric acid	200.3
Palmitic acid	256.4
Pivalic acid	102.1
Isobutyric acid	88.1
(S)-/(RS)-2-(4-(2-Methylpropyl)phenyl) propanoic acid	206.3
(-)-(R)-/(+)-(S)-2-(6-Methoxynaphthalen-2-yl) propionic acid	230.3
(S)-/(R)-/(RS)-2-Methyldecanoic acid	186.3

2. While flushing the *Schlenk*-flask with dry nitrogen, *p*-nitrophenol (1 eq, 1 mmol, 139 mg), 4-DMAP (0.1 eq, 0.1 mmol, 12.2 mg) and the chosen acid (1 eq, 1 mmol, *see* Table 2 for amount) are added to the inert flask and dissolved in dry dichloromethane (5 mL/mmol acid) (*see* **Note 16**).

3. Cool the reaction mixture to 0 °C using an ice bath.

4. While keeping the temperature at 0 °C, add EDC·HCl (1.2 eq, 1.2 mmol, 230 mg) to the reaction while flushing with dry nitrogen.

5. Stir the reaction for 1 h at 0 °C and further 12 h at room temperature (*see* **Note 17**).

6. Quench the reaction by addition of distilled water (in excess) and vigorous stirring (*see* **Note 18**).

7. Transfer the reaction mixture into a separation funnel and separate the aqueous phase from the organic phase. Wash the organic phase using first an aqueous sodium bicarbonate solution and secondly distilled water (*see* **Note 19**).

8. The organic phase is then dried by addition of anhydrous magnesium sulfate (*see* **Note 20**) and filtered through a funnel with Celite™ into a round bottom flask.

9. The solvent is evaporated under reduced pressure using a rotatory evaporator yielding the crude product.

10. Purify the crude product by column chromatography with silica gel using a mixture of PE–EA in a ratio of 90:10 (*see* **Note 21**). Fractions containing the desired ester are combined

and the solvent removed under reduced pressure resulting in either a colorless oil or solid as the final product.

11. For identification of the product and determination of the compounds purity, analytical data such as NMR, IR, and GC-MS should be measured. For exemplary data of 2-(R)-methyl-decanoic acid and literature refs. 41–47 for further compounds *see* **Note 22**.

3.2 Colourimetric Fingerprinting Assay

1. For calibration, a dilution series of 8–10 *p*NP solutions in acetonitrile in the range of 0.1–5 mM is created by diluting the stock solution. Vortex each solution for 1–2 min.

2. Pipette 135 μL assay buffer for each dilution into a MTP and add 15 μL of the calibration sample. Measure the absorption at $\lambda = 410$ nm in a microtiter plate reader. Plot each measured absorption [mAu] against the correlating concentration of each *p*NP solution [mM] in order to receive a calibration curve (*see* **Note 23**).

3. Create a master mix solution of assay buffer containing the substrate in a concentration of 0.25 mM and 10% acetonitrile using the substrate stock solution (*see* **Note 24**).

4. 120 μL master mix solution of each pNP-ester is pipetted into an MTP.

5. Add 30 μL of enzyme solution to each well containing the assay master mix and mix well by repeated pipetting (*see* **Note 25**). Blank reactions for determination of the autohydrolysis rate are prepared for each substrate by adding 30 μL assay buffer.

6. The absorption of the reaction is measured at 410 nm after 2–5 min reaction time in a microtiter plate reader (*see* **Note 26**) (Fig. 1).

7. For the activity fingerprint: Subtract the blank values (representing the autohydrolysis) from the measured absorption values of each reaction and calculate the enzymatic activity using the slope of the calibration curve (Volumetric activity: Eq. 1; Specific activity: Eq. 2). Each activity can then be assigned a color-intensity in order to create the fingerprint pixel diagram.

$$U_{\text{vol.}}\left[\frac{\mu\text{mol}}{\text{min} \times \text{mL}}\right] = \frac{\left(\frac{\text{Abs[mAu]}}{\varepsilon_{\text{Calib.}}\left[\frac{\text{mAu}}{\text{mM}}\right]}\right) \times 5}{t\,[\text{min}]} \tag{1}$$

Abs = Absorption [mAu] of reaction

$\varepsilon_{\text{Calib.}}$ = Slope of calibration curve [mAu/mM]

t = Time of enzymatic reaction

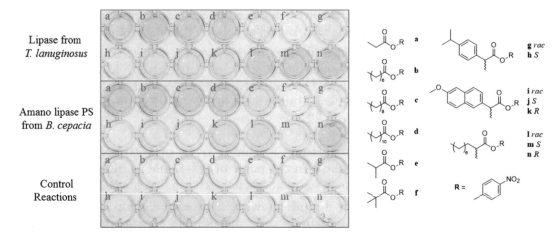

Fig. 1 Chemoselectivity and enantioselectivity screening of lipases from *T. lanuginosus* and *B. cepacia* using *p*-nitrophenyl ester *a–n*. The reaction was run for 5 min at room temperature (25 °C). Control reactions were carried out by adding buffer without enzyme in order to determine the rate of autohydrolysis

$$U_{\text{spec.}}\left[\frac{\mu\text{mol}}{\text{min} \times \text{mg}}\right] = \frac{U_{\text{vol.}}\left[\frac{\mu\text{mol}}{\text{min} \times \text{mL}}\right]}{c\left[\frac{\text{mg}}{\text{mL}}\right]} \qquad (2)$$

c = Protein concentration [mg/mL] of enzyme sample

8. For determination of apparent enantioselectivity E_{app}: Determine the hydrolytic reaction rates of the separate enantiomers of a single chiral compound (e.g., (*S*)- and (*R*)-2-methyldecanoic *p*-nitrophenyl ester) to determine E_{app} using Eq. 3 (*see* **Note 27**).

$$E_{\text{app}} = \frac{\text{Conversion rate}_{\text{fast enantiomer}}}{\text{Conversion rate}_{\text{slow enantiomer}}} \qquad (3)$$

3.3 HTS for Detergent Tolerance

3.3.1 Optimization of the Expression System

1. Transform commercially available *E. coli* strains BL21(DE3), Tuner(DE3), and JM109(DE3) with the plasmids encoding the wild-type hydrolase enzyme and incubate over night on selective LB agar plates containing either 50 μg/mL kanamycin (pET28 derivatives) or 100 μg/mL ampicillin (pET22 derivatives).

2. Prepare a sterile flat-bottom MTP with 150 μL selective LB medium in each well using a 8-channel pipette and a sterile reagent reservoir.

3. Pick eight clones from each agar plate into 150 μL LB medium using a tooth pick, close the lid and seal the MTP with tape. Incubate overnight at 37 °C and 900 rpm with 3 mm shaking diameter (*see* **Note 28**).

4. The following day prepare three sterile V-well MTPs with 120 µL selective TB autoinducing medium in each well using a 8-channel pipette.

5. Use a 96 pin steel replicator to transfer a few microliter of each culture and make three copies of the preculture in TB auto-inducing medium. Incubate one of each plate for 16 h at 25 °C, 30 °C, or 37 °C, respectively (see **Note 29**).

6. After incubation, centrifuge for 30 min at 1500 × g to pellet the cells.

7. Transfer 10 µL of the supernatant into a nonsterile flat-bottom MTP and dilute with 90 µL of 50 mM Tris–HCl pH 8 and add 100 µL of 2× pNPP Substrate solution

8. Monitor the increase of OD_{410} using a microtiter plate reader (see **Note 30**).

9. Use the slope of each well to calculate the amount of active lipase and the variation between the eight biological replicates. Data analysis will reveal the most suitable cultivation conditions (see **Note 31**).

3.3.2 Choosing Appropriate Detergent Concentrations

1. Prepare a benchmarking plate for the assay development. Pick 92 clones of the selected wild-type plasmid and expression strain combination into a sterile flat-bottom MTP together with four clones of cells transformed with the empty vector without the *lipA* gene to measure the experimental background (see **Note 32**).

2. Incubate the MTP containing the benchmarking variants overnight at 37 °C, 900 rpm in 150 µL of selective LB medium, add glycerol or DMSO, seal the plate with an adhesive foil and store at −80 °C (see **Note 33**).

3. Thaw the frozen benchmark microtiter plates (MTPs) and copy the plate into a sterile flat-bottom preculture MTP filled with 150 µL LB medium per well using a 96-pin steel replicator and incubate over night at 37 °C, 900 rpm. Store the benchmark MTP again at −80 °C in case you will need to repeat this process.

4. Transfer the fresh culture into a sterile V-well microtiter plate containing 120 µL autoinducing medium for target gene expression. Incubate 16 h at the previously determined optimal cultivation temperature, e.g., 30 °C in our case.

5. Collect the supernatant after 30 min centrifugation at 1500 × g and dilute if necessary.

6. Transfer 10 µL of the diluted culture supernatant to a nonsterile flat-bottom MTP and add 90 µL of various surfactants in concentrations ranging from 0% (w/v) to 0.1% (w/v), incubate for 2 h at room temperature (see **Note 34**).

7. After the incubation the activity is measured as described above, but using *p*NPB as substrate instead (*see* **Note 35**).

8. Calculate residual activity by first subtracting the activity measured in the empty vector controls to remove the background rate of activity. Then calculate the relative change in activity in each surfactant-containing reaction compared to the activity of the sample with buffer only. Based on the results, select the most appropriate surfactants (*see* **Note 36**).

9. Repeat the process and only use the chosen concentration(s) of the surfactants, instead of the gradient, to estimate the experimental error using all the available replicates on the MTP.

3.3.3 Screening of the Bacillus subtilis Lipase A Library

1. Prepare 60 mL of LB (100 µg/mL ampicillin for *bla* expressing *E. coli*) und transfer it into a sterile reagent reservoir for multichannel pipettes.

2. Thaw four microtiter plate libraries plates (−80 °C) on ice (*see* **Note 37**).

3. Fill four sterile flat-bottom MTPs with 150 µL of LB (100 µg/mL ampicillin) with a 8-channel automatic pipette (*see* **Note 38**).

4. Use a 96-pin steel replicator to copy the master plates into the plates from **step 3**, sterilize the replicator in 70% (v/v) ethanol between the different plates, and let it dry before inserting it into the next culture.

5. Close the plates and use tape to seal the plate by taping together the lid and the plate along the outside of the MTP.

6. Incubate overnight at 37 °C and 900 rpm.

7. Remove the preculture from the incubator and store at room temperature until inoculation of the main culture (*see* **Note 39**).

8. Prepare 60 mL of TB autoinduction medium (100 µg/mL ampicillin) and transfer it into a sterile reagent reservoir for multichannel pipettes.

9. Fill four sterile V-well bottom MTPs by adding 120 µL of TB autoinduction medium (100 µg/mL ampicillin).

10. Use a 96-pin steel replicator to copy the preculture into the plates from **step 9**, sterilize the replicator in 70% (v/v) ethanol between the different plates, let it dry before inserting it into the next culture.

11. Close the plates and use tape to seal the plates completely by taping together the lid and the plate along the outside of it (*see* **Note 40**).

12. Incubate for 16 h at 30 °C 900 rpm.

13. Separate the supernatant by centrifugation, 30 min 1500 × *g* (*see* **Note 41**).

14. Transfer 100 μL of the supernatant into a new nonsterile F-bottom MTP and dilute with 100 μL of 50 mM Tris–HCl pH 8.

15. Prepare five nonsterile F-bottom MTPs (four detergents and one buffer control) for each library plate (e.g., 20 plates if four library plates are screened in parallel with four different detergents in one concentration, or 52 to test at three different concentrations as in our screening) and label MTPs accordingly (*see* **Note 42**).

16. Prepare reagent reservoirs with the different detergents, set the 8-channel pipette to dispense 12 times 90 μL.

17. Transfer 10 μL of the diluted supernatant into each of the plates using a 12-channel pipette set to dispense 5 times with 10 μL, so that you will end up with MTPs which match the layout of your library (*see* **Note 43**).

18. Start the 2 h incubation by adding 90 μL of detergent solution into each well, set the start of each incubation apart by approximately 2 min and write the time on the plate (*see* **Note 44**).

19. After 2 h incubation, freshly prepare the *p*NPB substrate solution (*see* **Note 10**) by mixing 13.5 mL of assay buffer and 1.5 mL of *p*NPB in acetonitrile, briefly vortex, and pour it into a reagent reservoir. Pipette 100 μL of the substrate solution onto the 100 μL in the incubation MTP and measure the increase in OD_{410} and rinse the reagent reservoir. Repeat the process until all plates have been measured.

20. Export the data in a *.csv or *.xlx format and analyze the residual activity of each variant (Fig. 2) (*see* **Note 45**).

4 Notes

1. For stereoselective synthesis of this compound *see* ref. 16.

2. Beware; dichloromethane is toxic and suspected of causing cancer!

3. Caution when pipetting organic solvents. Due to the low density of acetonitrile the solvent can drip out of the tip of the pipette. Repeated pipetting of the solvent beforehand can minimize this effect.

4. The library in our study was composed of ~19,500 clones containing 3439 single amino acid variants [48]. We have found that a typical site-saturation mutagenesis (SSM) with a NNS codon (32 possible codons coding for all amino acids and one stop codon) on the lipA gene has a mean yield of 90% meaning that the library will contain 10% of wild-type clones or clones in which the Quikchange mutagenesis method failed

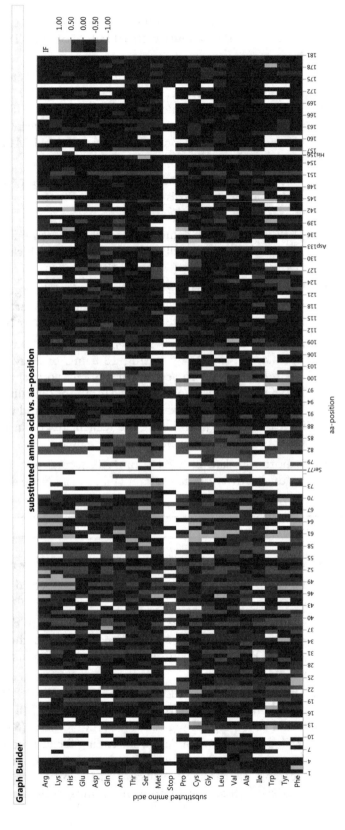

Fig. 2 Relative change of residual LipA activity compared to the wild-type enzyme in the presence of 0.35 mM SDS. The data are shown as a heat map representation. Amino acid positions beginning at the N-terminus are plotted along the X-axis (active site residues are labeled). The respective amino acid in the variant is given in the Y-axis. *Red* coloration indicates a decrease and *green* coloration indicates an increase in resistance as compared to the wild-type enzyme. *White* colored variants could not be measured due to loss of activity

and resulted in multiple insertions of the mutagenesis primer, as also observed in other studies [49]. We have devised a two stage method which is statistically the most efficient, in which we sequence 92 clones from a site-saturation mutagenesis and later add all missing variants by a second round of site-directed mutagenesis. The setup depends on oligonucleotide costs, yield of the SSM and costs of sequencing. The calculation of the optimal variant numbers for the SSM is described by Nov et al. [50]

5. In our study, the enzyme was secreted into the periplasm of *E. coli* BL21(DE3) and subsequently released into the culture supernatant thus avoiding a dedicated lysis step. To find the optimal conditions, the lipA gene was inserted into the multiple cloning sites of plasmids pET22b and pET28a (Novagen, USA) with and without fusion of the PelB signal to the N-terminus of the protein and compared in different *E. coli* expression host strains.

6. The MTPs for the frozen libraries contained 92 different variants, three wild-type clones in wells A1, E7, and H12, and one empty vector clone in D6 as internal control. In addition, our library creation approach resulted in about ten additional wild-type clones randomly distributed in the plates, which were also later used to measure the assay quality and experimental error.

7. To our experience, adding one NaOH pellet should give you consistent results. Refrain from using premixed ready to use media due to high variations from batch to batch.

8. Prepare separate stock solutions and combine afterwards, otherwise it will precipitate, store at 4 °C.

9. Add agar to individual blue cap flasks, then add LB medium and a stirring bar, take out of autoclave after sterilizing and mix briefly, this will greatly help when heating up again at a later time. Just before pouring the plates, add the required amount of the 1000× antibiotics stock solution.

10. Dissolve *p*NPB in acetonitrile and prepare the buffer separately. Combine only as much as needed for each individual measurement directly before the measurement. *p*NPB has a very high autohydrolysis rate and will turn yellow within a few minutes after dilution in buffer.

11. Weigh in *p*NPP in a 15 mL tube, resuspend in 2-propanol until completely dissolved. Weigh in sodium deoxycholate and gum arabic in a 50 mL tube, add Sørensen buffer, and vortex. Let both solutions agitate at 25 or 37 °C to facilitate solvation. Slowly add pNPP/2-propanol to the buffer to form the substrate emulsion. The solution should be turbid with a white to faintly yellow color. In case of a completely yellow color, most likely causes are a wrong pH of the buffer, or high water

content in your substrate stock from many freeze-thaw cycles, which can cause autohydrolysis of pNPP over time.

12. Weigh in surfactant on a very small disposable tray, fill a flask with the appropiate amount of buffer and add the whole tray and a magnetic stirring bar into the flask and stir. This ensures that the full amount is added; most of the surfactants may otherwise stick to the tray. Remove the tray later with tweezers. Measure and adjust the pH, since many surfactants will impact the pH. Solutions are used for up to a week, they can be used longer if they are stored in the dark and cool. However, formation of peroxides may be a problem over time.

13. Generic tips may also fit and are significantly cheaper; this is especially relevant for the tips used on the automated liquid handling platform.

14. Can also be reused, but the reservoirs used for the substrate should be marked and only used for this purpose to ensure that no residual enzymes cause a high background by hydrolysis in the reagent reservoir.

15. Do not, in any case, heat a closed flask due to the danger of explosion!

16. Make sure to always wear safety equipment when handling chemicals! 4-DMAP is toxic and may be fatal in contact with skin.

17. Check the reaction for full conversion before quenching by using thin-layer chromatography with a solvent mixture of 90:10 PE:EA.

18. Add enough equivalents of water to fully quench the EDC HCl in case of an incomplete conversion.

19. Caution when washing with sodium bicarbonate solution. The acidic reaction medium will cause the production of carbon dioxide. Release the pressure in the separation funnel after shaking.

20. If the solvent still contains traces of water, the magnesium sulfate will clump. Add to the organic phase until the magnesium sulfate keeps a powder form in the solvent.

21. This mixture should be adequate for purification of each ester described in this protocol. More polar mixtures up to 70:30 PE–EA are also possible. The separation of the compounds should always be checked by thin layer chromatography before purification.

22. This set of analytical data of the (R)-enantiomer is identical to the (S)-enantiomer. ^{1}H-NMR (600 MHz, CDCl$_3$) δ [ppm] = 0.88 (t, $^3J_{10,9}$ = 6.9 Hz, 3 H, 10-H), 1.31 (d, $^3J_{2\text{-Me}}$ = 7.0 Hz, 3 H, 2-CH$_3$), 1.22–1.46 (m, 6 × CH$_2$), 1.52–1.62 (m, 1 H, 3-H$_a$),

1.75–1.86 (m, 1 H, 3-H$_b$), 2.72 (ddq, $^3J_{2,3a}$ = 7.0 Hz, $^3J_{2,2\text{-Me}}$ = 7.0 Hz, 1 H, 2-H), 7.24–7.29 (m, 2 H), 8.25–8.29 (m, 2 H); ^{13}C NMR (151 MHz, CDCl$_3$) δ [ppm] = 14.10, 16.68, 22.67, 27.20, 29.24, 29.44, 29.48, 31.85, 33.63, 39.72, 122.42, 125.18, 145.24, 155.69, 174.39; GC-MS (EI, 70 eV): m/z (%) = 195 (18), 169 (80), 139 (25), 109 (43), 85 (90), 71 (75), 57 (100); IR (ATR film): ν = 3120, 3080, 2925, 2855, 1763, 1615, 1593, 1523, 1490, 1462, 1378, 1345, 1206, 1159, 1097, 1012, 882, 861, 746, 723, 689, 658. For analytical data of other p-nitrophenyl esters see the following literature: Propanoic-[41], butyric- [42], caprilyc- [43], lauric- [44], palmitic- [44], (RS)- 2-(4-(2-methylpropyl) phenyl) propanoic-[45], (+)-(S)-2-(6-methoxynaphthalen-2-yl) propanoic- [45], dimethylpropanoic [46], and 2-methylpropanoic-p-nitrophenyl ester [47].

23. The pNP stock solutions are diluted tenfold in the MTP wells. Beware of this when plotting the calibration curve.

24. The effect of cosolvents on the stability of the esters should always be tested before carrying out the enzymatic assay. In some cases, DMSO has been found to catalyze the autohydrolysis of the p-nitrophenyl esters.

25. When pipetting multiple rows of reactions make sure to use fresh pipette tips in order to avoid contamination with other substrates.

26. The reaction time depends on the activity of the enzyme. If the signal is too low or exceeds the detection limit, the reaction time should be lengthened or shortened, respectively. In case of very short reaction times, adapt the protein content of the enzyme solution by dilution.

27. The E-value is only an estimated value, due to the neglecting competition effects of the enantiomers when measuring these in separate reactions.

28. Completely seal by taping along the entire outside of the MTP to close any gaps between the lid and the plate. This is important to reduce edging effects.

29. Vary these parameters with your own system according to your needs, e.g., different cultivation lengths or precultures at 37 °C for 3 h with a subsequent shift to 15 °C. In our case, 16 h were chosen as incubation time as this was later more convenient during screening.

30. The remaining supernatant and cell pellet can optionally be used for SDS PAGE analysis. Further dilution may be necessary if the supernatants contain a lot of active enzyme.

31. In our case, incubation at 30 °C is chosen as the activity is still quite high. The supernatant was diluted by adding 100 µL KPi

buffer pH 8, which resulted in an activity of about 0.2 OD/ min at 410 nm using pNPB as the substrate. This moderate activity was chosen for easy handling since the screening requires the addition of substrate to 96 wells and very active clones might otherwise already reach an $OD_{410nm} > 1$ before the MTP is inserted into the microtiter plate reader.

32. Depending on the assay that is being developed, it might be useful to also replace some of the wild-type clones with additional references or already available variants to benchmark the overall performance.

33. The plate should be handled exactly the same way as the plates that contain the variants used for screening so that the results are comparable.

34. Variation of temperature and shaking of the plate during the incubation can be used to increase the stress if needed.

35. The pNPB substrate solution is not an emulsion and thus will not contain sodium deoxycholate and gum arabic, which could interfere with the surfactants present from the incubation.

36. We selected the most appropriate surfactants based on the charge of the hydrophilic head group, so that one detergent of each category (anionic, cationic, nonionic, and zwitterionic) is used for screening. The concentrations are chosen to result in a residual activity of about 30–50% compared to the incubation in buffer. This way, variants can later have an equal, higher or lower residual activity compared to the wild type.

37. Number of libraries plates that can be handled in parallel depends on the number of assays, available centrifuges, and microtiterplate readers.

38. Do not set the pipette to aspirate the maximum volume of the pipette to avoid carry over, instead set it to aspirate one less than the maximum, e.g., set to dispense 7 times 150 μL, and discard the first dispense.

39. Start this inoculation with 16 h incubation in mind, e.g., inoculate at 5 pm, plates will be ready the following day at 9 am.

40. The main reason to seal the plates is to avoid evaporation of the sample in the outer wells of the microtiter plate. If the plates are not sealed tightly, this leads to a concentration of the supernatant in the outer wells and subsequently to inconsistent results. A reduced cultivation temperature (30 °C) and a tight sealing solved this problem and resulted in a uniform performance in the assay reactions from inner and outer wells of the microtiterplate.

41. Depending on whether or not your target protein is released into the supernatant. If it is not released you will need to add additional steps for lysis of the cells.

42. During the screening, labeling can take a lot of time, so prepare the MTPs beforehand or print out labels.

43. This step can be done beforehand; the 10 µL supernatant will not dry out if you stack the plates and directly continue with the following step.

44. The timing depends on the number of plates, the length of the activity measurement and the number of available microplate readers that will be used for parallel measurements. It is often possible to optimize the assay for 384-well plates by cutting the volumes in half (5 µL supernatant, 45 µL surfactant solution, and 50 µL substrate per well). However, you will need to consider that in the 384-well reading mode the plate reader will only be able to measure each well once every 40–60 s depending on the speed of the MTP reader.

45. Use the mean or median value of the residual activity of the intentional wild-type clones, as well as the random wild-type clones, which are the result of the 90% yield of the mutagenesis. These residual activities can be used to calculate the true experimental error during the assay across different days using several thousand replicates. We have used three times the standard deviation (3 sigma) as a threshold to identify a significant increase of % relative activity caused by a single amino acid substitution. The measured residual activity of the wild-type clones in each plate can also be used to check the quality of the measurement. If these values are too far off, most likely there was an error somewhere and it is better to repeat the respective experiment with this plate.

References

1. Lehmann C, Sibilla F, Maugeri Z et al (2012) Reengineering CelA2 cellulase for hydrolysis in aqueous solutions of deep eutectic solvents and concentrated seawater. Green Chem 14:2719–2726

2. Wong TS, Roccatano D, Schwaneberg U (2007) Steering directed protein evolution: strategies to manage combinatorial complexity of mutant libraries. Environ Microbiol 9:2645–2659

3. Kumar V, Yedavalli P, Gupta V et al (2014) Engineering lipase A from mesophilic *Bacillus subtilis* for activity at low temperatures. Protein Eng Des Sel 27:73–82

4. Yedavalli P, Rao NM (2013) Engineering the loops in a lipase for stability in DMSO. Protein Eng Des Sel 26:317–324

5. Aharoni A, Thieme K, Chiu CP et al (2006) High-throughput screening methodology for the directed evolution of glycosyltransferases. Nat Methods 3:609–614

6. Becker S, Hobenreich H, Vogel A et al (2008) Single-cell high-throughput screening to identify enantioselective hydrolytic enzymes. Angew Chem Int Ed 47:5085–5088

7. Liebl W, Angelov A, Juergensen J et al (2014) Alternative hosts for functional (meta)genome analysis. Appl Microbiol Biotechnol 98:8099–8109

8. Chow J, Kovacic F, Dall Antonia Y et al (2012) The metagenome-derived enzymes LipS and LipT increase the diversity of known lipases. PLoS One 7:e47665

9. Cardenas F, Alvarez E, Castro-Alvarez MS et al (2001) Screening and catalytic activity in organic synthesis of novel fungal and yeast lipases. J Mol Catal B: Enzym 14:111–123

10. Pearson B, Wolf PL, Vazquez J (1963) A comparative study of a series of new indolyl compounds to localize beta-galactosidase in tissues. Lab Invest 12:1249–1259

11. Ben-David A, Shoham G, Shoham Y (2008) A universal screening assay for glycosynthases: directed evolution of glycosynthase XynB2 (E335G) suggests a general path to enhance activity. Chem Biol 15:546–551

12. Leis B, Angelov A, Mientus M et al (2015) Identification of novel esterase-active enzymes from hot environments by use of the host bacterium *Thermus thermophilus*. Front Microbiol 6:275

13. Fernandez-Alvaro E, Snajdrova R, Jochens H et al (2011) A combination of *in vivo* selection and cell sorting for the identification of enantioselective biocatalysts. Angew Chem Int Ed 50:8584–8587

14. Wahler D, Reymond JL (2001) High-throughput screening for biocatalysts. Curr Opin Biotechnol 12:535–544

15. Reymond JL, Fluxa VS, Maillard N (2009) Enzyme assays. Chem Commun 2009:34–46

16. Lauinger B (2016) Kolorimetrische und fluorimetrische Assays – auf der Jagd nach neuen Biokatalysatoren für die Synthesechemie. Dissertation, Heinrich-Heine-University, Düsseldorf

17. Schmidt M, Bornscheuer UT (2005) High-throughput assays for lipases and esterases. Biomol Eng 22:51–56

18. Tallman KR, Beatty KE (2015) Far-red fluorogenic probes for esterase and lipase detection. ChemBioChem 16:70–75

19. Leroy E, Bensel N, Reymond JL (2003) A low background high-throughput screening (HTS) fluorescence assay for lipases and esterases using acyloxymethylethers of umbelliferone. Bioorg Med Chem Lett 13:2105–2108

20. Lavis LD, Chao TY, Raines RT (2011) Synthesis and utility of fluorogenic acetoxymethyl ethers. Chem Sci 2:521–530

21. Zadlo A, Koszelewski D, Borys F et al (2015) Mixed carbonates as useful substrates for a fluorogenic assay for lipases and esterases. ChemBioChem 16:677–682

22. Nyfeler E, Grognux J, Wahler D et al (2016) A sensitive and selective high-throughput screening fluorescence assay for lipases and esterases. Helv Chim Acta 86:2919–2927

23. Babiak P, Reymond JL (2005) A high-throughput, low-volume enzyme assay on solid support. Anal Chem 77:373–377

24. Reymond JL (2008) Substrate arrays for fluorescence-based enzyme fingerprinting and high-throughput screening. Ann N Y Acad Sci 1130:12–20

25. Maillard N, Babiak P, Syed S et al (2011) Five-substrate cocktail as a sensor array for measuring enzyme activity fingerprints of lipases and esterases. Anal Chem 83:1437–1442

26. Wang B, Tang X, Ren G et al (2009) A new high-throughput screening method for determining active and enantioselective hydrolases. Biochem Eng J 46:345–349

27. Rachinskiy K, Schultze H, Boy M et al (2009) Enzyme test bench: a high-throughput enzyme characterization technique including the long-term stability. Biotechnol Bioeng 103:305–322

28. Liu AMF, Somers NA, Kazlauskas RJ et al (2001) Mapping the substrate selectivity of new hydrolases using colorimetric screening: lipases from *Bacillus thermocatenulatus* and *Ophiostoma piliferum*, esterases from *Pseudomonas fluorescens* and *Streptomyces diastatochromogenes*. Tetrahedron: Asymmetr 12:545–556

29. Franken B, Jaeger KE, Pietruszka J (2016) Screening for enantioselective enzymes. Springer, Berlin

30. Baumann M, Stürmer R, Bornscheuer UT (2001) A high-throughput-screening method for the identification of active and enantioselective hydrolases. Angew Chem Int Ed 40:4201–4204

31. Andersen RJ, Brask J (2016) Synthesis and evaluation of fluorogenic triglycerides as lipase assay substrates. Chem Phys Lipids 198:72–79

32. Trapp O (2007) Boosting the throughput of separation techniques by "multiplexing". Angew Chem Int Ed 46:5609–5613

33. Reetz M, Kühling KM, Wilensek S (2001) A GC-based method for high-throughput screening of enantioselective catalysts. Catal Today 67:389–396

34. Reetz MT, Kuhling KM, Hinrichs H (2000) Circular dichroism as a detection method in the screening of enantioselective catalysts. Chirality 12:479–482

35. Schrader W, Eipper A, Pugh DJ (2002) Second-generation MS-based high-throughput screening system for enantioselective catalysts and biocatalysts. Can J Chem 80:626–632

36. Reetz MT, Eipper A, Tielmann P et al (2016) A practical NMR-based high-throughput assay for screening enantioselective catalysts and biocatalysts. Adv Synth Catal 344:1008–1016

37. Henke E, Bornscheuer UT (1999) Directed evolution of an esterase from *Pseudomonas fluorescens*. Random mutagenesis by error-prone PCR or a mutator strain and identification of mutants showing enhanced enantioselectivity by a resorufin-based fluorescence assay. Biol Chem 380:1029–1033

38. Engstrom K, Nyhlen J, Sandstrom AG (2010) Directed evolution of an enantioselective lipase with broad substrate scope for hydrolysis of alpha- substituted esters. J Am Chem Soc 132:7038–7042

39. Janes LE, Kazlauskas RJ (1997) Quick E. A fast spectrophotometric method to measure the enantioselectivity of hydrolases. J Org Chem 62:4560–4561

40. Lima MLSO, Gonçalves CCS, Barreiro JC et al (2015) High-throughput enzymatic enantiomeric excess: Quick-ee. J Braz Chem Soc 26:319–324

41. Ke D, Zhan C, Li X et al (2009) The urea-dipeptides show stronger H-bonding propensity to nucleate β-sheetlike assembly than natural sequence. Tetrahedron 65:8269–8276

42. Qian L, Liu JY, Liu JY et al (2011) Fingerprint lipolytic enzymes with chromogenic *p*-nitrophenyl esters of structurally diverse carboxylic acids. J Mol Catal B: Enzym 73:22–26

43. Jin C, Li J, Su W (2009) Ytterbium triflate catalysed Friedel–Crafts reaction using carboxylic acids as acylating reagents under solvent-free conditions. J Chem Res 2009:607–611

44. Kobayashi S, Matsubara R, Nakamura Y et al (2003) Catalytic, asymmetric Mannich-type reactions of *N*-acylimino esters: reactivity, diastereo- and enantioselectivity, and application to synthesis of *N*-acylated amino acid derivatives. J Am Chem Soc 125:2507–2515

45. Bongen P (2014) Chemoenzymatische Synthesen - Hydrolasen in Methodik und Anwendung. Dissertation, Heinrich-Heine-University, Düsseldorf

46. Liu Z, Ma Q, Liu Y et al (2013) 4-(*N,N*-Dimethylamino)pyridine hydrochloride as a recyclable catalyst for acylation of inert alcohols: substrate scope and reaction mechanism. Org Lett 16:236–239

47. Lu N, Chang WH, Tu WH et al (2011) A salt made of 4-*N,N*-dimethylaminopyridine (DMAP) and saccharin as an efficient recyclable acylation catalyst: a new bridge between heterogeneous and homogeneous catalysis. Chem Commun 47:7227–7229

48. Fulton A, Frauenkron-Machedjou VJ, Skoczinski P et al (2015) Exploring the protein stability landscape: *Bacillus subtilis* lipase A as a model for detergent tolerance. ChemBioChem 16:930–936

49. Edelheit O, Hanukoglu A, Hanukoglu I (2009) Simple and efficient site-directed mutagenesis using two single-primer reactions in parallel to generate mutants for protein structure-function studies. BMC Biotechnol 9:61

50. Nov Y, Fulton A, Jaeger KE (2013) Optimal scanning of all single-point mutants of a protein. J Comput Biol 20:990–997

Chapter 13

Continuous High-Throughput Colorimetric Assays for α-Transaminases

Egon Heuson, Jean-Louis Petit, Franck Charmantray, Véronique de Bérardinis, and Thierry Gefflaut

Abstract

Transaminases are efficient tools for the stereoselective conversion of prochiral ketones into valuable chiral amines. Notably, the diversity of naturally occurring α-transaminases offers access to a wide range of L- and D-α-amino acids. We describe here two continuous colorimetric assays for the quantification of transamination activities between a keto acid and a standard donor substrate (L- or D-Glutamic acid or cysteine sulfinic acid). These assays are helpful for kinetic studies as well as for high-throughput screening of enzyme collections.

Key words Transaminase, Amino acids, Screening assay, Cysteine sulfinic acid, High-throughput screening

1 Introduction

Transaminases (TA) offer an efficient access to chiral amines, which are found in numerous pharmaceuticals, and therefore have gained considerable attention in the past few years [1–3]. Notably, α-transaminases (α-TA) catalyze the conversion of α-keto acids into α-amino acids of both L- or D-series. Many L- or D-selective TA have already proven useful for the preparation of a range of rare or nonnatural amino acids with high stereoselectivity, which are key building blocks for biologically active peptides or pseudopeptides. With the aim of mining biodiversity to identify new useful biocatalysts within the α-TA family, we have developed two continuous colorimetric assays to monitor transamination reactions [4]. As shown in Fig. 1, these assays are based on the use of L- or D-cysteine sulfinic acid (CSA) as irreversible amino donor. In the direct assay, L- or D-CSA is used as unique amino donor substrate and leads upon transamination to the release of sulfur dioxide, readily converted into sulfite ions by hydration. Subsequently, the nucleophilic

Uwe T. Bornscheuer and Matthias Höhne (eds.), *Protein Engineering: Methods and Protocols*, Methods in Molecular Biology, vol. 1685, DOI 10.1007/978-1-4939-7366-8_13, © Springer Science+Business Media LLC 2018

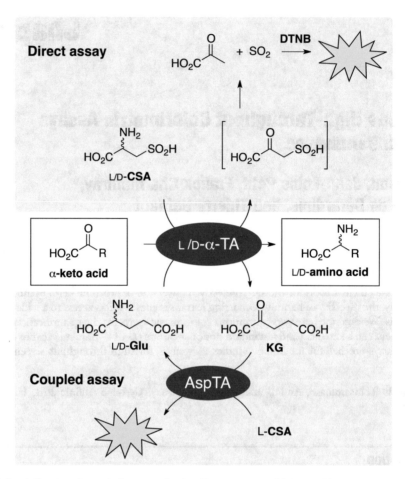

Fig. 1 Principle of direct and coupled assays. In the direct assay, L-CSA or D-CSA must be accepted by the transaminase of interest. CSA conversion yields the instable beta keto sulfonic acid. SO_2 is then formed in a spontaneous decomposition and detected via a dye-forming reaction. Coupled assay: If the transaminase doesn't convert CSA, Asp-TA is employed as a shuttling enzyme that facilitates CSA conversion

sulfite specifically reacts with Ellman's reagent (5,5'-dithiobis-(2-nitrobenzoic acid), DTNB) to give a yellow thiolate anion ($\lambda = 412$ nm, $\varepsilon = 14{,}150$ M^{-1} cm^{-1}) [5, 6]. Noteworthy, the use of L-or D-CSA offers the opportunity to identify TA with complementary stereoselectivities opening access to α-amino acids of the L- or D-series from a wide range of keto acids tested as acceptor substrates. In the coupled assay, L- or D-Glu is used as primary amino donor and gives upon transamination ketoglutaric acid (KG), which is converted back to L-Glu using aspartate transaminase (AspTA) as an auxiliary enzyme and L-CSA as secondary amino donor. The first enzymatic transamination reaction can therefore be monitored by colorimetric sulfite titration in the presence of Ellman's reagent DTNB, as described above for the direct assay. Considering that Glu is a preferred substrate for most α-TA, this coupled assay thus allows for identifying a large variety of valuable TA.

With dynamic ranges of 0.05–80 mU/mL and 0.25–80 mU/mL for the direct and coupled assays respectively [4], these two new complementary assays can be used for high-throughput screening of large collections of TA, as well as for kinetic studies of a variety of enzymes accepting L/D-CSA or L/D-Glu as amino donor substrate. We describe here, as application examples, the determination of kinetic parameters of the two *E. coli* enzymes AspTA and branched chain transaminase (BCTA) as well as the screening protocol of an α-TA library to detect new enzymes for the stereoselective synthesis of amino acids from the corresponding keto acids. Using these new assays for screening a library of 232 α-TA from biodiversity, we could identify new valuable biocatalysts for the synthesis of L- and D-homophenylalanine [4].

2   Materials

2.1   Chemical Synthesis of L- and D-CSA

Use chemicals and solvents of reagent or analytical grade.

1. 95% Formic acid.
2. 37% Hydrochloric acid.
3. 30% Hydrogen peroxide.
4. 30% Ammonia.
5. L-/D-Cystine ($C_6H_{12}N_2O_4S$).
6. 1 M Formic acid.
7. 1 M Acetic acid (AcOH).
8. Sulfonic acid resin column: pour 250 mL Amberlite® IR120Na in a glass column (4 cm diameter, 30 cm length) equipped with a Teflon stopcock and a cotton piece to retain the resin. Wash the column with 2 M HCl (1 L) and then with water, until neutrality (typically 2 L).
9. Basic resin column (AcO⁻ form): pour 200 mL Dowex® 1×8 (200–400 mesh, Cl⁻ form) in a glass column (4 cm diameter, 30 cm length) equipped with a Teflon stopcock and a cotton piece to retain the resin. Wash the column with 1 M NaOH (1 L), with water until neutrality of the effluent (approx. 400 mL), with 1 M AcOH until acidity (approx. 400 mL) and finally with H_2O until neutrality (approx. 400 mL).
10. Thin-layer chromatography (TLC): use silica gel 60F254 aluminum plates (Merck HX093822); elute with *n*-propanol–H_2O (7:3); reveal by immersion in 2 g/L ninhydrin in EtOH, followed by heating at 200 °C with a heat-gun (Rf of CSA = 0.5).

2.2 Enzyme Production and Purification

Use chemicals and solvents of reagent or analytical grade (>98%). Prepare all solutions using deionized water. Store buffers at 4 °C unless otherwise specified.

1. AspTA and BCTA from *E. coli* are produced from transformed *E. coli* BL21(DE3) strains. The plasmid pET22b_AspTA contains the AspC gene (Uniprot Acc. No. P00509) coding for AspTA. Plasmid pET22b_BCTA contains the ilvE gene (Uniprot Acc. No. P0AB80) coding for *E. coli* BCTA. Both plasmids also contain a gene for ampicillin resistance for selection of transformed cells. A sequence coding for a His-tag was added to the AspC and IlvE genes, for enzyme purification by immobilized metal affinity chromatography (IMAC). Store stocks of transformed cells at −80 °C in LB medium (3 mL) supplemented with glycerol (10%).

2. Ampicillin stock solution: 200 mg/mL ampicillin in water.

3. IPTG solution: 50 mM isopropyl β-D-1-thiogalactopyranoside.

4. Luria-Bertani (LB)-ampicillin medium: 10 g/L tryptone, 5 g/L yeast extract, 10 g/L NaCl, 100 mg/L ampicillin. In a 2 L beaker equipped with a stir bar, dissolve 50 g of solid LB mixture (Difco 244610) composed of tryptone (20 g), yeast extract (10 g) and NaCl (20 g) in 1.9 L water. Fill up to 2 L in a cylinder. Pour 200 mL of this solution in each of ten 0.5 L flasks equipped with a cotton stopper coated with an aluminum sheet and sterilize in an autoclave for 20 min at 121 °C. In each flask, add 0.1 mL ampicillin solution passed through 0.2 μM sterile syringe filters (Whatman FP30/0.2 CA-S).

5. 50 mM Phosphate buffer, pH 7.5: in a 0.5 L beaker, equipped with a stir bar, dissolve 6.80 g KH_2PO_4 in 900 mL water. Adjust pH with 5 M KOH (approx. 10 mL). Fill up to 1 L in a cylinder.

6. Lysis buffer: 50 mM potassium phosphate, 300 mM KCl, 10 mM imidazole, pH 8. In a 1 L beaker equipped with a stir bar, dissolve 6.80 g KH_2PO_4, 22.3 g KCl, and 0.68 g imidazole in 900 mL water. Adjust pH with 5 M KOH (approx. 10 mL). Fill up to 1 L in a cylinder.

7. Washing buffer: 50 mM potassium phosphate, 300 mM KCl, 20 mM imidazole, pH 8. Prepare as described for lysis buffer with 1.36 g imidazole.

8. Elution buffer: 50 mM potassium phosphate, 300 mM KCl, 250 mM imidazole, pH 8. Prepare as described for lysis buffer with 17.0 g imidazole.

9. Cleaning buffer: 50 mM potassium phosphate, 300 mM KCl, 500 mM imidazole, pH 9. Prepare as described for lysis buffer with 34.0 g imidazole.

10. Dialysis buffer: 50 mM potassium phosphate, 3 M $(NH_4)_2SO_4$, pH 7.5. In a 1 L beaker equipped with a stir bar, dissolve 6.80 g KH_2PO_4 and 396.4 g $(NH_4)_2SO_4$ in 600 mL water. Adjust pH with 5 M KOH (approx. 10 mL). Fill up to 1 L in a cylinder.

11. IMAC column: Ni-NTA Agarose (Qiagen). Prepare a 15 mL column and wash it with water (100 mL) and lysis buffer (100 mL). Between two purifications, use cleaning buffer (100 mL) and equilibrate with water (100 mL) and lysis buffer (100 mL). For storage, equilibrate with 20% EtOH (200 mL).

12. Dialysis membranes: 33 mm dialysis tubing cellulose membrane (Sigma-Aldrich).

2.3 Enzyme Assays, Kinetics, and Screening

Prepare all solutions using deionized water. Store buffers at 4 °C unless otherwise specified. Use chemicals and solvents of reagent or analytical grade (>98%).

1. Substrate and cofactor solutions: Prepare the various solutions listed below as follows: in a 50 mL beaker equipped with a stir bar, dissolve 0.34 g KH_2PO_4 and the appropriate amount of substrate (*see* below) in 40 mL water. Adjust pH to 7.5 with 1–5 M KOH (<5 mL) and bring to 50 mL in a cylinder. Concentrations and amounts of different substrates:

   KG solution, 20 mM: 0.15 g 2-ketoglutaric acid.

   MOPA solution, 20 mM: 0.15 g 4-methyl-2-oxopentanoic acid sodium salt.

   L-Glu solution, 200 mM: 1.48 g L-glutamic acid (*see* **Notes 1** and **2**).

   L-Asp solution, 200 mM: 1.33 g L-aspartic acid, (*see* **Note 1**).

   L-CSA solution, 200 mM: 1.53 g L-cysteine sulfinic acid (*see* **Notes 1** and **2**).

   Ammonium sulfate solution, 0.5 M: 3.30 g $(NH_4)_2SO_4$.

   DTNB solution, 10 mM: 0.198 g 5,5′-dithiobis(2-nitrobenzoic acid) previously dissolved in 5 mL EtOH.

   PLP stock solution, 10 mM: 0.133 g pyridoxal phosphate monohydrate.

   PLP solution, 0.5 mM: dilute 2.5 mL of 10 mM PLP-solution to 50 mL in a cylinder.

   NADH solution, 8.5 mM: 6 mg β-nicotinamide adenine dinucleotide reduced, disodium salt in 1 mL phosphate buffer just before use.

   Keto acid solution, 20 mM: For screening substrate specificity, prepare 20 mM solutions of various keto acids of your choice in phosphate buffer.

2. Glutamate dehydrogenase (GDH) from bovine liver (40 U/ mg) and malate dehydrogenase (MDH) from bovine heart (3000 U/mg) are commercially available as suspensions in 3 M ammonium sulfate (Sigma-Aldrich). Prepare enzyme solutions just before use: in a 1.5 mL microcentrifuge tube, centrifuge the appropriate volume of suspension at $14,000 \times g$ for 5 min. Discard the supernatant and gently dissolve the pellet in the appropriate volume of phosphate buffer, in order to reach the desired final concentration.

3. Bradford reagent (Bio-Rad).

4. BSA solutions, 0.01–0.1 mg/mL: prepare by serial dilution of bovine serum albumin (BSA) in phosphate buffer.

5. 96-Well microtiter plates, clear, flat bottom.

6. For screening tansaminase collections, enzyme solutions should be available as cell-free extracts, e.g., after cultivation and cell-disruption of recombinant *E. coli* clones in 96-well microtiter plates.

2.4 Equipment

1. Microplate reader.

3   Methods

3.1 Chemical Synthesis of L- or D-CSA

L- and D-CSA are prepared from commercially available L- and D-cystine following a modified procedure including oxidation of the disulfide bond of cystine with H_2O_2, followed by dismutation of the intermediate using ammonia [7]. A simple purification on an ion exchange column provides highly pure L- or D-CSA. Alternatively, L-CSA can be purchased from commercial sources.

1. In a 3-necked round bottom flask equipped with a stir bar and a thermometer, suspend cystine (20 g, 83.3 mmol) in 95% formic acid (417 mL).

2. Add to the suspension conc. HCl (15 mL, 174 mmol).

3. Add 30% H_2O_2 (20 mL, 200 mmol) dropwise, while keeping the temperature of the solution at 20 ± 3 °C using an ice bath and vigorous stirring. Stir then the mixture for 16 h at room temperature.

4. Concentrate the solution to dryness under reduced pressure (*see* **Note 3**) to obtain a yellow oil. Dissolve this oil in water (50 mL) and concentrate it again to dryness under reduced pressure. Repeat this step to completely remove formic acid by azeotropic distillation.

5. Dissolve the residue in water (200 mL) and adjust the pH value to 3.5 with 30% ammonia (approx. 5 mL). Cool the solution to 4 °C for 16 h. Filter the precipitate, and wash it three times with cold water (3 × 20 mL).

6. Suspend the precipitate in water (45 mL) and add 30% ammonia (45 mL). Stir the mixture for 2 h at room temperature.

7. Concentrate the mixture under reduced pressure to dryness. Dilute the residue in water (50 mL) and concentrate again to dryness under reduced pressure. Repeat this last step two times to completely remove ammonia. Suspend the residue in water (90 mL). Remove the precipitated cystine by filtration and wash it with cold water (10 mL). Concentrate the combined solution to approx. 20 mL under reduced pressure.

8. Pour the solution on the Amberlite® IR120 column (H⁺ form, 250 mL). Elute with water and collect 25 mL fractions. Follow the product elution by TLC. Combine the CSA containing fractions and reduce the solution to approx. 20 mL under reduced pressure.

9. Pour the solution on the Dowex® 1X8 column (AcO⁻ form, 200 mL). Wash successively with water (500 mL) and 1 M AcOH (500 mL), and elute with 1 M formic acid. Collect 25 mL fractions and follow CSA elution by TLC. Combine the CSA containing fractions and concentrate the solution under reduced pressure. Dilute the residue in H_2O (20 mL) and concentrate again to dryness under reduced pressure. Repeat this operation two times to completely remove formic acid. Typically, CSA is isolated as a white solid in 75% yield (*see* **Note 4**).

3.2 Production of AspTA from Escherichia coli

We describe here the cell culture conditions and the purification of AspTA (*see* **Note 5**).

1. Preculture: in a sterilized 0.5 L flask, add a 3 mL stock sample of AspC recombinant cells to 100 mL of LB-ampicillin medium. Stir at 200 rpm for 24 h at 37 °C.

2. In each of five sterilized 0.5 L flasks containing 200 mL of LB-ampicillin medium, add 4 mL of preculture suspension and stir at 200 rpm at 37 °C until OD at λ = 600 nm reaches approx. 0.7.

3. To each flask, add 2 mL of a 50 mM IPTG solution and stir at 200 rpm for 24 h at 30 °C.

4. Centrifuge the culture broth at 8000 × *g* for 15 min, combine and resuspend the cell pellets in lysis buffer (20 mL) and centrifuge at 8000 × *g* for 15 min. Repeat this washing step with lysis buffer three times. Usually, approx. 3 g of wet cells are obtained per liter of culture medium.

5. Suspend the cells in 20 mL of lysis buffer and disrupt cell walls by sonication at 4 °C (*see* **Note 6**). Centrifuge at 25,000 × *g* for 30 min at 4 °C to remove cell debris.

6. Adjust the lysate volume to 28.5 mL and add 1.5 mL of 10 mM PLP.

7. Measure AspTA activity (*see* **Note 7**): In a 0.3 mL well, add 20 μL of the lysate diluted 100 times (*see* **Note 8**), 20 μL of 10 U/mL MDH, 20 μL of PLP solution, 20 μL of NADH solution, 20 μL of KG solution, 50 μL of phosphate buffer, and initiate the reaction by adding 50 μL of L-asp solution. Record the OD linear decay at 340 nm. The activity (U/mL = μmol/(min·mL) of AspTA in the supernatant is calculated from the slope (min^{-1}) as follows: Activity = $((\text{slope}/6220)/0.59)\times 2\cdot10^{-4}\times50\times10^6$ (*see* **Note 9**). Usually a total activity of 25,000 U is found in the crude extract from 1 L culture.

8. Pour the cell lysate on a 15 mL IMAC column. Wash the column with the 150 mL of washing buffer and elute AspTA with 50 mL of elution buffer while collecting 5 mL fractions. Measure AspTA activity as described above and pool the AspTA-containing fractions. Usually the protein is obtained in fractions 3–5.

9. Fill a dialysis tubing with the AspTA solution, immerge in 200 mL dialysis buffer and stir slowly at 4 °C for 24 h. Refresh dialysis buffer (200 mL) two times after 8 and 16 h. After dialysis, remove the protein suspension from the tubing and store at 4 °C. Measure AspTA activity as described above. Usually around 10 mL of AspTA suspension is obtained with an activity of approx. 1500 U/mL. The suspension can be stored for months without loss of activity.

3.3 Production of BCTA from Escherichia coli

We describe here the recombinant production and purification of BCTA (*see* **Note 10**).

1. Prepare BCTA-containing cell lysate as described above for AspTA.

2. Measure BCTA activity as follows (*see* **Note 7**): In a 0.3 mL well, add 20 μL of cell lysate diluted ten times (*see* **Note 8**), 20 μL of 10 U/mL GluDH, 20 μL of PLP solution, 20 μL of NADH solution, 20 μL of ammonium sulfate solution, 20 μL of MOPA solution, 30 μL of phosphate buffer, and initiate the reaction with 50 μL of L-Glu solution. Record the OD linear decay. BCTA activity in the cell lysate is calculated as described above for AspTA. Usually a total activity of 2500 U is found in the crude extract from 1 L culture.

3. Purify BCTA by IMAC and dialyze as described for AspTA. Usually around 10 mL of BCTA suspension is obtained with an activity of approx. 100 U/mL. The suspension can be stored for months without loss of activity.

3.4 Enzyme Kinetics Using the Direct Assay

AspTA was chosen as a model of CSA-accepting enzyme to implement kinetic studies using the direct assay. All the measurements are done in triplicate for statistical reliability.

1. Measure the protein concentration of the purified AspTA suspension with the Bradford quantification method [8]. To get the calibration curve, add 20 µL of 0.01–0.1 mg/mL BSA solutions or phosphate buffer as a blank in different wells of a 96-well plate and add 180 µL of Bradford reagent. Incubate the plate at 30 °C for 15 min and measure OD at 595 nm. Plot OD as a function of BSA concentration. In a 1.5 mL microcentrifuge tube, centrifuge 20 µL of AspTA suspension in 3 M $(NH_4)_2SO_4$ at 14,000 × g for 5 min. Discard the supernatant and dissolve the pellet in phosphate buffer (1 mL). Dilute 100 µL of this solution to 1 mL. In a well, add the prepared AspTA solution (20 µL) and 180 µL of Bradford reagent (*see* **Note 11**). Incubate the mixture at 30 °C for 15 min and measure OD at 595 nm. Finally, determine AspTA concentration using the calibration curve obtained with BSA solutions.

2. Prepare a 0.1 U/mL AspTA solution: In a 1.5 mL microcentrifuge tube, centrifuge 10 µL of AspTA suspension produced in Subheading 3.2 at 14,000 × g during 5 min. Discard the supernatant and dissolve gently the pellet in 1 mL phosphate buffer. Dilute this solution to obtain a final activity of 0.1 U/ mL (*see* **Note 12**).

3. Prepare by serial dilutions 16, 8, 4, 2, and 1 mM KG and 64, 32, 16, 8, 4, and 1 mM L-CSA solutions from 20 mM KG and 200 mM L-CSA solutions respectively.

4. Measure the activity of AspTA at various substrate concentrations (*see* **Note 7**): In a 0.3 mL well, introduce 0.1 U/mL AspTA (20 µL), 20 µL of PLP solution, 20 of µL DTNB solution, 20 µL of KG solution (various concentrations), 70 µL of phosphate buffer, and initiate the reaction with 50 µL of L-CSA solution (various concentrations) (*see* **Note 13**). Record the OD variation at 412 nm over 30 min and use the slope (in min^{-1}) to calculate the initial velocity of AspTA from the linear part of the curve as follows: V_i (in mU) = ((slope/14150)/0.59)×2.10^{-4} (*see* **Note 9**).

5. Use a ping-pong bi-bi model to calculate the kinetic parameters of AspTA [9]. In our case, K_M values of 10 ± 0.5 mM and 1 ± 0.1 mM were found for L-CSA and KG, respectively, and a k_{cat} of 13,700 ± 700 min^{-1} was determined from measured V_{max} and protein concentration (*see* **Note 14**).

3.5 Enzyme Kinetics Using the Coupled Assay

BCTA is chosen as a model for enzymes accepting Glu as amino donor substrate and devoid of activity toward CSA. All the measurements are done in triplicate for statistical reliability.

1. Measure the protein concentration of the purified BCTA suspension as described above for AspTA (*see* **Note 15**).

2. Prepare as described above, a 10 U/mL AspTA solution and a 0.1 U/mL BCTA solution (*see* **Note 16**).

3. Prepare by serial dilutions 20, 15, 10, 8, 4, and 2 mM MOPA solutions from the 20 mM MOPA solution. Prepare 25 mM l-CSA by diluting 125 µL of 200 mM l-CSA to 1 mL.

4. Measure BCTA activity at various MOPA concentrations and 50 mM l-Glu (apparent K_M) (*see* **Note 7**): In a 0.3 mL well, add 20 µL of 0.1 U/mL BCTA, 20 µL of 10 U/mL AspTA, 20 µL of PLP solution, 20 µL of DTNB solution, 20 µL of MOPA solution (various concentrations), and 50 µL of 25 mM l-CSA, and initiate the reaction with 50 µL of l-Glu solution. Record the OD variation, calculate the initial velocities and determine the kinetic parameters as described above for AspTA. In our case an apparent K_M value of 0.33 ± 0.02 mM for MOPA and a kcat value of 268 ± 5 min^{-1} were found (*see* **Note 14**).

3.6 Screening of a TA Collection Using the Direct Assay

Both the direct and coupled assays are suitable for screening enzyme collections. In our study, we screened a library of putative transaminase genes from biodiversity, which were overexpressed in recombinant *E. coli* strains. 232 well-expressed TA were produced in 96-well microplates and screening with various keto acid substrates were performed on cell lysates after cell disruption by sonication [4].

1. Prepare the direct assay mixtures (Mix_D) for 100 assays: for each tested keto acid, mix 2 mL of l-CSA solution (*see* **Note 2**), 2 mL of DTNB solution, 100 µL of PLP stock solution, 2 mL of the desired keto acid solution, and 3.9 mL of phosphate buffer. The Mix_D can be stored at 4 °C for 12 h.

2. Dilute the cell lysates in order to allow measurement of TA activities within the assay dynamic range (*see* **Note 17**).

3. In each 0.3 mL well of a microtiter plate, add 20 µL of the various diluted cell-lysates and 80 µL of phosphate buffer, and initiate the reaction by adding 100 µL of Mix_D.

4. Monitor OD variation at 412 nm over 30 min (*see* **Note 7**), and calculate initial velocity from the slope (in min^{-1}) of the linear part of the curve: V_i (in mU) $= (slope/14{,}150)/0.59) \times 2.10^{-4}$.

3.7 Screening of a TA Collection Using the Coupled Assay

1. Prepare the coupled assay mixture (Mix_C) for 100 assays: for each tested keto acid, mix 2 mL of l-CSA solution, 2 mL of l-Glu solution (*see* **Note 2**), 2 mL of DTNB solution, 100 µL of PLP stock solution, 2 mL of keto acid solution and 1.9 mL of phosphate buffer. The Mix_C can be stored at 4 °C for 12 h.

2. Dilute the cell lysates in order to allow measurement of TA activities within the assay dynamic range (*see* **Note 17**).

3. In each 0.3 mL well of a 96-well microtiter plate, add 20 μL of various diluted cell-lysates, 80 μL of 0.25 U/mL AspTA (*see* **Note 18**), and initiate the reaction by adding 100 μL of Mix$_C$.

4. Monitor OD variation and calculate initial velocities as described above for the direct assay.

4   Notes

1. As acidic amino acids (Glu, Asp, and CSA) are poorly soluble in water in their neutral form, a suspension is first obtained. Solubilization is observed during pH adjustment and is complete at pH 7.5.

2. This chapter describes screening protocols for L-specific transaminases. If you want to assay D-specific transaminases, D-CSA must be used in the direct assay instead of L-CSA. In the coupled assay, d-glutamate has to be employed instead of L-glutamate, but keep in mind that the aspartate transaminase (reporter enzyme) needs L-CSA in the coupled reaction!

3. The absence of residual peroxides should be checked by using test sticks before the concentration. In our case we used Quantofix® peroxide 25 and never evidenced any remaining peroxide in the solution.

4. The structure and purity of CSA can be checked by the following analyses: Melting point: 160 °C; $[α]^{25}_D = +28°$ (c 1, 1 M HCl, L-enantiomer); ^1H and ^{13}C NMR (5 mg in 0.5 mL D$_2$O): 1H NMR (400 MHz) δ (ppm) 4.50 (1H, dd, J = 3.8, and 9.0 Hz), 2.99 (1H, dd, J = 3.8, and 14.2 Hz), 2.82 (1H, dd, J = 9.0, and 14.2 Hz); ^{13}C NMR (100 MHz, D$_2$O) δ (ppm) 179.6, 174.6, 138.0, 128.9, 128.3, 127.8, 59.9, 44.1, 40.3.

5. AspTA is also available from various organisms and commercial sources. Commercial enzyme preparations can also be used to implement the present colorimetric assays.

6. We used a Bandelin sonopuls sonicator, alternating 8 s on and 15 s off, for 60 min at 50% amplitude. The cell suspension is cooled by immersion in a water-ice bath during the sonication to avoid protein degradation.

7. Enzyme assays were automated on a Tecan Freedom EVO™ robotic platform. The workstation includes a multimode microplate reader (Safire™ II, Tecan) for UV/Vis absorbance measurement. All the experiments were performed in Greiner® 96-well plates incubated at 30 ± 2 °C and stirred 90 s before measurement.

8. If the measured activity is too high or too low, adjust the dilution of the supernatant to measure an enzyme activity of 1–4 mU/well.

9. The factor 0.59 is the calculated optical length of a well filled with 200 µL water.

10. As alternative to expression and protein purification, BCTA from various organisms and commercial sources can be used to implement the present colorimetric assay.

11. In our condition, the protein concentration of the purified AspTA suspension was about 16 mg/mL. The final protein concentration in the prepared solution (after centrifugation and 500-times dilution) was therefore around 0.03 mg/mL. This ensured that it was in the linear range of the calibration curve. If necessary, adjust the suspension dilution.

12. In our case, considering a starting activity of approx. 1500 U/mL for the AspTA suspension, 66 µL of AspTA solution were diluted to 1 mL to get a final mother solution of 0.1 U/mL.

13. All combinations of various KG and L-CSA concentrations are assayed.

14. Close values were measured using classical NADH assays based on reduction of pyruvate formed from CSA in the presence of lactate dehydrogenase, or on the reductive amination of KG formed from Glu in the presence of glutamate dehydrogenase.

15. In our condition, the protein concentration of the purified BCTA suspension was about 8 mg/mL. The final protein concentration in the prepared solution (after centrifugation and 200-times dilution) was then close to 0.04 mg/mL. If necessary, adjust the suspension dilution in order to be within the linear range of the calibration curve.

16. In our case, considering a starting activity of approx. 100 U/mL for the BCTA suspension, the pellet was dissolved in 0.5 mL before dilution of 50 µL of the solution to 1 mL.

17. In our case, lysates were diluted to get 0.1–0.5 mg/mL of protein.

18. The reference AspTA activity must be measured in the screening conditions (20 mM L-CSA, 2 mM KG, 1 mM DTNB, and 50 µM PLP) using the direct assay.

References

1. Koszelewski D, Tauber K, Faber K et al (2010) Omega-transaminases for the synthesis of non-racemic alpha-chiral primary amines. Trends Biotechnol 28:324–332

2. Hwang BY, Cho BK, Yun H et al (2005) Revisit of aminotransferase in the genomic era and its application to biocatalysis. J Mol Catal B Enzym 37:47–55

3. Taylor PP, Pantaleone DP, Senkpeil RF et al (1998) Novel biosynthetic approaches to the production of unnatural amino acids using trans-aminases. Trends Biotechnol 16:412–418

4. Heuson E, Petit JL, Debard A et al (2016) Continuous colorimetric screening assays for the detection of specific L- or D-α-amino acid trans-aminases in enzyme libraries. Appl Microbiol Biotechnol 100:387–408

5. Ellman GL (1959) Tissue sulfhydryl groups. Arch Biochem Biophys 82:70–77

6. Humphrey RE, Ward MH, Hinze W (1970) Spectrophotometric determination of sulfite with 4,4'-dithio-dipyridine and 5,5'-dithiobis (2-nitrobenzoic acid). Anal Chem 42:698–702

7. Emilliozzi R, Pichat L (1959) A simple method for the preparation of cysteinesulfinic acid. Acta Chem Scand 10:1887–1888

8. Bradford MM (1976) A rapid and sensitive method for the quantitation of microgram quantities of protein utilizing the principle of protein-dye binding. Anal Biochem 72:248–254

9. Segel IH (1975) Enzyme kinetics: behavior and analysis of rapid equilibrium and steady state enzyme systems. John Wiley & Sons, New York

Chapter 14

Colorimetric High-Throughput Screening Assays for the Directed Evolution of Fungal Laccases

Isabel Pardo and Susana Camarero

Abstract

In this chapter we describe several high-throughput screening assays for the evaluation of mutant libraries for the directed evolution of fungal laccases in the yeast *Saccharomyces cerevisiae*. The assays are based on the direct oxidation of three syringyl-type phenols derived from lignin (sinapic acid, acetosyringone, and syringaldehyde), an artificial laccase mediator (violuric acid), and three organic synthetic dyes (Methyl Orange, Evans Blue, and Remazol Brilliant Blue). While the assays with the natural phenols can be used for laccases with low redox potential, the rest are exclusive for high-redox potential laccases. In fact, the violuric acid assay is devised as a method to ascertain that the high-redox potential of laccase is not lost during directed evolution.

Key words Laccase, Directed evolution, High-throughput screening, Mutant libraries, Colorimetric assays, Syringyl-type phenols, Violuric acid, Synthetic organic dyes

1 Introduction

Laccases are multicopper oxidases capable of oxidizing a wide range of compounds, mainly substituted phenols and aromatic amines, using only oxygen from air. Their oxidative capabilities are limited by the redox potential at the catalytic T1 copper site, according to which laccases can be classified as of low- (up to +0.5 V), medium- (up to +0.7 V), and high-redox potential (up to +0.8 V). Laccases are considered green catalysts of great biotechnological potential, finding application in different industrial processes (pulp and paper, textile, biofuels production, organic synthesis, etc.) and in bioremediation (pollutants removal from soil and residual waters). For this reason, many efforts have been made to engineer these enzymes in order to obtain laccases with enhanced activity over different substrates and/or increased tolerance to adverse conditions, using both rational design and directed evolution approaches [1]. In this chapter we describe several colorimetric assays devised for the high-throughput screening of mutant laccase libraries,

Uwe T. Bornscheuer and Matthias Höhne (eds.), *Protein Engineering: Methods and Protocols*, Methods in Molecular Biology, vol. 1685, DOI 10.1007/978-1-4939-7366-8_14, © Springer Science+Business Media LLC 2018

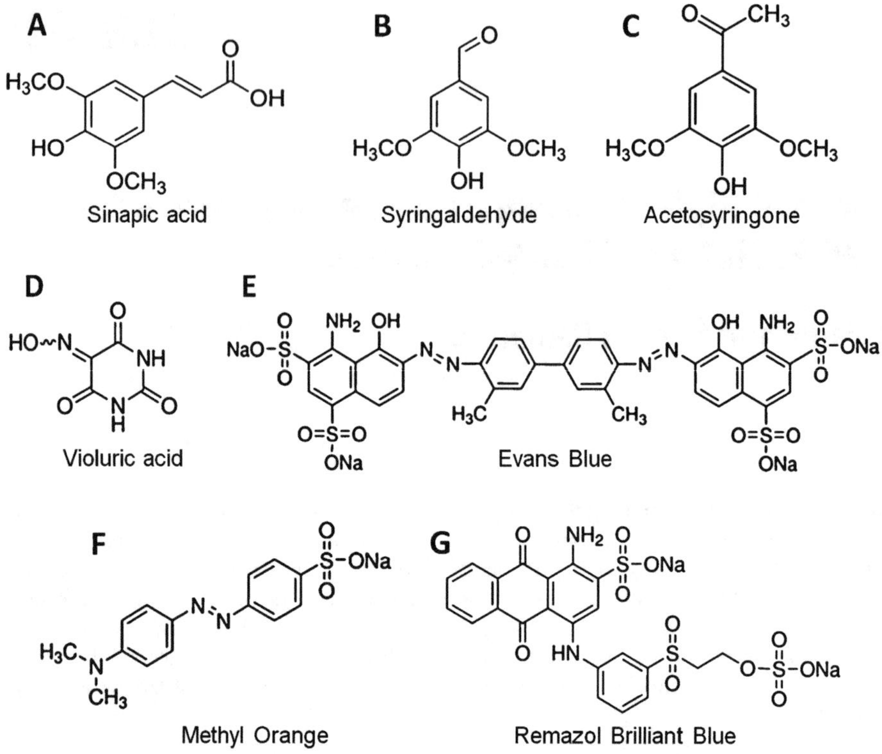

Fig. 1 Chemical structures of the substrates used in the assays: (**a**) sinapic acid, (**b**) syringaldehyde, (**c**) acetosyringone, (**d**) violuric acid, (**e**) Evans Blue, (**f**) Methyl Orange, (**g**) Remazol Brilliant Blue

which we have used for the directed evolution of high-redox potential laccases in the lab [2].

The first group of assays is based on the oxidation of sinapic acid, syringaldehyde, and acetosyringone (Fig. 1a–c), three syringyl-type phenols released during lignin biodegradation by wood-rotting fungi (the main producers of high-redox potential laccases in nature). These compounds have been described as efficient laccase redox mediators, acting as diffusible electron shuttles between the enzyme and the substrate and extending the oxidative capabilities of laccases [3]. As opposed to the more extensively studied artificial mediators such as 2,2′-azino-bis-(3-ethylbenzothiazoline-6-sulfonic acid) (ABTS), 1-hydroxybenzotriazole (HBT), and violuric acid, these S-type phenols have the advantages that (1) they can be easily obtained at low-cost from lignocellulosic residues; (2) they are environmentally friendly; and (3) due to their lower E^0 (~0.5 V), they can be oxidized by laccases of low and medium redox potential from bacteria and fungi.

Next, we describe an assay with violuric acid (Fig. 1d), which has been commonly used as a synthetic laccase mediator. With a redox potential of +1.1 V, this compound can only be directly oxidized by high-redox potential laccases. For this reason, we devised this assay as a tool to assure that the laccase redox potential was not diminished throughout the directed evolution pathway.

Finally, we describe another three activity assays for high-redox potential laccases based on the degradation of three synthetic dyes (Fig. 1e–g): two azoic dyes (Evans Blue and Methyl Orange) and one anthraquinoid (Remazol Brilliant Blue) dye. These dyes were selected for their chemical structures (azo dyes are the most commonly used in the textile industry) and because they gave a soluble and quantifiable response. These assays are useful for engineering laccases for the treatment of industrial dye-stuff effluents [4].

2   Materials

2.1   Biological Materials

1. *Saccharomyces cerevisiae* strain BJ5465 (protease deficient) for the heterologous expression of laccase.

2. Shuttle expression vector with the laccase coding sequence fused to an appropriate secretion signal peptide. In our case we used the pJRoC30 plasmid, with the GAL10 promoter and selection markers for *S. cerevisiae* (uracil) and *Escherichia coli* (ampicilin), containing a high-redox potential laccase CDS fused to an engineered alpha factor pre-proleader.

2.2   Culture Media, Buffers, and Solutions

1. Yeast transformation kit (Sigma).

2. Synthetic complete (SC) drop-out medium: 20 g/L glucose, 6.7 g/L Yeast Nitrogen Base (BD), 1.92 g/L yeast synthetic drop-out medium supplements without uracil (Sigma), 25 mg/L chloramphenicol, and 20 g/L agar.

3. Minimal medium: 20 g/L raffinose, 6.7 g/L Yeast Nitrogen Base, 1.92 g/L yeast synthetic drop-out medium supplements without uracil, 25 mg/L chloramphenicol.

4. Expression medium: 20 g/L galactose, 20 g/L peptone, 10 g/L yeast extract, 60 mM phosphate buffer (pH 6.0), 2 mM $CuSO_4$, 25 g/L ethanol, and 25 mg/mL chloramphenicol.

5. Sodium tartrate buffer, 100 mM, pH 4.0.

6. Sinapic acid stock solution: 25 mM in absolute ethanol.

7. Acetosyringone and syringaldehyde stock solutions: 20 mM in 20% ethanol.

8. Violuric acid stock solution: 200 mM in 20% ethanol.

9. Methyl Orange and Evans Blue stock solutions: 500 μM in water.

10. Remazol Brilliant Blue stock solution: 2 mM in water.

2.3   Equipment

1. Orbital shaker with humidity control.

2. 8-channel pipette.

3. 96-well flat-bottom polystyrene plates.

4. Automatic liquid handler, in the example Quadra 96–320 Liquid Handler (Tomtec).

5. UV-Vis plate reader, in the example SPECTRAMax Plus 384 (Molecular Devices).

3   Methods

3.1 Microcultures of S. cerevisiae Cells Expressing Laccase

The protocol used for the construction and expression of laccase mutant libraries in *S. cerevisiae* microcultures is described in detail in previous volumes [5, 6]. Here we briefly describe the conditions we used for cell growth and laccase expression for library screening, with slight modifications respecting those mentioned beforehand.

1. Transform an aliquot of competent yeast cells with the mutant DNA library and another with the parent gene. Grow on SC-dropout plates at 30 °C for 2–3 days.

2. Pick individual colonies and transfer to sterile 96-well plates containing 50 µL of minimal medium. In each plate, one well is left without inoculating as negative control, while a whole column is inoculated with yeast cells expressing the parent type as reference.

3. Cover plates, seal with Parafilm to avoid evaporation, and incubate for 2 days at 30 °C, with controlled humidity and agitation (200 rpm). Then, add 160 µL of expression medium to the 96-well plates and incubate for three more days (*see* **Note 1**).

4. Centrifuge plates at 2200 *g* for 15 min, at 4 °C. With the help of the automatic liquid-handler, transfer 30 µL of supernatant to new 96-well plates (not necessarily sterile). Up to three replica plates can be prepared for a multisubstrate screening with these volumes.

3.2 Laccase Activity High-Throughput Screening Assays

3.2.1 Oxidation of S-Type Phenols

1. In order to avoid spontaneous oxidation of the substrates, it is advisable to dilute the stock solutions in tartrate buffer right before performing the screening assay. The final concentration of the substrate solutions is 2 mM for acetosyringone and syringaldehyde, and 250 µM for sinapic acid (*see* **Note 2**).

2. With the help of the automatic liquid handler, add 220 µL of substrate solution to the 96-well plates containing the laccase sample.

3. As soon as the substrate is added, measure the initial absorbance in the plate reader at 370 nm for syringaldehyde, 520 nm for acetosyringone and 512 nm for sinapic acid.

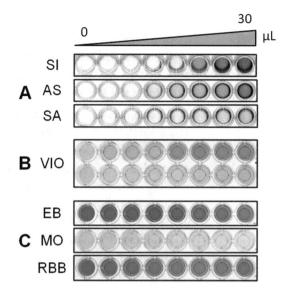

Fig. 2 Colored response of the activity assays with increasing volumes of supernatants from *S. cerevisiae* microcultures for (**a**) sinapic acid (SI), acetosyringone (AS), and syringaldehyde (SA); (**b**) violuric acid (VIO, the *bottom row* shows the color change due to copper chelation with a non-inoculated control); and (**c**) Evans Blue (EB), Methyl Orange (MO), and Remazol Brilliant Blue (RBB)

4. Incubate the plates at room temperature, preferably in darkness, and read the increase in absorbance regularly at the corresponding wavelength once the color starts to develop: syringaldehyde should turn yellow, acetosyringone will initially become yellow but then should turn red, and sinapic acid should turn pink (Fig. 2) (*see* **Notes 3** and **4**).

5. In our conditions, end-point is reached after 6–8 h for acetosyringone and syringaldehyde and 1–2 h for sinapic acid. Calculate relative activities with respect to the parent reference in each plate (*see* **Notes 5** and **6**).

3.2.2 Oxidation of Violuric Acid to Assess Redox Potential

1. Dilute violuric acid stock solution in tartrate buffer to a final concentration of 20 mM (*see* **Note 7**).

2. Add 220 μL of violuric acid working solution to the 96-well plates containing the laccase sample.

3. As soon as the substrate is added, measure the initial absorbance in the plate reader at 515 nm (*see* **Note 8**).

4. Incubate the plates at room temperature, preferably in darkness, and read increase in absorbance at 515 nm regularly once the color starts to change to red.

5. In our conditions, end-point is reached after 6–8 h (*see* **Note 5**). Calculate relative activities with respect to the parent reference in each plate.

3.2.3 Decolorization of Synthetic Organic Dyes

1. Dilute stock solutions in tartrate buffer to a final concentration of 50 μM for Methyl Orange and Evans Blue and 200 μM for Remazol Brilliant Blue (*see* **Note 9**).

2. Add 220 μL of substrate solutions to the 96-well plates containing the laccase sample.

3. As soon as the substrate is added, measure the initial absorbance in the plate reader at 470 nm for methyl orange, 605 nm for Evans Blue, and 640 nm for Remazol Brilliant Blue.

4. Incubate the plates at room temperature, preferably in darkness, and read decrease in absorbance regularly at the corresponding wavelength.

5. In our conditions, end-point is reached after 6–8 h for Evans Blue and Remazol Brilliant Blue and 20 h for Methyl Orange (*see* **Note 5**). Calculate relative activities with respect to the parent reference in each plate.

4 Notes

1. The optimal expression medium and incubation times can vary depending on the expressed laccase and should be adjusted for each case.

2. Due to the limited solubility in water of sinapic acid, syringaldehyde, and acetosyringone, stock solutions are prepared in absolute ethanol or in a 1:4 (ethanol:water) solution. Stock solutions of higher concentration may be prepared in absolute ethanol. The only important consideration is that the final concentration of ethanol in the reaction mix should be as low as possible (preferably <5%, in the conditions given it is below 2%), as the presence of organic solvents can affect laccase activity.

3. While the colored response is quite stable for syringaldehyde and acetosyringone, after some time the pink color of sinapic acid will change to yellow-orange, as the absorbance spectrum peak shifts from 512 nm to around 420 nm.

4. The development of the colored response from the oxidation of sinapic acid is pH-dependent. For this reason, the colorimetric assay should be validated if it is performed at a different pH.

5. Under the microculture conditions described, the high-redox potential laccase we used for the validation of the assays produces up to 120 mU/mL, measured with 3 mM ABTS ($\varepsilon_{418} = 36{,}000$ M^{-1} cm^{-1}). Using this activity as reference, the detection limit is of 0.15 mU in the well for all the assays at the indicated end-point times, except in the case of sinapic acid, which has a detection limit of 1 mU (*see* **Note 6**). Nevertheless, lower laccase activity can be detected with longer reaction times.

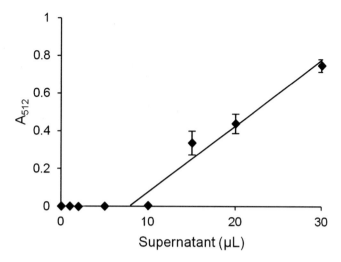

Fig. 3 Absorbance at 512 nm for the end-point assay with sinapic acid after 1 h of reaction with increasing volumes of microculture supernatant

6. Upon oxidation of sinapic acid by laccase, the phenoxy radicals undergo a fast β-β′ coupling rendering dimeric intermediates [7]. The colored response obtained in the assay is the result of the oxidation of the aforementioned intermediates. This produces an initial lag phase followed by a rapid increase at 512 nm. For this reason, it may be difficult to calculate relative activities accurately based on end-point measures when one clone has reached maximum absorbance at 512 nm and the parent reference has not yet developed any color. In our experience, 2 to 3-fold increases in activity can be measured with certainty (Fig. 3).

7. In order to dissolve the substrate, the stock solution should be prepared with mild heating ($<50\ ^\circ$C) immediately before diluting in the reaction buffer, as violuric acid will precipitate with time. Due to its low solubility, the violuric acid concentration used in this assay is not saturating for laccase activity. Nevertheless, it is sufficient to see substrate oxidation and to assure the high redox potential of laccase.

8. Violuric acid acts as a chelant of metal ions [8] and can therefore coordinate Cu(II) present in the expression medium. The violurate-Cu(II) complex presents an orange color with an absorbance peak at 420 nm that, however, does not interfere with the increase in absorbance measured at 515 nm (Fig. 4).

9. The high initial absorbance of the three dyes limited their use at higher concentrations, considering the plate reader's absorbance detection limit. Therefore, the concentrations used in the assays are not saturating for laccase activity, although they gave a measurable and linear response.

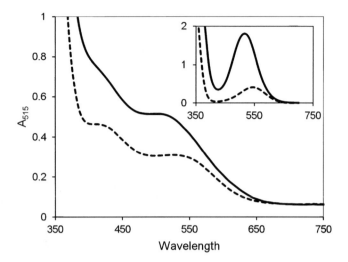

Fig. 4 Changes in the visible spectrum of violuric acid after 20 h oxidation with 30 μL of microculture supernatant. *Inset* shows the oxidation by a commercial high-redox potential laccase where the peak for the violurate complex of Cu(II) is not observed. Inital spectra are shown in *dashed lines*

Acknowledgments

This work was supported by the INDOX (KBBE-2013-613549) EU project and the NOESIS (BIO2014-56388-R) project of the Spanish Ministry of Economy and Competitiveness (MINECO).

References

1. Pardo I, Camarero S (2015) Laccase engineering by rational and evolutionary design. Cell Mol Life Sci 72:897–910

2. Pardo I, Chanagá X, Vicente AI et al (2013) New colorimetric screening assays for the directed evolution of fungal laccases to improve the conversion of plant biomass. BMC Biotechnol 13:90

3. Camarero S, Ibarra D, Martinez MJ et al (2005) Lignin-derived compounds as efficient laccase mediators for decolorization of different types of recalcitrant dyes. Appl Environ Microbiol 71:1775–1784

4. Husain Q (2007) Potential applications of the oxidoreductive enzymes in the decolorization and detoxification of textile and other synthetic dyes from polluted water: a review. Crit Rev Biotechnol 26:201–221

5. Bulter T, Sieber V, Alcalde M (2003) Screening mutant libraries in *Saccharomyces cerevisiae*. Methods Mol Biol 230:99–107

6. Bulter T, Alcalde M (2003) Preparing libraries in *Saccharomyces cerevisiae*. Methods Mol Biol 231:17–22

7. Lacki K, Duvnjak Z (1998) Transformation of 3,5-dimethoxy,4-hydroxy cinnamic acid by polyphenol oxidase from the fungus *Trametes versicolor*: product elucidation studies. Biotechnol Bioeng 57:694–703

8. Leermakers PA, Hoffman WA (1958) Chelates of violuric acid. J Am Chem Soc 80:5663–5667

Chapter 15

Directed Coevolution of Two Cellulosic Enzymes in *Escherichia coli* Based on Their Synergistic Reactions

Min Liu, Lidan Ye, and Hongwei Yu

Abstract

Directed evolution is a widely used technique for improving enzymatic properties. The development of an efficient high-throughput screening method is a key procedure, which is however often unavailable for many enzyme reactions, including the cellulase-catalyzed cellulose hydrolysis. Here, we describe a high-throughput screening assay for directed coevolution of two cellulases (an endoglucanase and a β-glucosidase) in form of a bicistronic operon based on their synergistic reactions. Insoluble filter paper is used as the real cellulose substrate to screen for positive enzyme variants, facilitated by the colorimetric assay coupled to glucose liberated from cellulose under catalysis of endoglucanase and β-glucosidase. Directed coevolution saves the labor and time required for two independent directed evolution cycles, which might provide reference for the engineering of other cellulosic enzymes or multienzyme systems.

Key words High-throughput screening, Directed evolution, Cellulose, Cellulase, Coexpression, Endoglucanase, β-glucosidase

1 Introduction

Cellulose is the most abundant renewable natural biological resource in the world. The production of biofuels and other biomaterials from cellulosic materials is critical for sustainable development. Generally, the widely accepted mechanism for hydrolyzing cellulose to the fermentable glucose involves synergistic actions of various cellulases including endoglucanase, exoglucanase, and β-glucosidase. Until recently, the cellulases still comprise a major cost in the economics of cellulosic industry [1–4]. Improving the cellulase activity by protein engineering and thus decreasing the cellulase amount required is one of the potential approaches to reduce production cost.

Directed evolution, independent of enzyme structure and the interaction between enzyme and substrate, is a widely used technique for improving enzymatic properties. An efficient high-throughput screening method is the key to successful directed

Uwe T. Bornscheuer and Matthias Höhne (eds.), *Protein Engineering: Methods and Protocols*, Methods in Molecular Biology, vol. 1685, DOI 10.1007/978-1-4939-7366-8_15, © Springer Science+Business Media LLC 2018

evolution, which is however often unavailable for enzyme reactions without visible color change or easily detectable features, with the cellulase-catalyzed cellulose degradation as an example. Developing efficient screening methods for cellulosic enzymes, especially using the exact substrate of cellulose rather than cellulose analogs or chromogenic and fluorogenic substrates, is imperative for their directed evolution, with the ultimate goal of "you get what you screen for" [4, 5].

Endoglucanase and β-glucosidase are two key enzymes in cellulose degradation. Endoglucanases catalyze the hydrolysis of internal β-1,4-glucosidic bonds of amorphous cellulose, generating oligosaccharides of various lengths, many of which can be hydrolyzed by β-glucosidase to liberate glucose [3]. Based on this principle, we constructed a bicistronic operon coexpressing a cocktail mixture containing both endoglucanase and β-glucosidase, the genes of which were connected by the internal ribosome binding site (IRBS) sequence originated from pET30a introduced upstream of the β-glucosidase encoding gene [6–8], as depicted in Fig. 1. This cocktail mixture would directly hydrolyze cellulose to glucose, the fermentable sugar readily detectable in a high-throughput manner [6]. The bicistronic operon was constructed based on the T7 promoter for transcriptional regulation and expressed in an *E. coli* strain harboring the DE3 prophage. For directed coevolution, the mutant library was constructed using error-prone PCR by treating the endoglucanase and β-glucosidase encoding genes together with the IRBS as an integral construct (Fig. 2). The 7-mm diameter Whatman No. 1 filter paper was used as a model cellulose substrate for screening and the glucose liberated from cellulose was detected colorimetrically by a coupled reaction of glucose oxidase and peroxidase [6, 9].

2   Materials

2.1   Biological and Chemical Materials

Prepare all solutions using analytical grade reagents and ultrapure water prepared by purifying deionized water to a sensitivity of 18 MΩ cm at 25 °C.

1. *Cellulomonas fimi* (ATCC 484).

2. *Trichoderma reesei* QM 9414 (ATCC 26921).

3. *E. coli* BL21 (DE3).

4. Plasmid pET30a.

5. Oligonucleotides:

   Primer 1: <u>ATTCGAA</u>GCTCCCGGCTGCCGCGTCGACTAC. (Underlined: *Nsp*V recognition sequence).

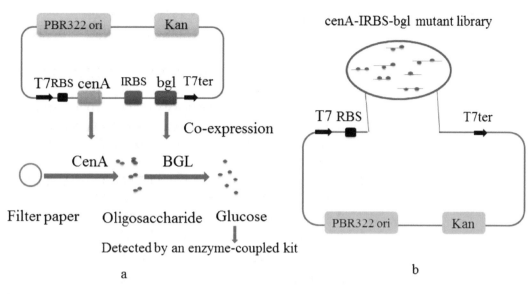

Fig. 1 Principle for the directed coevolution strategy of endoglucanase and β-glucosidase. (**a**) Construction of a bicistronic operon for coexpression of CenA and BGL based on the backbone of pET30a plasmid. BGL can hydrolyze oligosaccharides liberated by CenA to glucose, which can be easily detected by a glucose oxidase and peroxidase assay-coupled kit. (**b**) Targeting the genes of cenA, bgl, and the internal ribosome binding site as an integral structure for mutagenesis (named cenA-IRBS-bgl). The mutant library is constructed by introduction of the cenA-IRBS-bgl mutants generated by error-prone PCR into plasmid pET30a. RBS: ribosome binding site of pET30a; IRBS: internal ribosome binding site introduced by primer design

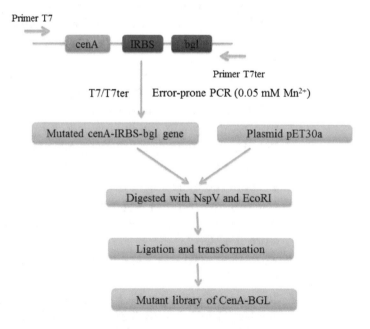

Fig. 2 Procedures for construction of the mutant library of CenA-BGL. Random mutations are introduced into the cenA-IRBS-bgl gene via error-prone PCR (with 0.05 mM Mn^{2+}) using primers T7 and T7ter

Primer 2: CGG<u>GGTACC</u>TCA*ATG*GTGGTGATGATGGT GC CACCTGGCGTTG. (Underlined: *Kpn*I recognition sequence).

Primer 3: GGAATTC<u>GGTACC</u>TTGCCCAAGGACTTTCAGT GGGGGTTCGCCACGGCTGCCTACCAGATCGA GGGCGCCGTC. (Underlined: *Kpn*I recognition sequence)

Primer 4: CCGG<u>AATTC</u>TCAATGGTGGTGATGATGGTGCG CCGCCGCAATCAGCT. (Underlined: *Eco*RI recognition sequence).

Primer 5: CGG<u>GGTACC</u>AATAATTTTGTTTAACTTTAA-GAAGGAGATATACATATGTTGCCCAAGGACTTT-CAG (Underlined: *Kpn*I recognition sequence).

Primer T7: TAATACGACTCACTATAGGG

Primer T7ter: TGCTAGTTATTGCTCAGCGG.

Prepare 10 µM solutions of each oligonucleotide.

6. LB medium: 10 g/L tryptone, 5 g/L yeast extract, 10 g/L NaCl. Adjust pH to 7.2 and sterilize by autoclaving at 121 °C for 20 min.

7. LB agar plates: 1.5% (w/v) agar in LB medium.

8. LB kanamycin plates: 1.5% (w/v) agar in LB medium, 50 µg/mL kanamycin.

9. Cellulose inducing medium: 3 g/L peptone, 2 g/L $(NH_4)_2SO_4$, 0.5 g/L yeast extract, 4 g/L KH_2SO_4, 0.3 g/L $CaCl_2 \cdot 2H_2O$, 0.3 g/L $MgSO_4 \cdot 7H_2O$, 0.02% (w/v) Tween 80, 20 g/L Avicel PH-101 (Sigma, USA). Sterilize by autoclaving.

10. Basal medium: 1 g/L $NaNO_3$, 1 g/L K_2HPO_4, 0.5 g/L KCl, 0.5 g/L yeast extract, 0.5 g/L $MgSO_4 \cdot 7H_2O$, 1 g/L Avicel PH-101. Sterilize by autoclaving.

11. PBS buffer: 8 g/L NaCl, 0.2 g/L KCl, 3.63 g/L $Na_2HPO_4 \cdot 12H_2O$, 0.24 g/L KH_2PO_4. Adjust pH to 7.4.

12. Tris–HCl buffer (50 mM, pH 8.0): 6.06 g/L Tris. Adjust pH to about 8.0 by HCl.

13. $CaCl_2$ buffer (60 mM, pH 7.2): 8.82 g/L $CaCl_2 \cdot 2H_2O$, 15% (v/v) glycerol. Adjust pH to 7.2. Sterilize by autoclaving and store at 4 °C.

14. NaAc buffer (50 mM, pH 4.8): 4.1 g/L sodium acetate. Adjust pH to 4.8.

15. Inoue buffer: Dissolve 15.1 g PIPES in 80 mL distilled water. Adjust pH to 6.7 using 5 M KOH and dilute to 100 mL with water (store at −20 °C). Mix 10.9 g $MnCl_2 \cdot 4H_2O$ (55 mM), 2.2 g $CaCl_2 \cdot 2H_2O$ (15 mM), and 18.6 g KCl (250 mM) with

20 mL 0.5 M PIPES buffer, and dilute to 1000 mL with water. Sterilize by filtration through 0.2 μm filter and store at 4 °C.

16. Cell lysis buffer: Add 0.48 g $MgCl_2 \cdot 6H_2O$ (5 mM), 250 mg lysozyme, 1000 U DNase I to 500 mL of Tris–HCl buffer, mix well and store at 4 °C.

17. Whatman No. 1 filter paper (Sigma, USA): Cut the Whatman No. 1 filter paper (φ = 9 cm) into plates with a diameter of 7.0 mm using an office paper punch for the 96-well microplate assays.

18. IPTG solution (1 mM): Dissolve 2.38 g IPTG in 10 mL water, sterilize by filtration through 0.2 μm Millipore filter and store at −20 °C.

19. Binding buffer: 7.6 g/L $Na_3PO_4 \cdot 12H_2O$, 29.2 g/L NaCl, 1.36 g/L imidazole. Adjust pH to 7.4. Sterilize by filtration.

20. Elution buffer: 7.6 g/L $Na_3PO_4 \cdot 12H_2O$, 29.2 g/L NaCl, 34 g/L imidazole. Adjust pH to 7.4. Sterilize by filtration.

21. Stripping buffer: 7.6 g/L $Na_3PO_4 \cdot 12H_2O$, 29.2 g/L NaCl, 14.6 g/L EDTA. Adjust pH to 7.4. Sterilize by filtration.

22. NaCl solution (1.5 M): 87.7 g/L NaCl. Sterilize by filtration.

23. NaOH solution (1 M): 40 g/L. Sterilize by filtration.

24. $NiSO_4$ solution (0.1 M): 26.3 g/L $NiSO_4$. Sterilize by filtration.

25. Bradford reagent.

26. Sterile 96-well microtiter plates with lid, flat bottom, clear.

27. Kits and enzymes:

    Genome extraction kits for bacteria and fungi.

    T4 DNA Ligase.

    FastDigest restriction enzymes (*NspV/KpnI/EcoRI*).

    Glucose detection kit (Rsbio, China).

    PrimeSTAR HS DNA polymerase (*TaKaRa*, Japan).

    TaKaRa LA Taq (*TaKaRa*, Japan).

2.2 Equipment

UV-VIS microplate reader.

Microplate shaking incubator.

Equipment for standard molecular biology operations (gel electrophoresis, incubators, PCR machine).

3 Methods

3.1 Cloning of cenA Gene

1. Cultivate *C. fimi* (ATCC 484) in the basal medium at 30 °C for 2 days. Extract the genome of *C. fimi* using a bacterial genome extraction kit.

2. Amplify the cenA (Uniprot: P07984) gene from *C. fimi* genomic DNA using primer 1 and primer 2 and PrimeSTAR HS DNA polymerase according to the manufacturer's instruction.

3. Digest the cenA gene with *NspV/Kpn*I in a 40 μL reaction system containing 34 μL of purified cenA gene, 1 μL of *Nsp*V, 1 μL of *Kpn*I, and 4 μL of 10 × FastDigest buffer at 37 °C for 1 h.

4. Introduce the *NspV/Kpn*I-digested cenA gene into pET30a digested with the same restriction enzymes to obtain plasmid pET30a-cenA by ligation using T4 DNA ligase at 22 °C for 30 min.

3.2 Cloning of bgl Gene

1. Cultivate *T. reesei* QM 9414 (ATCC 26921) in the cellulose inducing medium at 28 °C for 5 days. Extract the genome of *T. reesei* using a fungal genome extraction kit.

2. Amplify the bgl (Uniprot: O93785) gene from *T. reesei* genomic DNA using primer 3 and primer 4 and PrimeSTAR HS DNA polymerase according to manufacturer's instruction (*see* **Note 1**).

3. Digest the bgl gene with *Kpn*I/*Eco*RI in a 40 μL reaction system containing 34 μL of purified bgl gene, 1 μL of *Kpn*I, 1 μL of *Eco*RI, and 4 μL of 10 × FastDigest buffer at 37 °C for 1 h.

4. Introduce the *Kpn*I/*Eco*RI-digested bgl gene into pET30a digested with the same restriction enzymes to obtain plasmid pET30a-bgl by ligation using T4 DNA ligase at 22 °C for 30 min.

3.3 Construction of Bicistronic Plasmid pcenA-bgl

1. Introduce the IRBS sequence (AATAATTTTGTTTAACTT-TAAGAAGGAGATATACAT, 36 nt, *see* **Note 2**) upstream of the bgl gene using primer 5 and primer 4 to obtain the IRBS-*bgl* gene, using plasmid pET30a-bgl as the template and PrimeSTAR HS DNA polymerase according to the manufacturer's instruction.

2. Digest IRBS-bgl gene with KpnI/EcoRI in a 40 μL reaction system containing 34 μL of purified IRBS-bgl gene, 1 μL of KpnI, 1 μL of EcoRI, and 4 μL of 10 × FastDigest buffer at 37 °C for 1 h.

3. Introduce the *Kpn*I/*Eco*RI-digested IRBS-bgl into plasmid pET30a-cenA digested with the same restriction enzymes by ligation at 22 °C for 30 min using T4 DNA ligase to obtain the plasmid pcenA-bgl (*see* **Note 3**).

3.4 Error-Prone PCR for Generation of cenA-IRBS-bgl Mutants

1. Mix 25 μL of 2 × GC buffer I (Mg^{2+} plus), 8 μL of dNTPs (2.5 mM each), 0.5 μL of *TakaraLA Taq* (5 U/μL), 0.05 mM Mn^{2+} (*see* **Note 4**), 1 μL of primer T7, 1 μL primer T7ter, 10 ng template plasmid pcenA-bgl, and top up to a final volume of 50 μL using water.

2. Prepare the PCR product (cenA-IRBS-bgl) according to manufacturer's recommendations of *LA* Taq DNA polymerase using the following cycles: 1 min at 94 °C for denaturation, 30 cycles of 30 s at 94 °C for denaturation, 30 s at 54 °C for annealing and 2.5 min at 72 °C for extension.

3. Purify the PCR product of cenA-IRBS-bgl using a DNA cleanup kit according to the manufacturer's instruction.

3.5 Digestion for Mutant Library Construction

1. To digest the cenA-IRBS-bgl fragment, mix 34 μL of purified PCR product of cenA-IRBS-bgl with 4 μL of 10 × FastDigest buffer, 1 μL of *Nsp*V, and 1 μL of *Eco*RI and incubate at 37 °C for 1 h.

2. To digest the vector, mix 34 μL of pET30a plasmid with 4 μL of 10 × FastDigest buffer, 1 μL of *Nsp*V, and 1 μL of *Eco*RI and incubate at 37 °C for 1 h.

3. Purify the digested cenA-IRBS-bgl gene and pET30a plasmid using a DNA clean up kit according to the manufacturer's instruction.

3.6 Preparation of E. coli BL21 (DE3) Competent Cells

1. Pick a single colony from LB plate (cultured at 37 °C for 12–16 h) and inoculate it into 25 mL LB, followed by incubation at 37 °C for 6–8 h (250 rpm).

2. Transfer 0.5 mL, 1 mL, and 2 mL of the culture into 500 mL conical flasks containing 125 mL LB, respectively, and incubate at 18 °C overnight.

3. Measure the OD_{600} values of the three cultures the next morning. Cool the culture with an OD_{600} of 0.55 in an ice bath and discard the other two cultures.

4. Centrifuge the culture for 10 min at 4 °C ($2000 \times g$), and gently suspend the precipitate with 40 mL of cold Inoue buffer. Then centrifuge again and resuspend with 10 mL of cold Inoue buffer.

5. Add 0.75 mL DMSO, blend gently, and put on ice for 10 min.

6. Aliquot the cell suspension in volumes of 100 μL or 200 μL and store at −80 °C after briefly immersed in liquid nitrogen.

3.7 Construction of the Mutant Library

1. Ligate the *Nsp*V/*Eco*RI double-digested cenA-IRBS-bgl fragment and pET30a by mixing at a ratio of 3:1–5:1 (*see* **Note 5**), and incubating with T4 DNA ligase in a 20 μL reaction at 22 °C for 30 min.

2. Transform the ligation system into 200 μL *E. coli* BL21 (DE3) component cells using a heat pulse method (42 °C for 90 s) and incubate the cells on LB kanamycin plates at 37 °C overnight.

3. Create about 3000 colonies in each round of screening (Fig. 2).

3.8 High-Throughput Screening Protocol

1. Select and transfer the colonies to 96-well microplates (containing 200 μL of LB medium in each well) one by one with sterilized toothpicks. Incubate at 37 °C overnight in a microplate shaker. These plates are called the mother plates (*see* **Note 6**).

2. Transfer 10 μL of cultures (*see* **Note 7**) from each well of the mother plate to the corresponding fresh wells in another 96-well microplate (containing 200 μL of LB medium in each well, called daughter plate) and incubate at 37 °C in a microplate shaker for 2–3 h.

3. Add 20 μL of IPTG solution into each well of the daughter plate to induce protein expression, and continue growth for another 4 h.

4. Collect the cells by centrifugation at $4000 \times g$ for 10 min and freeze at −80 °C overnight (*see* **Note 8**), then lyse with lysis buffer at 37 °C for 1 h.

5. Centrifuge the microplates at 2000 g for 5 min (4 °C) to obtain the crude enzyme.

6. Store the mother plates at −80 °C by adding 15% glycerol.

3.9 Enzyme Activity Assay for High-Throughput Screening

1. Add appropriate amount of crude enzyme (*see* **Note 9**) from wells of the daughter plate into a new 96-well microplate with each well containing 0.05 M NaAc buffer and a disk ($\varphi = 7$ mm) of Whatman No. 1 filter paper in a 60 μL assay reaction.

2. Incubate the reaction mixture at 40 °C for 30 min.

3. Measure the final product of glucose with a coupled glucose oxidase and peroxidase assay kit (Rsbio, China). Transfer 100 μL (mixture of 50 μL buffer A and 50 μL buffer B) of reaction solution into each well of the reaction microplate above, incubate at 37 °C for 10 min, and assay at 505 nm using a microplate spectrophotometer (*see* **Note 10**).

3.10 Preparation of Ni-NTA Column for Enzyme Purification by In Situ Cleaning and Regeneration

1. Strip the resin by washing with 5–10 resin volumes of stripping buffer (*see* **Note 11**).

2. Wash the column with 5–10 resin volumes of 1.5 M NaCl and 10 resin volumes of water.

3. Wash the column with 1 M NaOH for 1–2 h, followed by 10 resin volumes of binding buffer and 10 resin volumes of water.

4. Wash the column with 5–10 resin volumes of 30% isopropanol and 10 resin volumes of water to complete in situ cleaning of the Ni-NTA column.

5. Strip the resin by washing with 5–10 resin volumes of stripping buffer.

6. Wash the column with 5–10 resin volumes of binding buffer and 10 resin volumes of water.

7. Recharge the water-washed column by loading 2.5 mL of 0.1 M NiSO$_4$.

8. Wash the column with 5 resin volumes of water and 5 resin volumes of binding buffer.

9. Wash the column with 5 resin volumes of 20% ethanol and store the column in 20% ethanol. The column is now ready for use.

3.11 Purification of the His-Tagged Protein

1. Install the Ni-NTA column on the ÄKTA purifier FPLC purification system.

2. Wash the column with 3–5 resin volumes of water.

3. Equilibrate the column with 5 resin volumes of binding buffer at a flow rate of 5 mL/min.

4. Charge the cell lysate to the column at a flow rate of 5 mL/min.

5. Wash the column with 6 resin volumes of binding buffer.

6. Wash the column with 4 resin volumes of elution buffer until the protein peak occurs and collect the protein.

7. Wash the column with 5 resin volumes of binding buffer and store the column with 20% ethanol.

3.12 Determination of Relative Activity and Specific Activity of the Wild-Type Strain and Positive Mutants

1. Inoculate the wild-type strain and positive mutants into 5 mL of LB medium (supplemented with 50 μg/mL kanamycin), and let grow overnight.

2. In the morning of the next day, dilute the culture into 50 mL of LB medium (containing 50 μg/mL kanamycin) and incubate at 37 °C until OD$_{600}$ reaches 0.6–0.8.

3. Add 0.1 mM IPTG and further incubate at 16 °C overnight (*see* **Note 12**).

4. Harvest the cells, wash with PBS buffer, resuspend in Tris–HCl buffer, and lyse by sonication.

5. Purify the supernatant using ÄKTA purifier FPLC purification system equipped with the HisTrap^{TM} HP column (GE Healthcare, Sweden).

6. Measure the protein concentration by the method of Bradford.

7. Determine the relative activity using the crude enzyme and the specific activity using the purified enzyme (*see* **Note 13**).

3.13 Directed Coevolution

Usually, several rounds of screenings are conducted to select positive mutants with higher activities than the wild-type enzyme. In our study, we conducted three rounds of mutagenesis and subsequently combined the most active variants of both enzymes to yield

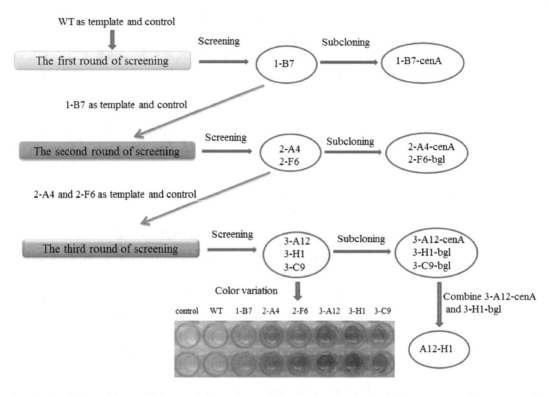

Fig. 3 Description of the multiple-round screening and the display of color variation among crude enzymes of the wild type (WT, CenA-BGL) and six positive variants

the final bicistronic plasmid (Fig. 3). At this point, the mutagenesis strategy (number of screening rounds, combination of beneficial mutations and genes) can be adapted to individual needs.

1. Use the wild-type plasmid of pcenA-bgl as the template and the wild-type strain as the positive control to perform the first round of error-prone PCR (*see* **Note 14**).

2. Select the best variant with improved activity, prepare plasmid, and send it for sequencing.

3. Subclone the genes from the improved variant into pET30a, send for sequencing and investigate its characteristics (Fig. 3).

4. Use the plasmid of the improved variant as the template and as the positive control in the second round of error-prone PCR.

5. Conduct subsequent rounds of error-prone PCR and screening. Use one or a few variants as starting points for further mutagenesis (Fig. 3) (*see* **Note 15**).

3.14 Combination of Positive Mutations

1. Combine the mutant genes of the most active variants to generate a final mutant bicistronic operon (Fig. 3).

2. Determine its activity and compare it to those of the wild-type and other mutant strains (*see* **Notes 16** and **17**).

4 Notes

1. There is only one intron in the bgl gene, thus over-lap extension PCR is more convenient than reverse transcription PCR (RT-PCR) for cloning of the bgl gene. For target genes with two or more introns, it would be better to use RT-PCR.

2. The ribosome binding site (RBS) and its surrounding mRNA sequences are important factors for determining the expression level of proteins [6], which are therefore often optimized and demonstrated with effectiveness in many commercial vectors, including pET30a. In this work, the sequence downstream of the T7 promoter and upstream of the initiation codon (ATG) was used as the internal RBS for expression of the downstream bgl gene.

3. The gene order significantly affects enzyme expression in the coexpression process. In general, the bicistronic construct under control of a single promoter often leads to poor expression of the downstream gene. Moreover, locating the gene with shorter coding sequence in the upstream and the gene with longer coding sequence in the downstream might result in higher expression level of the downstream protein than the reverse order. This might be ascribed to the more complex structure of the longer coding sequence which negatively affects gene expression [8].

4. Optimization of Mn^{2+} concentration (0.01 mM, 0.05 mM, 0.1 mM, and 0.15 mM were tested here) is often required for adapting the mutation frequency in the error-prone PCR. In our case, DNA sequencing analysis of nine randomly picked mutants generated with 0.05 mM Mn^{2+} revealed an average mutation frequency of 2.4 base pairs/cenA-bgl operon, indicating alteration of 1 or 2 amino acids in the CenA-BGL enzyme cascade in average, which was deemed as appropriate mutation rate for finding positive colonies.

5. A ratio of 3:1–5:1 between the mutant cenA-IRBS-bgl fragment and the plasmid can guarantee the efficiency of ligation and thus creation of effective colonies in the mutant library.

6. It is important to store the mother plates. In each round of screening, when positive colonies were found in the high-throughput screening process, the colonies were transferred from corresponding mother plates to new microplates for rescreening. If the enhancement of glucose production is repeatable, the corresponding colonies are cultivated in flasks and characterized.

7. In the high-throughput screening process, it is necessary to guarantee exactly the same inoculum concentration among each well in the microplate.

8. The cells from the daughter plate are frozen at −80 °C overnight before cell disruption with lysis buffer at 37 °C next day. Improvement of the lysis efficiency is expected after freezing at −80 °C overnight because of the thermal expansion phenomenon.

9. The color difference between different colonies would be difficult to distinguish if the glucose concentration is too high or too low. Thus, adjustment of the enzyme addition amount in the high-throughput reaction system is necessary.

10. The final product glucose was measured with a coupled glucose oxidase and peroxidase assay kit. The detection principle of this kit is as follows: Gluconic acid and hydrogen peroxide are generated from glucose under the catalysis of glucose oxidase; subsequently, the coexistence of hydrogen peroxide, 4-aminoantipyrine, and phenol leads to the formation of quinones that can be detected at 505 nm using a microplate spectrophotometer.

11. When significant contamination of the resin is observed, a cleaning procedure is required to restore column performance, whereas stripping and recharging of Ni-NTA column is not always necessary. However, if the performance is still not satisfactory after in situ cleaning, the Ni-NTA resin in the column should be stripped and recharged.

12. Low-temperature induction (at 16 °C overnight) with IPTG was used to enhance the soluble expression of CenA and BGL in flask cultures and characterization of positive colonies. In the high-throughput screening process, the CenA and BGL were induced at 37 °C for 4 h instead of the overnight low-temperature induction at 16 °C, in order to save the screening time.

13. The relative activity was determined with crude enzyme and expressed as unit/volume, in order to evaluate both the enzyme expression level and the enzyme activity of mutant colonies.

14. At least three wells should be incubated for each positive control to enhance the effectiveness of high-throughput screening in each round.

15. For further improvement of activity, mixed plasmids of 2-A4 and 2-F6 were used as templates in the third round of screening due to their comparable activity in glucose production. As effective controls, at least three wells for 2-A4 and three wells for 2-F6 should be set in each screening microplate.

16. All CenA and BGL mutants were separately subcloned and characterized from the CenA-BGL mutants obtained in each round of screening. After that, the CenA and BGL mutants with the highest activities (3-A12-CenA and 3-H1-BGL, Fig. 3) were combined to generate a new CenA-BGL mutant, which showed the highest activity in glucose production.

17. This coengineering approach based on bicistronic operon can also be applied for the directed evolution of an individual exoglucanase, which could be achieved by ligating the mutant exoglucanase gene with plasmid containing a wild-type gene of β-glucosidase to facilitate the screening. β-glucosidase can hydrolyze the cellobiose generated by exoglucanase to glucose. The catalytic efficiency of exoglucanase can be evaluated indirectly by measuring the glucose generated fom the synergetic action of β-glucosidase. Further, the engineered exoglucosidase together with the engineered endoglucanase and β-glucosidase can be combined into a single plasmid to construct a whole engineered cellulase system in *E. coli* for enhanced hydrolysis of cellulose.

Acknowledgment

This work was financially supported by the Natural Science Foundation of China (Grant No. 21176215), and National High Technology Research and Development Program of China (Grant No. SS2015AA020601).

References

1. Banerjee G, Scott-Craig JS, Walton JD (2010) Improving enzymes for biomass conversion: a basic research perspective. Bioenergy Res 3:82–92

2. Klein-Marcuschamer D, Oleskowicz-Popiel P, Simmons BA et al (2012) The challenge of enzyme cost in the production of lignocellulosic biofuels. Biotechnol Bioeng 109:1083–1087

3. Lynd LR, Weimer PJ, van Zyl WH et al (2002) Microbial cellulose utilization: fundamentals and biotechnology. Microbiol Mol Biol Rev 66:506–577

4. Zhang YHP, Himmel ME, Mielenz JR (2006) Outlook for cellulase improvement: screening and selection strategies. Biotechnol Adv 24:452–481

5. Cobb RE, Chao R, Zhao HM (2013) Directed evolution: past, present, and future. AICHE J 59:1432–1440

6. Liu M, Gu J, Xie W et al (2013) Directed coevolution of an endoglucanase and a beta-glucosidase in *Escherichia coli* by a novel high-throughput screening method. Chem Commun 49:7219–7221

7. Liu M, Yu HW (2012) Cocktail production of an endo-beta-xylanase and a beta-glucosidase from *Trichoderma reesei* QM 9414 in *Escherichia coli*. Biochem Eng J 68:1–6

8. Smolke CD, Keasling JD (2002) Effect of gene location, mRNA secondary structures, and RNase sites on expression of two genes in an engineered operon. Biotechnol Bioeng 80:762–776

9. Xiao ZZ, Storms R, Tsang A (2004) Microplate-based filter paper assay to measure total cellulase activity. Biotechnol Bioeng 88:832–837

Chapter 16

Program-Guided Design of High-Throughput Enzyme Screening Experiments and Automated Data Analysis/Evaluation

Mark Dörr and Uwe T. Bornscheuer

Abstract

The open source *LARA* software suite is designed for guiding manual or automated high-throughput screening experiments. Process planning, data reading, analysis, and visualization are herein explained in a step-by-step guide using exemplary dataset.

Key words High-throughput enzyme screening, Robotics, Process planning, Evaluation, Automated data analysis, Data visualization, Python, R

1 Introduction to the Challenges of High-Throughput Enzyme Screening

Manual and automated high-throughput enzyme screening [1–4] requires a well-planned and organized working schedule to (1) efficiently and reproducibly grow and harvest cells, (2) lyse cells to obtain a sufficient protein concentration, and finally (3) perform the most informative assays to determine enzyme activity, stability or other enzyme characteristics, e.g., kinetic constants, pH optimum, and solvent tolerance. For this purpose a widely applicable open source software suite *LARA* has been developed (*lara.uni-greifswald.de*) [5]. *LARA* supports its user with planning, designing, and scheduling the different phases of high-throughput screening experiments. A constantly updated overview of all required materials for running the process is visualized on a survey page during the design process. This includes chemicals; containers like microtiter plates, deep well blocks, and reagent reservoirs; and devices as well as an approximate schedule of the whole experiment. This helps to foresee and circumvent bottleneck steps and to predict the amounts of consumables and the time point at which they are required. The *LARA* suite uses only software that is freely and openly available.

Uwe T. Bornscheuer and Matthias Höhne (eds.), *Protein Engineering: Methods and Protocols*, Methods in Molecular Biology, vol. 1685, DOI 10.1007/978-1-4939-7366-8_16, © Springer Science+Business Media LLC 2018

The software is written platform-independent and should principally run under all common operating systems (*LINUX*, *Apple OSX* and also *MS-Windows*), but it has been developed and tested in a *LINUX* environment, so this is best supported (recommended distribution: Ubuntu 16.04 LTS)—for the database version a Linux-based server is recommended.

Please make sure that *Python* (currently version 2.7), *Python-Qt4*, and *R* (ver. >3.3) are installed on your computer. For convenience, we recommend R-Studio from rstudio.org.

2 Installation of the Components

The latest download instructions and release notes can be found at **lara.uni-greifswald.de (or the corresponding GitHub repository: https://github.com/LARAsuite)**.

The following summary of the installation procedure is described for a Linux system; for OSX and Microsoft Windows installation, please refer to the file INSTALL_OSX.rst or INSTALL_MSWIN.rst respectively—please note that software for closed source operation systems are not actively supported by the *LARA* developers. Alternatively, it is possible to use a virtual machine, like *VirtualBox* (*virtualbox.org*) to install a *LINUX* distribution and the *LARA* suite. Please always use the latest installation procedure described in the source code repositories.

The *LARA* software suite currently consists of three major parts: The *LARA* planning tool, which provides a graphical user interface (GUI) to plan screenings and robotic processes. This tool is a *Python* based application that needs to be started from a terminal (*LINUX* command line) as described in the README file. To use *LARA*, only basic knowledge of opening a command-line and changing to the right directory is necessary. For writing further process plug-ins, code generators for new robotic platforms etc., intermediate knowledge of *Python* and *QT4* is necessary. The second part is called *LARA-R* and consists of *R* packages for data reading, evaluation and visualization libraries. To use these packages basic knowledge of *R* is required (we recommend to additionally install *R-Studio* for easier handling). For expanding the code, intermediate to advanced *R* knowledge is needed. The third part, *LARA-django*, is a *Python* and *Python-django* (s. *www.djangoproject.com*) based collection of applications and libraries, which contain modules for data organization and database based storage (any *Python* supported database), and powerful web-based visualization. The evaluation modules use parts of the *LARA-R* packages. These need to be installed on the system to use all evaluation features. Since the source code of all modules is provided, the authors highly encourage the user to expand the given code according to their needs.

The *LARA* suite requires some additional software packages. To install these, please follow the instructions in the README file.

3   *LARA* Software Overview

The *LARA* suite is an integrated, open and flexible solution for planning and evaluation of for example high-throughput screening experiments. In contrast to proprietary reader software solutions and manual evaluation with spreadsheets it follows a different paradigm (Fig. 1): It accompanies all phases of the experiment, i.e., the planning phase, the performance of the experiments and the evaluation and visualization of the data generated (and if required, even data storage). All processes are implemented as modifiable and lightweight scripts to achieve highest flexibility and speed. The GUI based project planner was kept lean and easy to use. This is also a difference to large process evaluation packages like *KNIME* (*www.knime.org*) and *Orange* bioinformatics (*orange. biolab.si*). *LARA* is designed to use only a minimum of dependencies to special packages/libraries to avoid a sudden interruption of support if one of these software dependencies cannot be fulfilled.

Each step of the *LARA* support process is now described in the following chapters.

3.1   *Design and Planning-Phase*

LARA supports the scientist in the high-throughput experiment-planning phase with its intuitive graphical user interface. The *LARA* project planner follows the protein engineering work-flow paradigm, offering modules for example for cellular growth and protein expression, lysis, and analysis. Each of the modules can be graphically aligned in the process view (Fig. 2). New modules can be added as *Python/Qt4* plug-ins, using process abstraction classes and classes for GUI elements (a minimal sample code for a plug-in is provided in the repository). Every change in the process view will affect the process summary view. This process summary preview (Fig. 3) informs the user about the required chemicals, consumables and gives a rough estimate of the required durations of a certain step. Each module has an optional field where the user can set the starting time manually (currently this feature is used for planning only and will not affect a possible robotic process). This already helps to find bottlenecks of a process.

3.2   *Example Procedure*

A very common process in protein engineering is to grow cells, induce protein expression with, e.g., IPTG, harvest and lyse the cells and perform some assay with the crude cell extract. How can such a process be planned with the *LARA* software?

1. Open a terminal and change to the *LARA* directory. Then start *LARA* by typing. */lara.py.*

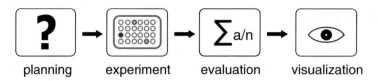

planning   experiment   evaluation   visualization

Fig. 1 *LARA* workflow paradigm

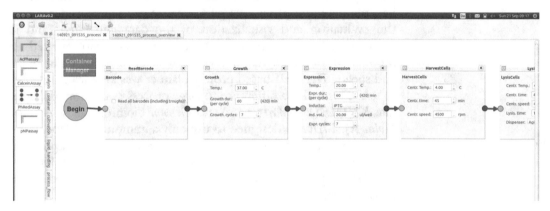

Fig. 2 *LARA* process planner

2. Press the "New Process" button and begin to drag your process modules onto the process editor surface—this will generate a new process view tab, a process overview tab and the corresponding *LARA* xml (.lml) file, containing the process steps.

3. Provide a process name (when a name is provided, the application can be closed at any time of design without the need of an explicit saving step—every change is saved when the application is closed and additionally after every 10 min).

4. Add more modules if required.

5. Adjust parameters like growth temperature and duration for each module, if required, in the project modules.

6. Finally, make sure that the last step is connected to the end process symbol (or double click on the end process symbol).

7. Inspect the results in the related process overview tab.

Each individual process step has a default value of containers (e.g., microtiter plates) that are associated with this step. This number can be changed by double-clicking the module icon. On a change of container numbers, the total amount of required containers will automatically be reevaluated. It should be kept in mind that some devices, e.g., centrifuges, require a defined number of containers to operate properly or have restrictions on the amount and type of containers that can be used.

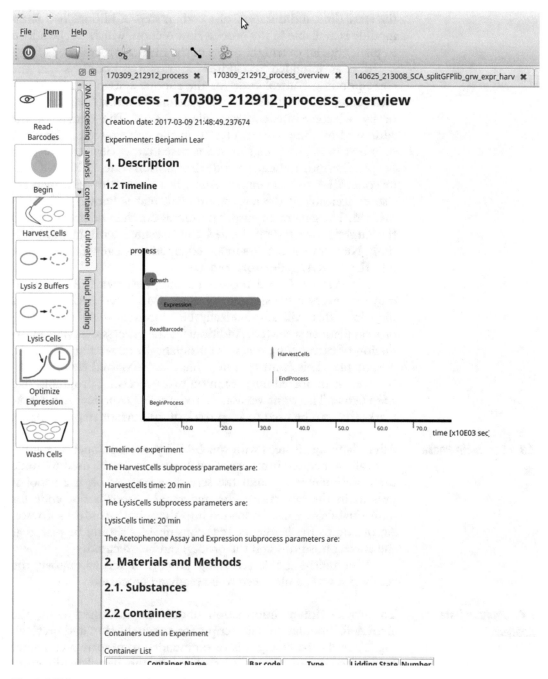

Fig. 3 *LARA* process summary preview

During the whole design process one can always switch to the corresponding overview tab of the process to get an idea about the tentative scheduling and amount of consumables required (see Fig. 3).

For manual processing, user-defined timing can be set in the *settings* and *timing* tab of each module, which contain an input for the start time and duration of a certain step. Additionally a *pause* module is available in the process flow section, which can be added to plan, e.g., an overnight pause of the manual process.

If it is intended to address a certain robotic system, the corresponding control code for the robotic system can be generated (compiled) by pressing the "generate" button (gears symbol) or by selecting the corresponding menu). The details of the addressed machine are manifested in the process steps. Currently only text based outputs (for script based target devices, like JavaScript, ThermoMomentum code, EpiMotion code, TECAN Gemini code, SiLA xml) are implemented, but binary output formats are also conceivable if the target output format is known and documented. The generated output programs can then be transferred to the target robot system, loaded and executed without modifications. New devices can be supported, by adding a generator module into the *LARA*/generators folder.

Leaving the *LARA* process planner/designer will automatically save every value and setting as an XML based *LARA* process file (.lml) that will automatically be reopened when the *LARA* process planner is started. Additionally, the process overview information of each open process is automatically generated as an html output file. This html overview file can be viewed by any webbrowser or imported into common text processing programs, like *LibreOffice*. The print version of the process overview has checkmarks that can be used to keep track of the current step processed.

3.3 Screening Phase

After planning all steps with the *LARA* process planner, the automatically generated html overview document can be used to guide the experimenter through the screening process or, if a robot is present in the laboratory, the generated process script code for individual devices can be loaded into the robot's control software. According to the list generated, consumables should be placed at the correct positions and the process can be executed.

After finishing this screening process in the laboratory, the received raw data files need to be read and processed.

3.4 Primary Data Analysis

To provide initial information about the measured data, the *LARA-R* modules contain scripts for fast evaluation and previewing. For this processing it is recommended to prepare a container layout description file as exemplified in the template directory (*LARA-R/laraDataReader/plate_layout_templates*). This layout description file has a simple, but versatile comma separated values (*csv*) structure that is used to describe each well and provides further details of the experiments, e.g., per well volume, added substrate, substrate concentration, and cofactor and/or enzyme concentration. The geometry of the container or plate is arbitrarily definable in the header line of the layout file. The information provided here are then automatically read during the evaluation process. In the layout file each sample is assigned a sample class, e.g., sample (S), reference (R), positive control (pCtrl), negative control (nCtrl), blank (BL), and a small description separated by a colon. The sample class is used to define which wells are used, e.g., for blank subtraction or comparison of the samples to an internal reference. The short description is used to tag and memorize each individual sample. This description can be plotted in each output graph to label the graph and generates the automatic graph file name. The definition of the container layout is therefore a key step for the evaluation, since the evaluation module uses this information about the samples to group and relate the measurements. This also allows averaging multiple repeats of an experiment just by using the same description. The syntax details are explained in the layout templates. Important: please currently avoid special characters, like umlauts and commas in your entries.

After defining the layout, the raw data can be read by the laraDataReader *R* modules. For manual evaluation it is recommended to install *R* and *R-studio* as a working environment. For larger data volumes an automatic data processing pipeline with a database connection and web presentation is also provided. The file reader modules (located in *laraDataReader/R/*) are written in R and support a set of plate readers (e.g., from BMG, TECAN and Thermo) and HPLC ASCII outputs. More file formats can be specified in the import data module and can be added with intermediate programming skills. The files "import modules" are all derived from the LA_ImportData base class and share a common parameter set. The class of the file name (as assigned by the structure method) defines which module is chosen to read the data (*see* demo session in Subheading 4). The file name can either be a single file name or a regular expression pattern to select multiple file names.

For long-term data storage, proprietary binary file formats and databases should be avoided. Simple text outputs in structured formats like *JSON, XML* or *csv* are preferable. It is highly recommended to store information about the used measurement conditions, like wavelengths, number of data points, temperature, number of repeated reads, time units, starting time, recording duration, and well information within these files. Commonly, the method editor of a measuring device allows for adjusting the output format and information saved. These generated files should have names according to the *LARA* file name convention (*see* Subheading 5).

The next step after reading the raw data is its processing. For this purpose the following modules currently exist:

- Combining multiple measurement files to one single kinetics for each well.
- Baseline subtraction.

- Detection of the longest linear region with calculation of slope and intercept.

- Calculation of the initial slope.

- Best n evaluation (depending on given criterion and number n).

- splitGFP evaluation (combination of linear activity measurements with single-point fluorescence measurements) [6].

- Michaelis–Menten kinetics.

For special evaluations of enzyme assays, please find individual modules in the /*LARA*_dataworkflow/*LARA*EvalVis/R folder.

Finally, the processed data can be visualized by laraEvalVis-modules (/*LARA-R*/laraEvalVis/) for common plots, like kinetic plots, chromatogram plots, histogram plots, well overview plots and spectral plots. Once the data is read, their visualization can be achieved through one common plot command (LA_plot) with a standard set of parameters, but depending on the class of the plotting data LA_plot triggers the right plotting routine. This makes the plotting routines very powerful since they rely only on one common data structure; this is illustrated in the examples section below.

3.5   Higher Order Data Analysis

Utilizing the *LARA*-database (part *of LARA-django*), it is also possible to extend the evaluation of only one single experiment. *LARA-django* provides mechanisms to combine data from several experiments, e.g., different measurements of the same sample, or even multiple projects—to extract for example the ten best mutants of multiple, independent screening rounds. Please refer to the *LARA-django* demo data to explore these features.

---

4   Demo Session for Primary Data Analysis

How can these libraries now be used practically? This small demo session should illustrate the simplicity and power of the *LARA* libraries. With four simple commands, it reads kinetic UV-vis absorption data of a transaminase acetophenone assay performed in a standard SBS 96-well microtiter plate, generated by a plate reader and plots the best linear fit for each individual well into one single high-resolution image. This and more demos can also be found in *laraEvalVis/demo*. After installing the R libraries as described in the *LARA* documentation, one can open a new, empty *R* file, e.g., in *R-studio*, and enter (the #-sign denotes comments that are not executed):

```
library("LARADataReader")
library("LARAEvalVis")
setwd("./")
# reading multiple single point absorption measurement files generated
# by a Thermo Varioskan microtiter plate reader with two wavelengths (600nm and
# 660nm)
# and corresponding plate layout information (layout=TRUE) and path length
# correction (PLC)
# the import method further specifies, how the data is interpreted during the
# reading phase.
vs_assay_df = LA_ImportData(structure("BC_4480_20150611.*varioskan_SPabs.*",
      class="varioskan"), method='KINabs', layout=TRUE, PLC=TRUE)
# find best linear models for each individual well
best_lin_mod_df = selectBestLinModel(kin_data_df=vs_assay_df, wavelength=340)
# plotting all the data as kinetics data
class(vs_assay_df) <- "kinetics"
LA_Plot( plot_data_df=vs_assay_df, slopes_df=best_lin_mod_df, ylim=c(0.3,1.1),
      xlim=c(0, 10.0), markBest=TRUE, bestNegSlope=TRUE, plotBestLinMod=TRUE,
      description=TRUE)
```

This short demo script consists of four lines of code to read, evaluate and plot the data. If no parameters are supplied, default values are assigned as shown in the documentation. For each module the documentation of its purpose and parameters can be viewed by writing help("[command-name]") into the *R* console, or very user friendly in *R-studio* by moving the cursor on the method name an pressing *F1*.

4.1 Examples of Evaluation and Data Visualisation

To visualize the growth curves of *E. coli* cells measured at approx. 2 h time intervals by individual absorption measurements at 600 nm (*see* Fig. 4).

```
# growth plot demo
growth_df = LA_ImportData(structure("BC_4693.*omega_SPabs.*", class="omega"),
method='SPabs', layout=TRUE, PLC=TRUE)
class(growth_df) <- "kineticsPlate"
LA_Plot( plot_data_df=growth_df, ylim=c(0.0,5.5), xlim=c(0, 180000.0),
xlab="time / s", description=T )
```

A kinetic plot can be obtained from the information in the above plots by the following four commands (Fig. 5):

```
# kinetic plot demo
tecan_assay_df = LA_ImportData(structure("0316_racemase_daao_vanillicacid.xlsx",
  class="tecanXLSX"), method='KINabs', barcode="0316", layout=TRUE,
PLC=TRUE)
# fast best model
best_lin_mod_df = selectBestLinModel(kin_data_df=tecan_assay_df)
```

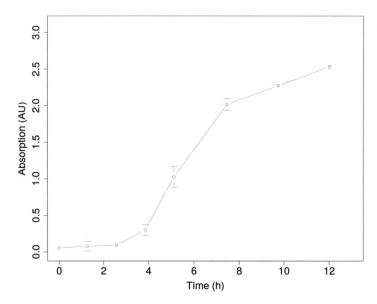

Fig. 4 Bacterial growth curve. *Blue circles* denote absorption during growth phase; *red circles* represent absorption after inducer addition

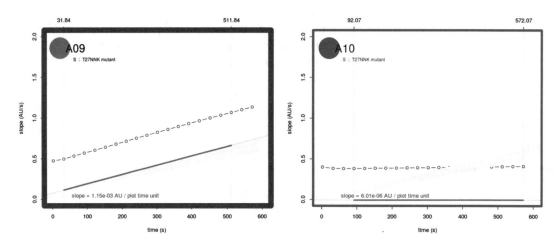

Fig. 5 Exemplary activity plots from two wells of an MTP showing an active (*left*) and inactive (*right*) variant in the screening. The well number, sample type, and description are given in the left upper corner. A *thick red frame* highlights the best variant. The *colored circle* graphically represents overall relative activity. Original raw data (*black circles*); best linear fit (*red solid line*); activity range of wild-type control reactions (*rose area*); average wild-type activity (*white line in rose area*); two sigma deviation (*grey area*); best linear range (*perpendicular dotted lines*)

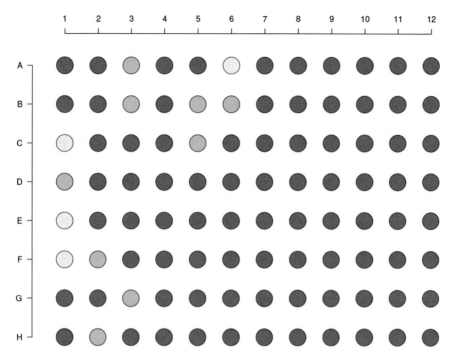

Fig. 6 Per-well activity plot of the same data

```
class(tecan_assay_df) <- "kinetics"
LA_Plot(plot_data_df=tecan_assay_df, slopes_df=lin_mod, ylim=c(0.0,6.5),
xlim=c(0, 3000.0), description=TRUE, filename="_kinPlot_lin_mod_1")
```

The same data can be plotted as color encoded plate overview of the activities in well plot format (Fig. 6) and as a "top 15" plot (Fig. 7) by the following commands:

```
class(tecan_assay_df) <- "wellplot"
LA_Plot(plot_data_df=tecan_assay_df, slopes_df=lin_mod, ylim=c(0.0,6.5),
xlim=c(0, 3000.0), description=TRUE, filename="_kinPlot_lin_mod_1")
# or as top 15 plot (s. Figure 7)

class(tecan_assay_df) <- "top15"
LA_Plot(plot_data_df=tecan_assay_df, slopes_df=lin_mod, ylim=c(0.0,6.5),
xlim=c(0, 3000.0), description=TRUE, filename="_kinPlot_lin_mod_1")
```

For more complex examples, like splitGFP evaluation [6] and Michaelis–Menten kinetic plots, please refer to the package demo files. It is also possible to read and plot chromatograms from HPLC or GC data. Examples of complete automated protein activity screenings are described in [5] and at *lara.uni-greifswald.de*.

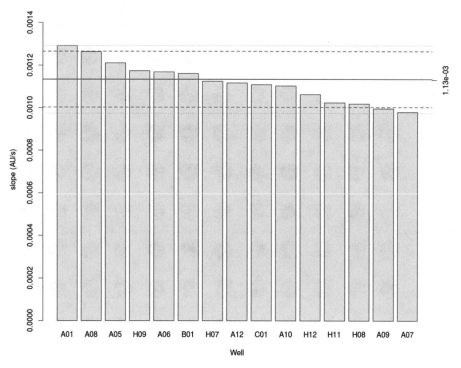

Fig. 7 Plot of 15 best candidates. The *red solid line* represents the average wild-type activity, *dashed lines* the minimum and the maximum of the wild-type activity, the *grey lines* two standard deviations of the wild-type references

4.2 User Adjustments to the Scripts

There are two common cases where adjustments to the scripts are required. The first case is the addition or fine tuning of an assay evaluation, the second the addition of a new data input format from an unsupported device.

A new assay evaluation can be defined within modules in the laraEvalVis/R/folder. To add a new assay evaluation, see the code in, e.g., *enzymeAssays.R*.

In case a new, unknown output format needs to be read, such as a new microtiter plate reader or a high-throughput GC/MS device etc., a new data input method needs to be written. The folder *laraDataReader/R/* contains a collection of input methods for common device classes. All import methods are derived from the base class LA_ImportData, which is defined in the module import-Data.R. To write a new module it is in most cases sufficient to select and copy the most similar one, adjust its name and change the method according to the requirements of the format that should be imported by the module.

Important: Whenever a change in one of the modules was made, the packages need to be recompiled and installed by the following commands to make them available for the user processing script:

```
# Change to the root directory of the scripts
# (the directory that contains the folders LARADataReader and LARAEvalVis),
# start R and type (remember that the devtools R package is installed):
library("devtools") # this will provide to tools for package compilation
document("LARADataReader") # this will compile and document the module
install("LARADataReader") # this will install the new module
```

For further adjustments, refer to the *LARA* web page or look inside the well-documented script code.

5   *LARA* File Naming Scheme Convention

On many lab device computers in the world it is not uncommon to find files named "test measurement1.dat", "prot1 kinetics.txt" etc. Within weeks it is impossible to assign such names to a certain experiment. Therefore it is highly recommended to use a standardized output file name scheme, including a unique identifier, like a container barcode to connect the measured data with the measured container, the date and time of the measurement, the device, the method, and some user defined tags to make the file human identifiable. *LARA*, especially the automatic evaluation part, relies on standardized file names.

LARA file name scheme:

```
UID_YYMMDD_HHmmss_DEVICE_METHOD_USERSTRING.file_type
```

Example:

```
1234_160528_122457_varioskan_SPabs_kemp_activity.csv
```

6   Summary

The *LARA* software suite supports high-throughput screening efforts beyond common spreadsheet evaluation. It guides the researcher in planning, evaluation, and visualization. The evaluations can be performed on many levels from the initial data processing (pathlength correction, baseline subtraction, smoothing) to project wide interpretation of results. Every module is designed to be very lean, but powerful. Its application ranges from manual screening to robot-assisted processes and data evaluation of medium-sized robotic platforms. The code is designed to be quickly adjustable to the individual needs of a scientist, given that basic programming skills are present.

References

1. Romero PA, Arnold FH (2009) Exploring protein fitness landscapes by directed evolution. Nat Rev Mol Cell Biol 10:866–876
2. Hertzberg RP, Pope AJ (2000) High-throughput screening: new technology for the 21st century. Curr Opin Chem Biol 4:445–451
3. Goddard JP, Reymond JL (2004) Enzyme assays for high-throughput screening. Curr Opin Biotechnol 15:314–322
4. Bisswanger H (2014) Enzyme assays. Perspect Sci 1:41–55
5. Dörr M, Fibinger MPC, Last D et al (2016) Fully automatized high-throughput enzyme library screening using a robotic platform. Biotechnol Bioeng 113:1421–1432
6. Santos-Aberturas J, Dörr M, Waldo GS et al (2015) In-depth high-throughput screening of protein engineering libraries by split-GFP direct crude cell extract data normalization. Chem Biol 22:1406–1414

Chapter 17

Solid-Phase Agar Plate Assay for Screening Amine Transaminases

Martin S. Weiß, Uwe T. Bornscheuer, and Matthias Höhne

Abstract

Agar plate assays represent a useful method for high-throughput prescreening of larger enzyme libraries derived from for example error-prone PCR or multiple site-saturation mutagenesis to decrease screening effort by separating promising variants from less active, inactive, or neutral variants. In order to do so, colonies are directly applied for enzyme expression and screening on adsorbent and microporous membranes instead of elaborately preparing cell lysates in 96-well plates. This way, 400–800 enzyme variants can be prescreened on a single membrane, 10,000–20,000 variants per week and per single researcher respectively (25 membranes per week).

The following chapter gives a detailed protocol of how to screen transaminase libraries in solid phase, but it also intends to provide inspiration to establish a direct or coupled agar plate assay for screening variable enzymatic activities by interchanging assay enzymes and adapting assay conditions to individual needs.

Key words Agar plate assay, Solid-phase assay, High-throughput screening assay, Directed evolution, Transaminase, Horseradish peroxidase, Glycine oxidase

1 Introduction

Random mutagenesis and multiple site-saturation mutagenesis still represent key methods for evolving biocatalysts to fit the desired process conditions such as high solvent, substrate and product titers or elevated temperatures [1]. In most of the cases random mutagenesis methods are applied to improve an already existing enzyme property by mimicking nature's evolution process and simply screening a large number of random variants [2]. To reduce the screening effort by sorting out less active or inactive variants and to focus only on the promising ones, various assay systems with different throughput levels are available. Among those, 96-well microtiter plate assays are commonly used, and allow for growth of the expression host, and expression and screening of the variants. In this way the screening throughput is more or less restricted to

Uwe T. Bornscheuer and Matthias Höhne (eds.), *Protein Engineering: Methods and Protocols*, Methods in Molecular Biology, vol. 1685, DOI 10.1007/978-1-4939-7366-8_17, © Springer Science+Business Media LLC 2018

screening not significantly more than 1000 variants per week per researcher (without any robotic support) [2].

Solid-phase assay screening intends to increase the throughput by circumventing the laborious inoculation of single variants from agar plates in microtiter plates. Instead, colonies carrying the plasmids with the mutated genes are directly transferred from the master agar plate to microporous membranes. Placing the membranes—colonies facing up—on different agar plates (containing appropriate inducer and assay substances) allows expression and screening of the enzyme variants by diffusion of the assay compounds from the agar into the membrane. In general, agar plate assays comprise the same principle steps as microtiter plate assays do: (1) Transfer of the colonies to a microporous membrane, (2) enzyme expression on an induction plate, (3) cell permeabilization, e.g., by chloroform treatment, and (4) finally assaying the desired reaction by placing the membrane on an assay plate containing all the reagents necessary for the detection of the desired activity. Thus, up to ~800 colonies can be prescreened on a single membrane and as handling of the membranes does not involve any time-consuming and elaborate processing steps, such as centrifugation of microtiter plates after cell disruption, it is easily feasible to prescreen 25 membranes representing 10,000–20,000 different variants per week per researcher. However, not only the throughput is relevant for efficient catalyst evolution, but also a well-prepared library, which should provide a suitable mutation frequency without comprising too many wild-type sequences. Using an agar plate assay allows preparation and investigation of several different libraries in parallel. Observation of the ratio of active to inactive variants on a membrane can help to adjust the mutation frequency to the individual's needs or can help to select one of the investigated libraries for further screening.

There are various interchangeable assay reactions available that allow translation of the desired enzyme activity into a detectable signal, most frequently by formation of a dye to visualize colonies containing the desired enzyme variants on agar plates or membranes (*see* Table 1) [14].

For coupled assay systems, coexpression of one or more assay enzymes together with the variant of interest allows efficient screening in solid-phase without any need to apply purified enzymes and taking advantage of very close proximity of all involved assay enzymes [12, 13]. Application of externally applied purified assay enzymes may lead to diffuse color formation due to diffusion of the intermediate products before its local concentrations are high enough to allow fast subsequent conversion by the assay enzymes in situ, if the activity of the assay enzymes is insufficient [12, 13]. For coexpression of multiple enzymes on different plasmids, subcloning in compatible plasmids exhibiting independently regulated plasmid replication origins is required [15].

Table 1
Selected exemplary literature using the most commonly used assay principles in agar plate screening

Enzyme activity	Principle	References
Hydrolases, decarboxylases, kinases, and glycosyltransferases	Detection of pH change	[3]
Transaminases	Direct formation of a colored product	[4–6]
Glycosynthase	Coupled assay using chromogenic substrates	[7]
Oxidases	Direct detection of hydrogen peroxide	[8–11]
Racemases and transaminases	Coupled assays detecting hydrogen peroxide	[12, 13]

Table 2
Suitable horseradish peroxidase substrates for agar plate screenings at different pH working ranges

Compound name	pH-working range	Dye solubility
3-Amino-9-ethylcarbazole (AEC)	<6.0	Insoluble
4-Chloro-1-naphthol (4-CN)	7.0–8.0	Insoluble
3,3′-Diaminobenzidine (DAB)	7.0–7.6	Insoluble
4-Aminoantipyrine (4-AAP) and vanillic acid or phenol	>8.0	Soluble

Horseradish peroxidase-coupled assays allow screening of a variety of different enzymatic activities by detection of hydrogen peroxide and subsequent formation of a dye [8–13]. There are several chromogenic substrates for horseradish peroxidases that lead to soluble or insoluble products. However, the pH working range is different for each substrate due to changes in the absorbance spectra, altered stability or specific reactivities leading to the proper formation of the dye at different pH values (*see* Table 2).

The following chapter intends to provide general inspiration to establish a direct or coupled agar plate assay for screening variable enzymatic activities. Thus, we exemplarily describe a glycine oxidase and horseradish peroxidase double-coupled agar plate assay for detection of transaminase activities toward amines by following the formation of the by-product glycine, which is oxidized in situ, generating hydrogen peroxide and finally leading to the formation of a red quinone imine dye (*see* Fig. 1) [12].

The solid-phase assay screening procedure consists of several consecutive steps depicted in Fig. 2. *E. coli* BL21 (DE3) carrying the plasmid coding for glycine oxidase (Fig. 2a) is transformed with the plasmid mixture obtained from the mutagenesis experiment.

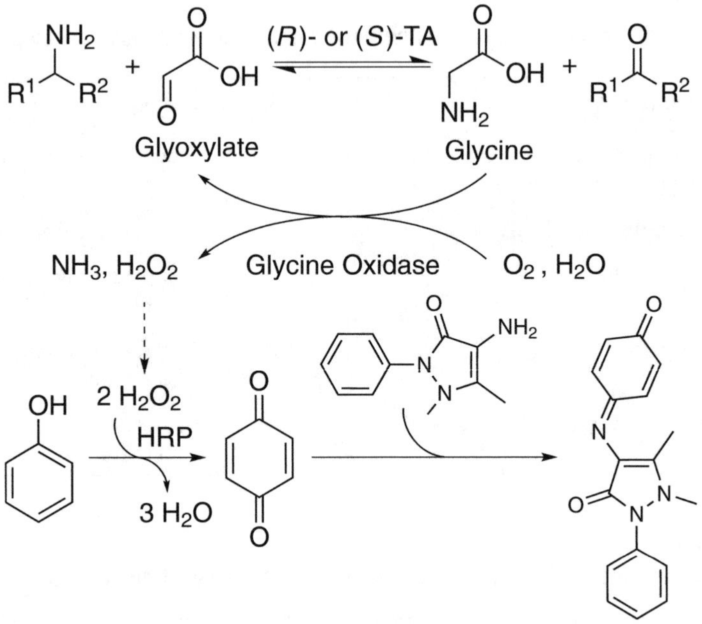

Fig. 1 Application of glyoxylate as amino acceptor substrate allows screening for activity toward different amines by either (R)- or (S)-selective transaminases. Production of achiral glycine is then followed by oxidation and hydrogen peroxide formation leading to the formation of a red quinone imine dye [12]. Toxic phenol can be substituted by vanillic acid [12]. Figure reproduced from [12], http:/pubs. acs.org/doi/pdf/10.1021/ac503445y

Transformed cells are spread out on agar plates and are incubated overnight for colony growth on the plates (Fig. 2b). Transfer of the colonies on a microporous membrane (Fig. 2c) allows induction of the colonies on agar plates containing suitable inducers (Fig. 2d), while the original plate is kept to serve as a master plate. After expression, the membrane is treated with chloroform for cell permeabilization (Fig. 2e). To reduce possible background reactions the membrane is placed on dialysis plates overnight allowing low molecular weight compounds to diffuse into the dialysis agar (Fig. 2e). Finally, incubation of the membranes on assay plates containing all the substrates and reagents for the assay reaction allows screening of the library of interest by selecting most colored colonies for further investigation (Fig. 2f). In Table 3, we suggest a possible timetable for the assay procedure for screening 25 agar plates per week.

2   Materials

1. *E. coli* BL21 (DE3) expression strain harboring the plasmid pCDF-1b encoding the glycine oxidase gene (*see* **Note 1**).

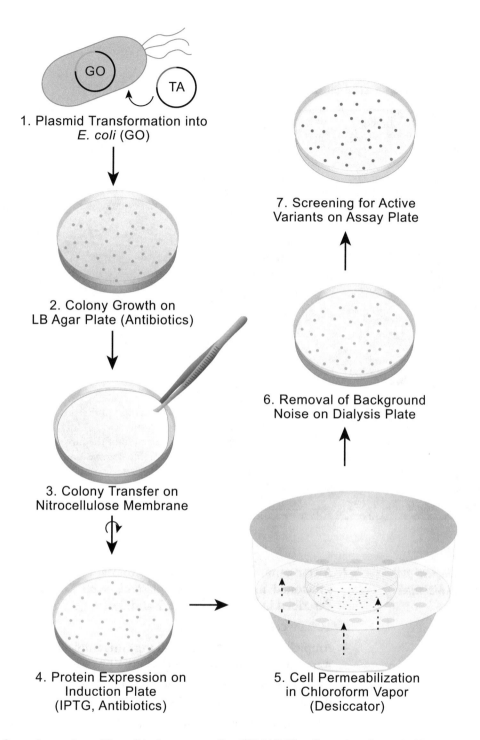

1. Plasmid Transformation into
E. coli (GO)

2. Colony Growth on
LB Agar Plate (Antibiotics)

3. Colony Transfer on
Nitrocellulose Membrane

4. Protein Expression on
Induction Plate
(IPTG, Antibiotics)

5. Cell Permeabilization
in Chloroform Vapor
(Desiccator)

6. Removal of Background
Noise on Dialysis Plate

7. Screening for Active
Variants on Assay Plate

Fig. 2 General procedure of the solid-phase assay. *E. coli* BL21(DE3) cells are transformed with both plasmids coding for glycine oxidase and the transaminase library of interest (**a**) and colonies are grown on dual selection LB agar plates (**b**). Afterward, colonies are transferred to nitrocellulose membranes (**c**) that are placed—colonies facing up—on induction plates containing IPTG for expression of the proteins (**d**). Then, cells are permeabilized by chloroform treatment (**e**). To eliminate false-positive background color formation, permeabilized cell colonies are dialyzed overnight by placing the membranes on dialysis plates (**f**). Finally, screening is conducted by incubation of the membranes on assay plates (**g**). Figure reproduced from [12], http:/pubs. acs.org/doi/pdf/10.1021/ac503445y

Table 3
Suggested schedule for solid-phase assaying

Monday	Tuesday	Wednesday	Thursday	Friday
Preparation of agar plates, induction plates, dialysis plates, buffers and stock solutions	Transformation of the libraries and plating out 75 plates for incubation overnight	If no suitable colony density on the plates was achieved, repeat the program of yesterday	Preparation of fresh assay plates	Reserved for repetitions and organization of the next week's experiments
Preparation of 1% agarose in screening buffer and storage at 60 °C		Labeling of 25 membranes, transfer of the colonies to the membranes, induction and regrowth of the colonies on the master plates	Screening of the membranes: 45 min time-delayed for every five or ten membranes	
		After at least 6 h of expression on induction plates chloroform treatment and overnight dialysis	Following color formation of the colonies and selection of the most colored colonies on the master plates	
			Arrangement of a hit plate	

2.1 Biological and Chemical Materials

2. A transaminase of interest and a transaminase mutant library subcloned in a compatible plasmid, such as pET22b(+) as described in [16] (*see* **Notes 1** and **2**).

3. 10% glycerol (autoclaved).

4. Ampicillin stock solution: 100 mg/mL ampicillin in deionized water. Sterilize by filtration (0.2 μm pore size filter).

5. Spectinomycin stock solution: 50 mg/mL spectinomycin in deionized water. Sterilize by filtration.

6. IPTG stock solution: 238 mg/mL IPTG in deionized water. Sterilize by filtration.

7. Tris–HCl buffer, 30 mM, pH 8.5.

8. CHES buffer, 50 mM, pH 9.5. Adjust the pH with NaOH.

9. 1 M Glyoxylate stock solution: 92 g/L glyoxylic acid in deionized water. Store at 4 °C.

10. 4-AAP stock solution: 0.3 g/mL 4-aminoantipyrine in pure ethanol. Store at room temperature.

11. Vanillic acid stock solution: 0.13 g/mL vanillic acid in methanol. Store at room temperature.

12. Amine donor stock solution: 1 M (S)-1-phenylethylamine in pure ethanol. Pipette 263 μL of (S)-$(-)$-1-phenylethylamine, add 1737 μL of pure ethanol, mix and store at room temperature.

13. HRP solution: Carefully dissolve 3000 U horseradish peroxidase lyophilisate in 2.5 mL ultrapure water. Keep on ice and do not store the solution longer than necessary. Always prepare it fresh directly before use.

14. LB medium: 0.5% (w/v) yeast extract, 1.0% (w/v) NaCl, 1.0% (w/v) tryptone. Sterilize by autoclaving.

15. SOC medium: 20 g/L tryptone, 5 g/L yeast extract, 10 mM NaCl, 2.5 mM KCl, 20 mM glucose. Add glucose as sterile-filtered 1 M stock solution to the autoclaved solution containing the remaining components.

16. Selection plates: 0.5% (w/v) yeast extract, 1.0% (w/v) NaCl, 1.0% (w/v) tryptone, 1.5% (w/v) agar-agar, 100 mg/L ampicillin, 35 mg/L spectinomycin. Add 12 g of agar-agar to 800 mL of LB medium and autoclave the mixture. Cool down to ≈45 °C while stirring. For the simultaneous selection of cells that harbor both plasmids (encoding the glycine oxidase and the transaminase variant) add 800 μL of ampicillin stock solution and 800 μL of spectinomycin stock solution (*see* **Note 1**). After mixing, aliquot the solution in sterile Petri dishes (30–35 plates at 20–25 mL) and store them at 4 °C after solidification.

17. Induction plates: 0.5% (w/v) yeast extract, 1.0% (w/v) NaCl, 1.0% (w/v) tryptone, 1.5% (w/v) agar-agar, 100 mg/L ampicillin, 35 mg/L spectinomycin, 1 mM IPTG. Prepare in the same way as the selection plates, but additionally add 800 μL of IPTG stock solution before pouring the plates.

18. Dialysis plates: 30 mM Tris–HCl, pH 8.5, 0.4% (w/v) agarose, 0.1 mM PLP, 10 μM FAD. Suspend 3.2 g of agarose in 800 mL of Tris–HCl buffer and boil the suspension by microwave heating or on a hot plate until the agarose is completely dissolved. Cool the solution down to 38 °C while stirring. Add 19.77 mg PLP and 6.28 mg FAD. Aliquot the homogeneous solution into petri dishes (30–35 plates at 20–25 mL) and (*see* **Note 2**).

19. Assay plates: 50 mM CHES pH 9.5, 1% (w/v) agarose, 3 g/L 4-aminoantipyrine, 0.65 g/L vanillic acid, 0.92 g/L glyoxylic acid, 10 mM amino donor. Suspend 2 g of agarose in 200 mL

CHES buffer and boil the suspension until the agarose is completely dissolved. Cool down the solution to 40 °C while stirring. Add 160 μL of 4-AAP stock solution, 1 mL of vanillic acid stock solution, 2 mL of glyoxylic acid stock solution and 2 mL of amine donor stock solution (*see* **Note 3**). Horseradish peroxidase application on the assay plates takes place right before the membranes are applied as described in the methods section.

20. Control plates: Same ingredients as assay plates but glyoxylic acid is omitted.

21. Nitrocellulose membranes: Binding capacity for proteins >200 μg/cm^2, pore size 0.2 μm, membrane strength 0.15 mm ± 0.05 mm.

2.2 Equipment

1. Electroporator.

2. Desiccator: Fill up a desiccator with chloroform to 1 cm height, close the lid and store it for at least 5 h at room temperature.

3. Sieve: A usual steel strainer used in the kitchen, remove the handle in order to fit it in the desiccator.

3 Methods

1. Thaw a sample of electrocompetent cells containing plasmids necessary for the assay on ice for 5–10 min. Add 0.5 μL of plasmid DNA from your library of interest having a plasmid DNA concentration of around 20 ng/μL (*see* **Note 4**). Add the aliquot containing the DNA in a precooled 0.2 mm electroporation cuvette and pulse with 1.8 kV (*see* **Note 5**).

2. Immediately afterward, add 1 mL of LB-SOC medium and carefully transfer the suspension in a sterile 1.5 mL tube. Cure the cells for 1 h at 37 °C and 180 rpm. To achieve suitable colony numbers on the plates, plate out 10, 20, 30, 40, and 50 μL of the suspension on selection plates (room temperature) containing ampicillin and spectinomycin. Incubate the plates at 30 °C overnight.

3. In the same way, transform and plate cells with the plasmid harboring the gene of the wild-type transaminase (or the starting transaminase mutant used for the mutagenesis). This plate will be used as a control to investigate background color formation.

4. The next morning, incubate the plates at 37 °C, if the colonies are not big enough, until the desired size is obtained (*see* **Note 6**).

5. Use a waterproof pen to label the nitrocellulose membranes with a membrane number and the current date. Additionally,

put three different check marks on the margin of each membrane (ideally in a triangle), to be able to document the orientation of the membrane on the agar plate.

6. Place one labeled nitrocellulose membrane, labeling facing down, on an agar plate covered with colonies of reasonable size (*see* **Note 7**). Make sure the membrane has contact to the assay agar over the whole plate. Use a waterproof pen to label the agar plate with the same check marks in the same orientation, plate number and date. One nitrocellulose membrane with colonies is required for a negative control.

7. Take the membrane off the agar plate using tweezers and place it, colonies facing up, on an induction plate. Make sure that the membrane is in contact with the induction medium over the whole plate. For coexpression of the variants of interest and the assay enzymes, incubate the membrane for 6–7 h at 30 °C (*see* **Note 8**).

8. The selection plates, where the colonies were transferred from, are incubated for 6 h at 37 °C for regrowth of the colonies. By doing so, an exact copy of the colonies on the membranes is generated that can serve as master plates for further investigation of the interesting variants. Seal the master plates and store them at 4 °C.

9. After expression, the membranes are placed on a mesh in the desiccator (saturated with chloroform vapor) for 60 s at room temperature (*see* **Note 9**) to permeabilize the cells (Figs. 3 and 4).

10. After cell permeabilization place the membranes on precooled dialysis plates and store them overnight at 4 °C (*see* **Note 10**).

11. Spread 100 μL of horseradish peroxidase solution on an assay plate (*see* **Note 11**). Immediately afterward place a membrane covered with chloroform-treated and dialyzed colonies, colonies facing up, on the assay plate with horseradish peroxidase. If necessary strip remaining liquid of the membrane at the lid of the petri dish. Make sure the whole membrane is in contact with the assay agarose. Incubate the assay plates at 37 °C and monitor color formation for up to 6 h (*see* **Note 12**).

12. Select the most colored colonies on each membrane and locate them on the corresponding master plate for regrowth on a hit-agar plate or for rescreening in microtiter plates, depending on the number of interesting variants for further investigation (*see* **Note 13**).

Fig. 3 Arrangement of the membranes on a mesh prior to cell permeabilization by chloroform vapor

Fig. 4 Chloroform vapor treatment of membranes in a desiccator filled with chloroform to 1 cm of height

4   Notes

1. The transaminase gene can be subcloned in pET22b as described in [16], but in principle, other plasmids can be used too. However, it is important that the plasmid encoding the glycine oxidase is compatible with the one containing the transaminase. The two plasmids should contain different antibiotic selection markers. In our hands, a codon-optimized glycine oxidase gene [12] subcloned in the compatible vector pCDF-1b worked well. In this case, selection for two plasmids is carried out by ampicillin (transaminase of interest) and spectinomycin (glycine oxidase, assay enzyme). Substitute these

antibiotics according to your needs. For three or more different plasmids, reduce the amount of each antibiotic.

To create the mutant library expression strain harboring both plasmids, transform competent *E. coli* cells with pCDF-1b (glycine oxidase). The obtained transformants are made competent again to achieve high transformation efficiencies for multiple plasmid transformations. Transform again. Repeat the protocol for the preparation of electrocompetent cells. Do not forget to add the corresponding antibiotics to the culture media.

2. Do not dry the dialysis plates after aliquoting. They are supposed to be wet enough to facilitate diffusion during the dialysis step.

3. Do not stir the assay agar too strongly, to avoid air intake and air bubble formation. The amount of substrate for the enzyme of interest (here transaminases) depends on the activity of your template and the cost of the amine substrates. We recommend concentrations of 1–10 mM. For higher amine substrate concentrations it is important to check, whether the buffer capacity is sufficient or whether the pH has to be adjusted. For other assay systems substitute all reagents to assay your reaction of interest. Be advised: Depending on both the substrates and your protein of interest the choice of buffer compound and the adjusted pH value is crucial. For transaminases even pH 9.0 instead of pH 9.5 will dramatically affect the signal strength. Make sure that all assay enzymes (here glycine oxidase and horseradish peroxidase) at least tolerate or are compatible the conditions required by your enzyme of interest. Make sure to use chromogenic dyes that are suitable for the desired pH value of the assay agar. Besides the desired assay reaction, a background color formation might occur, which often also has a certain pH optimum. Do not store the assay plates for more than a few hours to avoid auto-oxidation or other side reactions.

4. Please consult further literature for preparation of random mutagenesis libraries or saturation libraries [17, 18], as it would go beyond the scope of this chapter to discuss this in detail.

5. In our hands, transformation by electroporation provided the highest and quite reproducible transformation efficiencies. Both properties are crucial for efficient screening of large libraries. However, despite all efforts, transformation efficiency varies. Less than 800 colonies should be plated out per membrane, in order to get separated colonies on the membrane and to be able to identify hit colonies on the master plates. Too few colonies per membrane are not worth the screening effort. The

amount of colonies obtained is very dependent on the amount of DNA added. Try to always add the same amount of DNA (e.g., 20 ng) and determine the individual transformation efficiency for each batch of competent cells. With some experience it is easily possible to get the ideal numbers of colonies.

6. Do not grow the plates too long, as too large colonies cannot be discriminated form each other. This will decrease the screening efficiency. Too small colonies are difficult to transfer to the membrane and later on are more difficult to locate on the master plate, when interesting colonies are supposed to be selected for further investigation. Consider that during expression the colonies still continue to grow. For these reasons colonies should not be larger than the head of a pin (<1 mm diameter) when they are transferred to the nitrocellulose membrane.

7. Whenever possible try to implement a positive control on each membrane. In best cases the template already shows initial activity against the target substrate. This way properly working assay reagents and procedure can be monitored for each membrane. It is also possible to apply a positive control from a different plate. To do so cut out a piece of agar from a plate covered with colonies carrying wild-type or reference plasmids and transfer the colonies to the membrane before placing the membrane on the plate with the target library. However, this positive control will not exactly behave as the colonies obtained from placing the membrane on the agar containing the colonies of the library of interest as cell material will remain on the master plate.

8. Expression conditions such as temperature, duration, kind of inducer and inducer concentration need to be adapted to the needs of both your assay enzymes and your protein of interest, as well as to your expression system. We recommend elaborating the expression conditions for the assay enzymes in a first step (in this case expression of cells containing only glycine oxidase and incubation on glycine containing assay agar). In parallel, expression conditions for the protein of interest can be investigated in shake flasks to get an idea where to start. In every case a good compromise needs to be made for all enzymes involved in the assay reaction. Investigation of the whole enzymatic setup is possible by application of characterized wild-type proteins (here transamiases) and application of different substrates being accepted with different activities. Thus, determine the limit of detection for your individual assay conditions.

9. In our hands, cell permeabilization using chloroform was most straightforward. Close and incubate the desiccator for a few

minutes to ensure a chloroform-saturated atmosphere. Freeze–thaw cycles or liquid nitrogen might be possible as well, but we observed that the nitrocellulose membranes broke into pieces and we did not put further effort in optimizing these approaches. Membranes from different suppliers might behave differently.

10. In every case a negative control experiment employing assay plates without the substrate of interest (in this case 1-phenylethylamine) is required to investigate if any compounds in the cell lysate lead to a false-positive signal in the assay. In our exemplary case a background signal could arise from intracellular glycine, or other amino acids that are converted to glycine by the transaminase. Incubating the membranes in a dialysis step helped to remove most of the background signal.

11. Make sure to apply the membrane immediately after application of horseradish peroxidase solution in order to avoid diffusion into the assay agarose and to allow the membrane to soak in the solution. Make sure that there is not too much liquid between the membrane and the assay agar. The amount of humidity should be enough for the substrates to diffuse into the nitrocellulose membrane when placed on the agar but low enough to avoid any easy diffusion of reaction products.

12. Too long incubation of the plates leads to color formation that is potentially not correlated to the desired enzyme activity. Also, diffusion of the dye plays a bigger role for longer incubations. Investigate the maximum duration of incubation by application of positive and negative controls. We recommend application of insoluble dyes for long-term incubations.

13. According to our experience, variants with more than fivefold increased activity compared to the reference can be discriminated in solid phase on a hit plate. For more accurate investigation, most colored colonies should be expressed in deep-well blocks followed by the screening of the crude lysates. For multiple plasmids in the cells, add all antibiotics and express all cascade enzymes. As a reference, apply *E. coli* BL21 (DE3) cells transformed as well with all the plasmids necessary for the assay cascade and the protein of interest to guarantee comparability.

References

1. Bornscheuer UT, Huisman GW, Kazlauskas RJ et al (2012) Engineering the third wave of biocatalysis. Nature 485:185–194

2. Leemhuis H, Kelly RM, Dijkhuizen L (2009) Directed evolution of enzymes: library screening strategies. IUBMB Life 61:222–228

3. Bornscheuer UT, Altenbuchner J, Meyer HH (1998) Directed evolution of an esterase for the stereoselective resolution of a key intermediate in the synthesis of epothilones. Biotechnol Bioeng 58:554–559

4. Green AP, Turner NJ, O'Reilly E (2014) Chiral amine synthesis using ω-transaminases: an amine donor that displaces equilibria and enables high-throughput screening. Angew Chem Int Ed 53:10714–10717

5. Martin AR, DiSanto R, Plotnikov I et al (2007) Improved activity and thermostability of (S)-aminotransferase by error-prone polymerase chain reaction for the production of a chiral amine. Biochem Eng J 37:246–255

6. Hailes H, Baud D, Ladkau N et al (2015) A rapid, sensitive colorimetric assay for the high-throughput screening of transaminases in liquid or solid-phase. Chem Commun 51:17225–17228

7. Mayer C, Jakeman DL, Mah M et al (2001) Directed evolution of new glycosynthases from *Agrobacterium* β-glucosidase: a general screen to detect enzymes for oligosaccharide synthesis. Chem Biol 8:437–443

8. Delagrave S, Murphy DJ, Pruss JL et al (2001) Application of a very high-throughput digital imaging screen to evolve the enzyme galactose oxidase. Protein Eng 14:261–267

9. Escalettes F, Turner NJ (2008) Directed evolution of galactose oxidase: generation of enantioselective secondary alcohol oxidases. ChemBioChem 9:857–860

10. Alexeeva M, Enright A, Dawson MJ (2002) Deracemization of α-methylbenzylamine using an enzyme obtained by in vitro evolution. Angew Chem Int Ed 41:3177–3180

11. Rowles I, Malone KJ, Etchells LL et al (2012) Directed evolution of the enzyme monoamineoxidase (MAO-N): highly efficient chemoenzymatic deracemisation of the alkaloid (±)-crispine A. ChemCatChem 4:1259–1261

12. Weiß MS, Pavlidis IV, Vickers C et al (2014) Glycine oxidase based high-throughput solid-phase assay for substrate profiling and directed evolution of (R)- and (S)-selective amine transaminases. Anal Chem 86:11847–11853

13. Willies SC, White JL, Turner NJ (2012) Development of a high-throughput screening method for racemase activity and its application to the identification of alanine racemase variants with activity towards L-arginine. Tetrahedron 68:7564–7567

14. Reymond JL (2006) Enzyme assays. Wiley-VCH, Weinheim

15. Tolia NH, Joshua-Tor L (2006) Strategies for protein coexpression in *Escherichia coli*. Nat Methods 3:55–64

16. Steffen-Munsberg F, Vickers C, Thontowi A et al (2013) Connecting unexplored protein crystal structures to enzymatic function. ChemCatChem 5:150–153

17. Tee KL, Wong TS (2013) Polishing the craft of genetic diversity creation in directed evolution. Biotechnol Adv 31:1707–1721

18. Gillam EMJ, Copp JN, Ackerley DF (eds) (2014) Directed evolution library creation: methods and protocols. Springer, New York

Chapter 18

Ultrahigh-Throughput Screening of Single-Cell Lysates for Directed Evolution and Functional Metagenomics

Fabrice Gielen, Pierre-Yves Colin, Philip Mair, and Florian Hollfelder

Abstract

The success of ultrahigh-throughput screening experiments in directed evolution or functional metagenomics strongly depends on the availability of efficient technologies for the quantitative testing of a large number of variants. With advanced robotics, libraries of up to 10^5 clones can be screened per day as colonies on agar plates or cell lysates in microwell plates, albeit at high cost of capital, manpower and consumables. These cost considerations and the general need for high-throughput make miniaturization of assay volumes attractive. To provide a general solution to maintain genotype–phenotype linkage, biochemical assays have been compartmentalized into water-in-oil droplets. This chapter presents a microfluidic workflow that translates a frequently used screening procedure consisting of cytoplasmic/periplasmic protein expression and cell lysis to the single cell level in water-in-oil droplet compartments. These droplets are sorted based on reaction progress by fluorescence measurements at the picoliter scale.

Key words Directed evolution, Ultrahigh-throughput screening, Single-cell, Microfluidic droplets, Assay miniaturization, In vitro compartmentalization, Hydrolase, Microfluidics, Functional metagenomics

1 Introduction

The selection of enzyme variants, e.g., in directed evolution experiments or from metagenomic libraries, is crucially dependent on screening large numbers of library members in high-quality quantitative enzyme assays [1]. Screening at the single cell level by fluorescence-activated cell sorting (FACS) allows analysis of up to 10^8 mutants per day, but is only possible for reactions in which a link between the genotype, i.e., the gene coding for the enzyme of interest, and the phenotype, e.g., the reaction product, is maintained [2–4]. Biomimetic compartments [5] provide a general solution for genotype–phenotype linkage and enable the recovery of the identity of enzyme variants. Additionally, these compartments drive down experimental cost a million fold over classical formats with reaction volumes at the picoliter to femtoliter scale

Uwe T. Bornscheuer and Matthias Höhne (eds.), *Protein Engineering: Methods and Protocols*, Methods in Molecular Biology, vol. 1685, DOI 10.1007/978-1-4939-7366-8_18, © Springer Science+Business Media LLC 2018

[6]. Microfluidic devices allow production of large numbers of such compartments in monodisperse form [7, 8].

Several enzymatic assays have been successfully miniaturized in droplet compartments [1, 9]. Here, we focus on the single cell lysate assay [10], a well-established format for the high-throughput screening of enzymes. This assay can be carried out in different compartmentalized formats such as agarose beads [11] or double emulsions [12], but this chapter outlines procedures employed for work in water-in-oil droplets [1]. Instead of expressing the protein of interest in *E. coli*, in vitro expression can also be used [13–16].

The miniaturized single cell lysate assay requires the use of three microfluidic operation units [17]. Picoliter droplets are generated in a flow-focusing device, incubated ("in line" or off-chip), and subsequently sorted on chip based on a fluorescence readout indicating reaction progress. This approach proved highly effective in recovering improved variants of a promiscuous sulfatase/phosphonate hydrolase from 10^7 sequences, which were randomly diversified [10]. It was also instrumental in identifying rare enzymes via their promiscuous activity in the screening of a million-membered metagenomic library [18]. In both cases, 10^6–10^7 compartments containing single variants were screened, overcoming the screening capacity of microtiter plates by two- to three orders of magnitude, thanks to the sorting of droplets at kHz frequencies on chip. Figure 1 summarizes the different operations of a platform technology in which miniaturized single cell lysate assays are employed for the screening of enzymes in directed evolution rounds or in functional metagenomics.

The following protocol has been implemented for droplets with a typical diameter of 15 μm (corresponding to a volume of 2 pL) using substrates that use fluorescein as a leaving group which serves as an optical reporter of reaction progress.

2   Materials

2.1   Preparation of Microfluidic Devices

1. SU-8 photoresist (*see* **Note 1**).
2. Polydimethylsiloxane (PDMS) Sylgard 184 kit (Dow Corning).
3. Desiccator.
4. Oven.
5. Biopsy punch of 1 mm diameter.
6. Glass coverslips with 1 mm and 0.13 mm thickness.
7. An oxygen plasma generator (Femto, Diener Electronics).
8. Trichloro(1H,1H,2H,2H–perfluorooctyl)silane (PFOTS) 1% v/v in the fluorous oil HFE 7500 (3 M).
9. Indium wire (51 In/32.5 Bi/16.5 Sn; Indium Corporation and electrical wiring (multicore, diameter 0.75 mm RS).

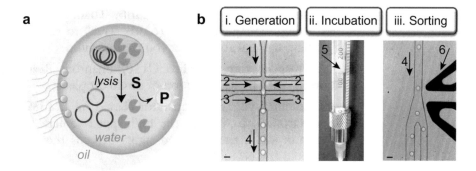

Fig. 1 Miniaturized cell lysate assay using microfluidic droplets. (**a**) Single cell lysis in a droplet compartment. Single *E. coli* cells are encapsulated in monodisperse water-in-oil droplets using a microfluidic chip together with a fluorogenic substrate (S) and lysis reagents. After droplet formation, the cell is lysed and the cytoplasmic enzymes are released inside the compartment and can now encounter and process the substrate. Hydrolysis of the fluorogenic substrate generates a fluorescent product P. The droplet maintains the link between the genotype (i.e., plasmids encoding the enzyme) and the phenotype (i.e., the fluorescent product). (**b**) The unit operations used for the miniaturized single cell lysate assay. (i) Cells transformed by the gene library (1) are mixed with substrate and lysis reagents (2) and emulsified by fluorous oil containing surfactant (3). The *arrow 4* shows the direction of the droplets. (ii) Droplets are incubated off-chip in a syringe (shown by *arrow 5*). (iii) Droplets are reinjected into the sorting chip (the *arrow 4* shows the direction of the flow). Positive droplets are selected by dielectrophoresis induced by the electrodes designated by the *arrow 6*. Scale bars: 40 μm

2.2 Preparation of Bacterial Suspensions Prior to Droplet Encapsulation

1. Competent *E. coli* cells to transform the DNA library (e.g., E. cloni 10G Elite; Lucigen).

2. Library cloned into a high-copy plasmid (e.g., pZero-2, Invitrogen, or pRSFDuet-1, Novagen).

3. Screening buffer: 100 mM MOPS, 115 mM NaCl, 100 μg/mL kanamycin (or other suitable antibiotic depending on the vector used), complete EDTA-free protease inhibitor: one tablet/ 50 mL, Percoll (25%, v/v; Sigma-Aldrich). Adjust pH to 8.0.

4. Syringe filters (pore size: 0.22 μm and 5 μm; Sartorius).

2.3 Generation, Incubation, and Sorting of Microfluidic Droplets

1. Fluorinated oil HFE-7500 (3 M) with fluorosurfactant: 1% (w/w) PicoSurf1 (Dolomite) in HFE-7500 (3 M) (*see* **Note 2**).

2. Lysis buffer: 20% (v/v) BugBuster®, 30 kU/mL lysozyme, and the fluorogenic substrate (typically in the μM range), 50 mM MOPS, pH 7.5.

3. Syringe pumps (Nemesys, Cetoni), syringes (BD plastic and gas tight glass syringes (SGE)) and PTE tubing (Smiths Medical) to operate the microfluidic chip.

4. A fluorescence microscope, e.g., Olympus IX73 equipped with a 40× objective (NA 0.60), 488 nm laser source, dichroic filter, and photodetector (PMT, Hamamatsu).

5. A pulse generator (TGP110, Thurlby Thandar Instruments) and a high-voltage amplifier (610E, Trek) to select droplets that have a signal above a set voltage threshold.

6. A high-speed camera (Miro C110, Phantom Research) to monitor the physical selection of single droplets.

7. A differential voltage comparator (based on LM339, STMicroelectronics) to trigger the high-speed camera and pulse generator when the observed signal exceeds the set voltage threshold (*see* **Note 3**).

2.4 Recovery of DNA and Transformation

- 1H,1H,2H,2H–perfluorooctanol (Sigma-Aldrich) to break the droplet emulsion prior to recovering the DNA.

- DNA Clean & Concentrator-5 (Zymo Research) to purify the plasmid DNA.

- Nontransformable DNA (e.g., linear salmon sperm DNA, Life Technologies).

- Competent *E. coli* cells to transform the DNA library (e.g., E. cloni 10G Elite; Lucigen).

3 Methods

3.1 Preparation of Microfluidic Devices [19]

1. PDMS is used to cast the fluidic master and produce cheap replicates (*see* **Note 1**). Mix the PDMS elastomer and the curing agent thoroughly in a 10:1 (v:v) ratio.

2. Degas the mixture in a desiccator under vacuum for about 30 min until all air bubbles have disappeared.

3. Pour PDMS onto SU-8 mould and cure in the convection oven at 65 °C for at least 4 h.

4. Peel the PDMS off the SU-8 mould and punch holes to form the inlets/outlets with a biopsy punch (*see* Fig. 2).

5. Bond the PDMS chips to a glass coverslip (thickness: 1 mm for flow-focusing chip, 0.13 mm for sorting chip) using the oxygen plasma generator [20].

6. Treat the channels by flushing a freshly prepared solution of 1% v/v PFOTS in HFE-7500 through the channels.

7. Cure the devices in the convection oven at 65 °C for another 20 min, remove from the oven, and store at room temperature.

8. Place the chip on a hot plate (set at 145 °C) and insert low melting-point indium composite solder into the electrode channel entrance holes to melt the metal (*see* **Note 4**).

9. Connect electrical wiring (length, 2 cm) by introducing it into the still molten indium composite solder.

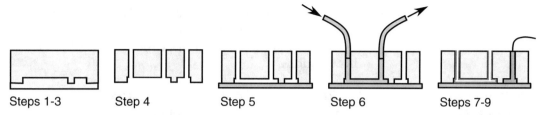

Steps 1-3   Step 4   Step 5   Step 6   Steps 7-9

Fig. 2 Preparation of microfluidic devices. Cartoon representation of the cross section of a microfluidic device at the different steps described in Subheading 3.1. Color coding: *light-grey*—SU-8 mould; *light-blue*—PDMS; *dark-blue*—glass-cover slide; *orange*—PFOTS solution and resulting hydrophobic coating; *dark-grey*—metal electrode

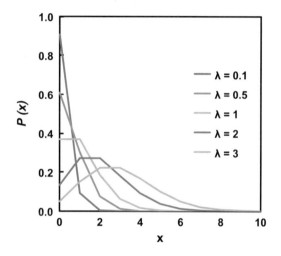

Fig. 3 Poisson Distribution. Probability distribution *P(x)* to encapsulate *x* entities in a droplet as a function of occupancy λ, assuming a droplet size of 2 pL (Ø:15 μm). The figure shows the percentage of droplets containing *x* entities at different occupancies λ

3.2 Preparation of the Cell Suspension and Poisson Distribution

1. Transform competent *E. coli* cells with the plasmid library and plate on agar plates containing the appropriate antibiotic.

2. Grow bacterial cells at 37 °C and incubate in adequate conditions to achieve optimal protein expression.

3. Centrifuge the culture (4 min, 2000 × *g*), resuspend the pellet in buffer, repeat twice, filter through a syringe filter (pore size: 5 μm) and record the optical density OD_{600} of the filtrate at 600 nm.

4. To achieve a given cell occupancy, dilute the cells in buffer to achieve the corresponding initial cell concentration (Fig. 3).

   Given that an OD_{600} of 1 corresponds to a concentration of about $5 \cdot 10^8$ cells/mL and using droplets with a diameter of 15 μm, for a cell occupancy of 0.1, a value of OD_{600} of 0.15 will result in 9% of droplets containing a single cell and 1% containing 2 or more cells.

5. Filter all solutions before compartmentalization using syringe filters to avoid clogging of microfluidic channels. Use a filter pore size of 5 μm for the cell suspension and 0.22 μm for all other solutions.

3.3 Generation, Incubation, and Sorting of the Microdroplets

1. Place the microfluidic device on the stage of the inverted microscope. Fill two 1 mL gas-tight syringes with the cell suspension and the solution containing the substrate and lysis agent, respectively. Fill a 5 mL syringe with the oil–surfactant mix (*see* **Note 2**). Connect the syringes to PTE tubes, which are sufficiently long to reach the microfluidic inlets. Fix the syringes tightly on the syringe pumps and connect the tubes to the appropriate inlets of the chip (Fig. 4a, flow-focusing design).

2. Start the syringe pumps in order to initiate the generation of monodisperse droplets. Set flow rates for the aqueous and the oil phases to 50 μL/h and 500 μL/h, respectively. Wait for

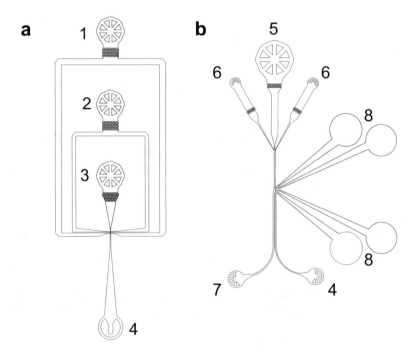

Fig. 4 Designs of microfluidic devices. (**a**) Flow-focusing device for the generation of monodisperse water-in-oil droplets. The two aqueous solutions, containing the substrate and the cells are injected from inlet 2 and 3, respectively. Inlet 1 is used for the oil–surfactant mixture. Droplets are collected from the outlet 4. (**b**) Sorting device for the fluorescence-based screening of droplets. The emulsion is reinjected into the device using the inlet 5. Droplets are respaced by extra fluorous oil injected by the two inlets 6. Laser-induced fluorescence is recorded and used as a selection signal to sort droplets into the outlet 4. Positive droplets are deflected by dielectrophoresis induced by the electrodes 8. Droplets that do not satisfy the user-defined fluorescence threshold flow into the waste channel 7

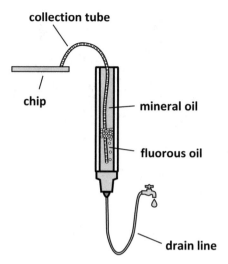

Fig. 5 Collection of microdroplets in a syringe. The outlet tubing of the droplet-making chip is inserted into the syringe where droplets accumulate at the interface between mineral oil and the fluorous oil used for the emulsion. During collection, the drain line is closed so that the oil does not drip

droplet formation to stabilize (~3 min). This setup results in the formation of droplets with 15 μm (2 pL) in diameter at a rate of 4 kHz.

3. Once the droplet formation is stable, start to collect droplets by inserting the outlet tube into the back of a glass syringe (the plunger is removed so that the tube reaches inside the barrel of the syringe). This syringe is connected to a long PTE tube, prefilled with HFE-7500 and the tube clamped to avoid dripping of oil during collection. Droplets will accumulate at the interface (Fig. 5) (*see* **Note 5**).

4. After collection, add mineral oil on top of the droplets until the barrel is completely filled. Remove the clamp and gently reinsert the plunger back into the barrel.

5. Incubate the droplets according to previously determined conditions (*see* **Notes 6–8**).

6. After incubation, reinject droplets into the sorting device (Fig. 4, sorting design). Connect two gas-tight syringes filled with the oil–surfactant mixture and set flow rates to 300 μL/h. The extra oil increases the distance between droplets and prevents the sorting of multiple droplets with a single electric pulse. Reinject droplets at a rate of 30 μL/h, this setup results in a sorting rate of around 2 kHz. The asymmetric Y-shape of the sorting junction ensures that all droplets flow automatically into the wider waste channel, having a lower flow resistance.

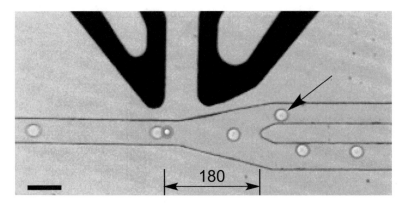

Fig. 6 Snapshot of the droplet sorting device. In the absence of electric pulsing, droplets flow to the bottom (waste) channel. Whenever a fluorescence read-out exceeds a freely set threshold, the electric pulse is activated, dielectrophoretically attracting a single droplet (indicated by an *arrow*) to the sorting channel. Scale bar: 60 μm

7. Focus the laser beam about 180 μm upstream of the sorting junction through a 40× microscope objective (refer to Fig. 6).

8. Connect the photomultiplier tube (PMT) to the pulse generator. The differential voltage comparator inputs are the PMT signal (V_{sig}) and a user-defined reference voltage (V_{ref}) that corresponds to the threshold product concentration (*see* **Note 7**). Wire the output of the comparator to the trigger input of the pulse generator. When V_{sig} exceeds V_{ref}, a square pulse (0.5 ms) is triggered. Amplify the generated pulse 1000-fold to between 0.6 and 0.8 kV (*see* **Note 9**) with the high-voltage amplifier and apply to the electrodes of the sorting device (Fig. 7) (*see* **Note 10**).

9. Use an Eppendorf tube to collect sorted droplets. Being lighter than the fluorous oil carrier phase, the sorted droplets accumulate at the air–oil interface of the collection vessel and can be conveniently harvested despite their small volume.

3.4 Recovery of DNA and Transformation

1. Add 100 μL water into the collecting Eppendorf tube after having rinsed the connecting tubing (from outlet to tube) with HFE-7500 oil (to make sure no droplets are left inside the tubing).

2. Gently add 200 μL of perfluorooctanol to the Eppendorf tube containing the microdroplets, making sure it rinses the tube's walls.

3. Remove the aqueous phase and add ~100 ng of non-transformable DNA (typically ~1 nM final concentration) to prevent loss of selected plasmids during DNA recovery. Purify and concentrate using a spin column. Elute into 7 μL milliQ water.

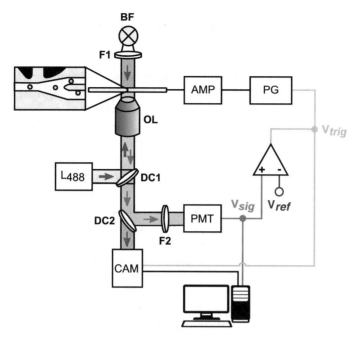

Fig. 7 Microscopy and sorting setup. The microfluidic chip is illuminated by the microscope light source (BF) using a band-pass filter (F1, 593/25 nm) to prevent green illumination light from reaching the detector. The laser beam (488 nm) passes a dichroic mirror (DC1, 495 nm) and is focused through a 40× objective (OL), about 180 μm upstream of the sorting junction (inset and Fig. 6). The induced fluorescence is directed to a second dichroic mirror (DC2, 555 nm), which allows the red light to be captured by a high-speed camera (CAM) for imaging of the chip. Lower wavelengths are directed to a band-pass filter (F2, 516–530 nm) before reaching the photomultiplier tube (PMT) for detection. The PMT output voltage (V_{sig}) is connected to a voltage comparator. If an operator-defined threshold (V_{ref}) is exceeded, the comparator sets its output voltage (V_{trig}) to high, thereby triggering the pulse generator (PG) and image acquisition by the high-speed camera (CAM). The generated pulse is amplified (AMP) and applied to the electrodes, which are embedded in the microfluidic chip finally causing the fluorescent droplet to be dielectrophoretically pulled into the collection channel

4. Transform plasmid DNA into electro-competent *E. coli* cells.

5. Plate out transformed bacteria on petri dishes containing LB-agar and appropriate antibiotics. Incubate plates overnight at 37 °C.

6. Samples from individual colonies are re-grown in deep 96-well plates and their activity confirmed in cell lysates (*see* **Notes 11** and **12**).

7. Colonies are re-grown and their DNA is extracted (using a plasmid DNA extraction kit) and sequenced.

4 Notes

1. The master mould of the microfluidic device is made out of SU-8 2015 photoresist (MicroChem) using standard

photolithographic techniques [21] with the flow-focusing and droplet sorting designs (e.g., those deposited at http://openwetware.org/wiki/DropBase).

2. Alternatively, the surfactant 008-fluorosurfactant (Ran Biotechnologies) can be employed. Similar surfactants can be synthesized following published protocols [12, 19]. The oil–surfactant mixture has to be optimized for several aspects: stability of the droplets to prevent progressive coarsening of the emulsion, small molecule transfers between compartments or uptake of these small molecules by the oil and also protein stability and activity in the presence of a two-phase system [22]. A given reaction time frame will depend on how long the reaction products can be retained in the droplets. Addition of BSA [23] or small molecules (e.g., cyclodextrins) [24] to the aqueous phase, adjustment of pH, variation of the surfactant concentrations and choice of the fluorous carrier oil are parameters that can improve the retention of fluorophores by several orders of magnitude. Alternatively, chemical modification of the substrate, e.g., by introduction of a polar functional group, can increase product retention [25]. A simple method enabling the quantification of fluorescent molecules in microdroplets facilitates the analysis of the effects of oil–surfactant mixtures on enzyme activity as well as those of additives on the retention of fluorescent molecules within the compartment [22].

3. The implementation of the electronic trigger module using a differential comparator is the simplest and cheapest option. Other methods use microprocessors to make comparisons between measured optical signal and trigger threshold (e.g., an open source Arduino board [26] or more sophisticated Field Programmable Gate Array logic [27]).

4. Alternatively, sodium chloride electrodes have recently been used [28]: prepare two syringes with a solution of 5 M NaCl in water, connect them to a metallic needle and fit a plastic tubing onto the needles. Then, fill the channels with the NaCl solution.

5. Droplets collected immediately after their formation form a packed emulsion layer within the collection syringe. In order to avoid possible coalescence during passage through a metallic needle, flow rates have to be kept low during this transit (i.e., around 100 µL/h). Coalescence within the tubing can also occur whenever the pressure is too high within the microfluidic chip: for instance, as a consequence of the inlet becoming clogged by dust or debris. Monitoring whether the chip remains in a pristine state over time helps to address problems quickly by immediate intervention for troubleshooting. The

fluidics operation is otherwise well optimized and should, if filtered solutions are used, consistently run for several hours. Alternative setups for storing droplets include glass vials [27] or long tubing [26].

6. The sensitivity of the optical setup has to be determined for each analyte to assess the minimum number of enzymatic turnovers needed to detect a signal. Typically, a few thousand molecules can be readily detected using laser sources and sensitive detectors. Current FACS instruments can detect 4000 fluorescein molecules. The minimal incubation time for the droplets can then be inferred assuming catalytic rates and expression levels are known.

7. For directed evolution, the fluorescence signal brought about by the parental or wild-type enzyme has to be determined in order to set up a selection threshold for screening of the library.

8. Background reaction resulting from other enzymes contained in *E. coli* and buffer-catalyzed hydrolysis have to be evaluated to make sure fluorescent signals arise from the protein of interest. This threshold can be empirically measured by measuring the mean fluorescence of cells bearing a plasmid expressing an unrelated protein without relevant activity.

9. To drive the selected droplets into the collection channel, the required pulse settings may be varied (e.g., length or amplitude), depending on droplet frequency, flow rates, and droplet size: the bigger the droplets and higher flow rates, the stronger the pulses required. However, the higher flow rates also require shorter pulses (at least below the period between two droplets) to ensure that only single droplets are selected. Typical settings for the procedures described in this chapter are a single pulse at 600 V of 500 μs width. The settings are adjusted before each sorting experiment to achieve optimal performance, which is verified by monitoring the sorting process by visual inspection of a movie taken by the high-speed camera.

10. When using salt electrodes, connect the output of the amplifier to the metallic needles of the syringes loaded with the salt solution.

11. Encapsulation of multiple cells and phenotypic variation will influence the apparent product formation rates, so that the tail of a fluorescence histogram will include false positives. Depending on the width of the selection window (and the sensitivity of the threshold set by the operator) between less than 1% and 90% false positives have been typically observed.

12. If hits are rare, we recommend accepting even a majority of false positives (e.g., as a consequence of high background that makes distinguishing positives difficult). True positives can be

verified in a subsequent microtiter plate screen, provided it is available, or by repeated droplet selections. We also recommend oversampling the library screened by a factor of 10 so that all hits can be recovered.

References

1. Colin PY, Zinchenko A, Hollfelder F (2015) Enzyme engineering in biomimetic compartments. Curr Opin Struct Biol 33:42–51

2. Leemhuis H, Stein V, Griffiths AD et al (2005) New genotype-phenotype linkages for directed evolution of functional proteins. Curr Opin Struct Biol 15:472–478

3. Yang G, Withers SG (2009) Ultrahigh-throughput FACS-based screening for directed enzyme evolution. ChemBioChem 10:2704–2715

4. Schaerli Y, Wootton RC, Robinson T et al (2009) Continuous-flow polymerase chain reaction of single-copy DNA in microfluidic microdroplets. Anal Chem 81:302–306

5. van Vliet LD, Colin PY, Hollfelder F (2015) Bioinspired genotype-phenotype linkages: mimicking cellular compartmentalization for the engineering of functional proteins. Interface Focus 5:20150035

6. Agresti JJ, Antipov E, Abate AR et al (2010) Ultrahigh-throughput screening in drop-based microfluidics for directed evolution. Proc Natl Acad Sci U S A 107:4004–4009

7. Theberge AB, Courtois F, Schaerli Y et al (2010) Microdroplets in microfluidics: an evolving platform for discoveries in chemistry and biology. Angew Chem Int Ed 49:5846–5868

8. Huebner A, Sharma S, Srisa-Art M et al (2008) Microdroplets: a sea of applications? Lab Chip 8:1244–1254

9. Mair P, Gielen F, Hollfelder F (2017) Exploring sequence space in search of functional enzymes using microfluidic droplets. Curr Opin Chem Biol 37:137–144. doi:10.1016/j. cbpa.2017.02.018

10. Kintses B, Hein C, Mohamed MF et al (2012) Picoliter cell lysate assays in microfluidic droplet compartments for directed enzyme evolution. Chem Biol 19:1001–1009

11. Fischlechner M, Schaerli Y, Mohamed MF et al (2014) Evolution of enzyme catalysts caged in biomimetic gel-shell beads. Nat Chem 6:791–796

12. Zinchenko A, Devenish SRA, Kintses B et al (2014) One in a million: flow cytometric sorting of single cell-lysate assays in monodisperse picolitre double emulsion droplets for directed evolution. Anal Chem 86:2526–2533

13. Courtois F, Olguin LF, Whyte G et al (2008) An integrated device for monitoring time-dependent in vitro expression from single genes in picolitre droplets. ChemBioChem 9:439–446

14. Houlihan G, Gatti-Lafranconi P, Kaltenbach M et al (2014) An experimental framework for improved selection of binding proteins using SNAP display. J Immunol Methods 405:47–56

15. Diamante L, Gatti-Lafranconi P, Schaerli Y et al (2013) *In vitro* affinity screening of protein and peptide binders by megavalent bead surface display. Protein Eng Des Sel 26:713–724

16. Mankowska SA, Gatti-Lafranconi P, Chodorge M et al (2016) A shorter route to antibody binders via quantitative *in vitro* bead-display screening and consensus analysis. Sci Rep 6:36391

17. Kintses B, van Vliet LD, Devenish SRA et al (2010) Microfluidic droplets: new integrated workflows for biological experiments. Curr Opin Chem Biol 14:548–555

18. Colin PY, Kintses B, Gielen F et al (2015) Ultrahigh-throughput discovery of promiscuous enzymes by picodroplet functional metagenomics. Nat Commun 6:10008

19. Kotz K, Cheng X, Toner M (2007) PDMS device fabrication and surface modification. J Vis Exp 8:319

20. Eddings MA, Johnson MA, Gale BK (2008) Determining the optimal PDMS-PDMS bonding technique for microfluidic devices. J Micromech Microeng 18:067001

21. McDonald JC, Duffy DC, Anderson JR et al (2000) Fabrication of microfluidic systems in poly(dimethylsiloxane). Electrophoresis 21:27–40

22. Kaltenbach M, Devenish SR, Hollfelder F (2012) A simple method to evaluate the biochemical compatibility of oil/surfactant mixtures for experiments in microdroplets. Lab Chip 12:4185–4192

23. Courtois F, Olguin LF, Whyte G et al (2009) Controlling the retention of small molecules in emulsion microdroplets for use in cell-based assays. Anal Chem 81:3008–3016

24. Zinchenko A, Devenish SR, Hollfelder F (2017) Rapid quantitative assessment of leakage of assay components from microdroplet to test the suitability of oil/surfactant combinations. Submitted

25. Woronoff G, El Harrak A, Mayot E et al (2011) New generation of amino coumarin methyl sulfonate-based fluorogenic substrates for amidase assays in droplet-based microfluidic applications. Anal Chem 83:2852–2857

26. Gielen F, Hours R, Emond S et al (2016) Ultrahigh-throughput-directed enzyme evolution by absorbance-activated droplet sorting (AADS). Proc Natl Acad Sci U S A 113:E7383–E7389

27. Beneyton T, Coldren F, Baret JC et al (2014) CotA laccase: high-throughput manipulation and analysis of recombinant enzyme libraries expressed in *E. coli* using droplet-based microfluidics. Analyst 139:3314–3323

28. Sciambi A, Abate AR (2015) Accurate microfluidic sorting of droplets at 30 kHz. Lab Chip 15:47–51

Chapter 19

Isolation of pH-Sensitive Antibody Fragments by Fluorescence-Activated Cell Sorting and Yeast Surface Display

Christian Schröter, Simon Krah, Jan Beck, Doreen Könning, Julius Grzeschik, Bernhard Valldorf, Stefan Zielonka, and Harald Kolmar

Abstract

Fluorescence-activated cell sorting (FACS) in combination with yeast surface display (YSD) has proven to be a valuable tool for the engineering of antibodies. It enables the fast and robust identification and isolation of candidates with prescribed characteristics from combinatorial libraries. A novel application for FACS and YSD that has recently evolved addresses the engineering of antibodies toward pH-switchable antigen binding, aiming at reduced binding at acidic pH, compared to neutral pH. Therefore, we give guidance for the incorporation of such pH switches into antibody variable domains using combinatorial histidine scanning libraries. The protocol describes a flow cytometric sorting technique for the enrichment of antigen-specific molecules. Moreover, we provide information on how to screen the obtained antibody pools from initial sorting to isolate and characterize pH-sensitive variants.

Key words Fluorescence-activated cell sorting (FACS), Yeast surface display, Fab display, Antibody engineering, Protein engineering, pH-dependent antigen binding, pH-switch engineering

1 Introduction

Monoclonal antibodies have emerged as one of the most promising and hence rapidly growing class of therapeutics for various indications, such as inflammatory diseases and cancer [1, 2]. The generation of antibodies comprising desired properties is yet challenging and often requires antibody engineering toward affinity, specificity, or enhanced biophysical stability [3].

In order to improve the performance of therapeutic antibodies, engineering approaches have emerged to incorporate pH sensitivity into the antigen-binding site of antibodies exhibiting reduced

Christian Schröter, Simon Krah and Jan Beck contributed equally to this work.

Uwe T. Bornscheuer and Matthias Höhne (eds.), *Protein Engineering: Methods and Protocols*, Methods in Molecular Biology, vol. 1685, DOI 10.1007/978-1-4939-7366-8_19, © Springer Science+Business Media LLC 2018

antigen binding at acidic pH (pH 6.0), compared to physiological pH (pH 7.4). Upon internalization, antibody–antigen complexes can enter the endosomal pathway, where pH-dependent antigen binding allows dissociation of the complex in the acidified endosome (pH ~6.0). As a consequence, released antigen can enter the degradative pathway whereas the free antibody is recycled to the cell surface via its interaction with the neonatal Fc-receptor (FcRn) [4]. The reuse of antibodies can result in enhanced antigen clearance that may enable less frequent or lower antibody dosing [5]. Decrease in affinity over a modest pH drop from pH 7.4 to pH 6.0 can be achieved through insertion of histidine residues into the complementary determining regions of the variable domains of an antibody. It has been shown that the number of histidine residues, the synergistic interplay of multiple histidine residues as well as the magnitude of pK_a changes upon antibody binding affect pH sensitivity [6].

Fluorescence-activated cell sorting (FACS) in combination with yeast surface display (YSD) (Fig. 1) has become one of the most valuable tools for the engineering of antibodies affording the benefit of rapid isolation of hit candidates with predefined characteristics from combinatorial libraries [7–9]. One of the main advantageous features of YSD lies in the fact that the eukaryotic expression machinery of yeast ensures proper protein folding of human antibody-derived fragments. In combination with FACS, a sensitive and quantitative analysis of each library variant can be performed by real-time and online data analysis [10]. Furthermore, libraries exceeding a diversity of 10^9 transformants can easily be created by homologous recombinations and antibody heavy and light chain repertoires can be combined by yeast mating technology [11, 12]. These features have been acknowledged by several successful protein engineering approaches, wherein protein variants with pH-dependent antigen binding were isolated from combinatorial libraries using FACS and YSD [13, 14]. In this regard, we recently published a novel approach that comprehensively considered the combinatorial effects of multiple histidine substitutions [14]. Synthesis of the variable regions of a parental antibody Fab fragment allowed incorporation of histidines within all three complementarity determining regions (CDRs) of the variable regions of heavy and light chain (VH and VL) with average mutation rates of 2–3 histidine per variant. Moreover, a novel screening strategy enabled the efficient screening of the antibody Fab histidine scanning libraries and isolation of antibody variants that revealed sharp, reversible pH-dependent binding profiles. A general outline of the procedure for isolating pH-dependent antibodies is given in Fig. 2.

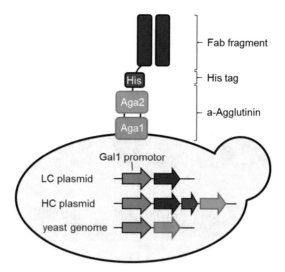

Saccharomyces cerevisiae

Fig. 1 Schematic illustration of the yeast surface display system as described by Boder and Wittrup for antibody Fab fragment expression [15]. The heavy chain of the Fab fragment is fused to the a-agglutinin yeast adhesion receptor protein Aga2p, which assembles with the coexpressed Aga1p forming two disulfide bonds. Aga1p in turn is covalently linked to β-glucan of the extracellular yeast cell matrix and thereby anchors the protein assembly to the yeast cell wall. Two single copy *E. coli* yeast shuttle plasmids encode for heavy and light chains, which include TRP or LEU auxotrophic markers for selection in yeast and ampicillin or kanamycin markers for *E. coli* selection, respectively. The heavy chain fusion protein includes a His$_6$-tag for detection of Fab expression. The system requires the chromosomally encoded AGA1 in yeast (e.g., EBY100). Expression of AGA1 as well as shuttle plasmid encoded proteins are under control of the inducible galactose 1 promotors

2   Materials

2.1 Strains, Plasmids, and Libraries

1. *S. cerevisiae* strain EBY100 (MATa) (Invitrogen)

   (URA3-52 trp1 leu2Δ1 his3Δ200 pep4::HIS3 prb1Δ1.6R can1 GAL) (pIU211:URA3).

2. *S. cerevisiae* strain BJ5464 (MATα) (American Type Culture Collection, ATCC)

   (URA3-52 trp1 leu2Δ1his3Δ200 pep4::HIS3 prb1Δ1.6R can1 GAL).

3. *E. coli* strain Top10 (Invitrogen)

   (F- *mcr*A Δ(*mrr-hsd*RMS-*mcr*BC) Φ80*lac*ZΔM15 Δ*lac*X74 *rec*A1 *ara*D139 Δ(*ara leu*) 7697 *gal*U *gal*K *rps*L (StrR) *end*A1 *nup*G).

4. Plasmids: Two derivatives of pYD1 vectors (Invitrogen) are used for cell surface display of heterodimeric antibody Fab

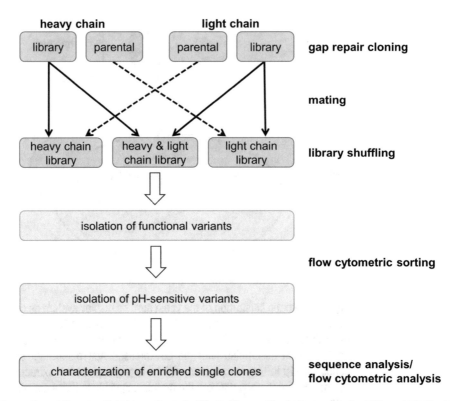

Fig. 2 General workflow for the generation of pH-sensitive antibody fragments by YSD and FACS. Gap repair cloning is applied to generate separate heavy chain and light chain combinatorial histidine-scanning libraries via homologous recombination that are derived from a parental antibody. Subsequently, heavy and light chain diversities can be combined with corresponding parental chains or shuffled in order to introduce additional diversity by mating. Resulting libraries are initially screened by FACS to isolate functional antigen-specific variants, followed by 3–5 consecutive rounds of FACS for identification and isolation of pH-sensitive candidates. After successful enrichment of pH-sensitive variants, isolated single clones are sequenced, allowing for the identification of antibody domains that are further characterized by flow cytometry

fragments encoding for the heavy chain fusion protein (Xpress-tag-(G4S)-VH-CH1-(G4S)-His$_6$-(G4S)$_3$-Aga2p) and the light chain protein. In case of the light chain vector, the aMFpp8 leader sequence is followed by VL-Cλ or VK-Cκ which results in soluble expression of the light chain [16]. In addition, both plasmids encode different auxotrophic markers allowing yeast growth in tryptophan or leucine deficient media and ampicillin or kanamycin resistance genes for selection in *E. coli*.

5. A detailed protocol for the generation of yeast libraries is not given at this point, but has been described in detail elsewhere [11]. However, we provide a brief outline of the histidine mutagenesis procedure in this section: Combinatorial histidine substitution libraries can be commercially obtained (GeneArt, Life Technologies or EllaBiotech). Gene synthesis of the variable regions of the parental Fab fragment allows combinatorial

histidine substitutions within all three CDRs of the VH and VL regions (CDR H1-H3 and CDR L1-L3) (*see* **Note 1**). The occurrence of a histidine codon at defined positions is adjusted to 10%. Theoretical diversities are calculated as previously described [17]. VH and VL gene libraries are amplified with specific oligonucleotides using PCR and cloned into yeast shuttle vectors by gap repair cloning following the protocol of Benatuil and colleagues to generate two separate libraries in EBY100 and BJ5464 cells, respectively [11]. Considering the diversities that can be covered by magnetic activated cell sorting (MACS; *see* **Note 2**) and FACS in combination with YSD, heavy and light chain repertoires are either combined or paired with corresponding parental chains using yeast mating technology [12]. After mating, cells are cultivated in Trp and Leu deficient media. These libraries represent the starting material for flow cytometric sorting (Fig. 2).

2.2 Media and Reagents

1. SD −Trp/−Leu medium: Dissolve 26.7 g minimal SD-Base (Clontech) in deionized H_2O and adjust volume to 900 mL. Sterilize by autoclaving. Dissolve 8.56 g $NaH_2PO_4 \cdot H_2O$, 5.4 g Na_2HPO_4 and 0.64 g Dropout-mix −Trp/−Leu (Clontech) in deionized H_2O and adjust the volume to 100 mL. Sterilize by autoclaving. Combine both solutions, add 10 mL of Penicillin–Streptomycin (Gibco, 10,000 U/mL) and remove any particles by filter sterilization using a 0.22 μm bottle top filter.

2. SD −Trp/−Leu plates: Dissolve 26.7 g of minimal SD-Agar (Clontech) in deionized H_2O and adjust volume to 900 mL. Sterilize by autoclaving. Dissolve 8.56 g of $NaH_2PO_4 \cdot H_2O$, 5.4 g of Na_2HPO_4 and 0.64 g Dropout-mix −Trp/−Leu (Clontech) in deionized H_2O and adjust the volume to 100 mL. Sterilize by autoclaving. Combine both solutions and prepare plates.

3. SG −Trp/−Leu medium: Dissolve 37 g of minimal SD-Base + Gal/Raf (Clontech) in deionized H_2O and adjust volume to 500 mL. Dissolve 8.56 g of $NaH_2PO_4 \cdot H_2O$, 5.4 g of Na_2HPO_4 and 0.64 g Dropout-mix −Trp/−Leu (Clontech) in deionized H_2O and adjust the volume to 100 mL. Dissolve 110 g of PEG8000 in deionized H_2O and adjust the volume to 400 mL. Sterilize by autoclaving and combine all three solutions. Add 10 mL of Penicillin-Streptomycin (Gibco, 10,000 U/mL) and remove any particles by filter sterilization using a 0.22 μm bottle top filter (Merck Millipore).

4. SD Low −Trp/−Leu medium: Dissolve 5 g dextrose, 6.7 g yeast nitrogen base in deionized H_2O and adjust volume to 900 mL. Sterilize by autoclaving. Dissolve 8.56 g $NaH_2PO_4 \cdot H_2O$, 5.4 g Na_2HPO_4, and 0.64 g Dropout-mix

−Trp/−Leu (Clontech) in deionized H_2O and adjust the volume to 100 mL. Sterilize by autoclaving. Combine both solutions, add 10 mL of Penicillin-Streptomycin (Gibco, 10,000 U/mL) and remove any particles by filter sterilization using a 0.22 μm bottle top filter (Merck Millipore).

5. LB Amp medium: Dissolve 10 g NaCl, 10 g tryptone, and 5 g yeast extract in 1 L deionized H_2O. Sterilize by autoclaving. Once the medium has chilled (to approximately 50 °C), add 1 mL of sterile filtered ampicillin solution (100 mg/mL in deionized H_2O).

6. LB Amp plates: Dissolve 10 g NaCl, 10 g tryptone, 5 g yeast extract, and 15 g agar to a volume of 1 L in deionized water. Sterilize by autoclaving. When medium is chilled off (to approximately 50 °C), add 1 mL of sterile filtrated ampicillin solution (100 mg/mL in deionized H_2O) and prepare plates.

7. LB Kana medium: Dissolve 10 g NaCl, 10 g tryptone, and 5 g yeast extract in 1 L deionized water. Sterilize by autoclaving. Once the medium has cooled down (to approximately 50 °C), add 1 mL of sterile filtered kanamycin sulfate solution (30 mg/mL in deionized H_2O).

8. LB Kana plates: Dissolve 10 g NaCl, 10 g tryptone, 5 g yeast extract, and 15 g agar in 1 L deionized water. Sterilize by autoclaving. Once the medium has cooled down (to approximately 50 °C), add 1 mL of sterile filtered kanamycin sulfate solution (30 mg/mL in deionized H_2O) and prepare plates.

9. Yeast library freezing solution: Dissolve 2 g of glycerol and 0.67 g of yeast nitrogen base in a volume of 100 mL *Dulbecco's Phosphate-Buffered Saline* (DPBS). Sterile-filter the solution.

2.3 Selection Reagents for FACS

1. Alexa Fluor 647 labeled mouse anti-His IgG antibody, 0.2 mg/mL (Qiagen).

2. 1 mg/mL Streptavidin, R-phycoerythrin conjugate (Invitrogen).

3. Target proteins devoid of a His-tag (His-tag is present in the YSD heavy chain fusion protein).

4. EZ-Link Sulfo-NHS-Biotin (Thermo Scientific).

5. Pierce™ Biotin Quantitation Kit (Life technologies).

6. *Dulbecco's Phosphate-Buffered Saline* (DPBS) (Gibco).

7. PBS-1: adjust DPBS to pH 7.4 with 1 M NaOH.

8. PBS-2: adjust DPBS to pH 6.0 with 1 M HCl.

2.4 Equipment

1. Cryogenic vials.

2. Freezing container.

3. 0.22 μM Steriflip and Steritop filtration units.

4. Yeast plasmid preparation kit.

5. *E. coli* plasmid miniprep kit.

6. Shaking incubator (20 and 30 °C).

7. Flow cytometry device.

3 Methods

In the following sections, all centrifugation steps to pellet yeast cells are performed at $3000 \times g$ for 3 min. Ensure that fluorophores are shielded from light and perform incubation steps on ice.

3.1 Target Protein Preparation

For labeling of antigen, a plethora of different tags and fluorophores are commercially available (*see* **Note 3**). Nevertheless, we recommend the application of the EZ-Link™Sulfo-NHS-Biotin (Thermo Scientific) according to the manufacturer's protocol. Herein, biotin is linked to N-terminally accessible primary amines (NH_2) or to surface exposed lysine residues. Since antigen labeling with biotin may affect antigen functionality, consider assessing the quality of conjugated antigen by yeast surface display (Subheading 3.4.1) (*see* **Notes 4** and **5**). Furthermore, biotinylated antigens require secondary labeling using fluorophore-conjugated streptavidin that allows signal amplification.

3.2 Yeast Library Preparation for the Expression and Display of Antibody Fragments

1. Thaw an aliquot of the frozen library at room temperature and resuspend cells in SD −Trp/−Leu medium, yielding a final OD_{600} of 0.1–0.5. An absorbance value of 1 at 600 nm corresponds to approximately 1×10^7 cells/mL. The total number of cells in the starting culture should exceed the calculated library diversity at least ten times.

2. Expand the culture overnight at 30 °C and 225 rpm until stationary phase is reached (OD_{600} of approximately 4–6). At this point, storage of cells is possible for up to 4 weeks at 4 °C.

3. Harvest at least the number of cells corresponding to a tenfold excess of the library diversity by centrifugation and resuspend cells in SG −Trp/−Leu medium to an initiate OD_{600} of 1. Cultivate cells for 24–48 h at 20 °C and 225 rpm for expression of antibody fragments.

3.3 Yeast Library Cryoconservation for Long-Term Storage

1. Harvest cells from a freshly grown SD −Trp/−Leu culture and pellet the cells by centrifugation.

2. Inoculate SD Low −Trp/−Leu medium to an initiate OD_{600} of 0.5–1 and cultivate the cells for 48–72 h at 30 °C and 225 rpm. Library diversity should be oversampled at least ten times.

3. Harvest cells by centrifugation and remove the supernatant

4. Resuspend the cells in yeast library freezing solution with final cell concentrations of approximately 1×10^{10} cells/mL and transfer suspensions into cryogenic vials, followed by incubation at room temperature for 5–10 min.

5. Freeze vials at $-70\,^\circ$C.

3.4 Labeling of Library Yeast Cells for Flow Cytometric Analysis

All samples should be handled in parallel reactions. Cell washing is performed by centrifugation of cell suspension, followed by aspiration of the supernatant and resuspension of the cell pellet in 0.5 mL PBS-1 per 1×10^7 cells. A second centrifugation step yields the washed cell pellet.

3.4.1 Assessment of Saturating Antigen Concentrations

For determination of appropriate antigen concentrations and to evaluate the quality of biotinylated antigen, binding signals on yeast cells are analyzed by FACS using streptavidin, R-phycoerythrin conjugate (SA-PE) (*see* **Note 6**).

1. For this, yeast cells (displaying antigen-specific antibody fragments as control and/or library cells) are induced for antibody expression. Next, split 5×10^7 cells into 1.5 mL tubes with 1×10^7 cells per tube.

2. Wash cells and resuspend the pellets in 20 µL PBS-1 containing different antigen concentrations of biotinylated antigen (typical serial dilutions range from 62.5 nM to 1 µM) in PBS-1 ($20\,\mu\text{L}/1 \times 10^7$ cells) and incubate the cells on ice for 30 min (*see* **Note 7**).

3. Afterward, wash the cells, followed by secondary labeling ($20\,\mu\text{L}/1 \times 10^7$ cells) with SA-PE (diluted 1:20 in PBS-1) and subsequent FACS analysis.

4. In FACS histograms, antigen binding signals can be quantified and the sample that exhibits a relative fluorescence intensity converging with the measured value obtained from cells stained with a higher antigen concentration can be considered as saturated. If the above outlined concentrations do not yield saturating antigen binding signals, use higher antigen concentrations.

3.4.2 Isolation of Displayed Antibody Fragments that Exhibit Antigen Binding

The initial round of flow cytometry sorting allows for the isolation of binders that successfully display the antibody fragments and simultaneously bind the antigen of interest. Therefore, cells are incubated with biotinylated antigen (Subheading 3.1), followed by labeling with SA-PE as wells as Alexa Fluor 647 conjugated anti-Penta-His antibody for detection surface expressed antibody Fab fragments.

Overview of samples to be prepared:

Sort Sample (**steps 1–3**): Primary labeling with biotinylated anti-gen, secondary labeling with SA-PE and Alexa Fluor 647 con-jugated anti-Penta-His antibody.

Target Binding Control (**steps 4–6**): Primary labeling with bioti-nylated antigen, secondary labeling with SA-PE.

Fab Display Control (**steps 7** and **8**): Alexa Fluor 647 conjugated anti-Penta-His antibody and SA-PE.

The following protocol is given for handling 10^7 cells and can be up-scaled according to the cell number used (*see* **Note 8**). The number of cells for sorting should exceed the theoretical diversity by at least ten times. By increasing the numbers of cells, volumes and amounts of labeling reagents should be increased proportionally.

1. For preparing the *Sort Sample*, harvest the induced yeast cells by centrifugation (1×10^7 cells) and remove the supernatant by aspiration. Wash the cells and resuspend the cell pellet in 20 μL PBS-1 containing a saturating antigen concentration (*see* Subheading 3.4.1) and incubate the cells on ice for 30 min.

2. Wash and resuspend the cells in 20 μL PBS-1 containing Alexa Fluor 647 conjugated anti-Penta-His antibody (diluted 1:20) and SA-PE (diluted 1:20), followed by incubation on ice for 30 min.

3. Wash the cells and resuspend the cell pellet in 0.5 mL PBS-1 and keep on ice shielded from light until sorting.

4. For preparing the *Target Binding Control*, pellet cell suspen-sion (1×10^7 cells) by centrifugation and remove the superna-tant by aspiration. Wash the cells and resuspend the cell pellet in 20 μL PBS-1 containing a saturating antigen concentration (*see* Subheading 3.4.1) and incubate the cells on ice for 30 min.

5. Wash the cells and perform secondary labeling by resuspension of the cell pellet in 20 μL PBS-1 containing SA-PE (diluted 1:20), followed by incubation on ice for 30 min.

6. Wash the cells and resuspend the cell pellet in 0.5 mL PBS-1 and keep on ice shielded from light until FACS analysis.

7. For preparing the Fab Display Control, pellet cell suspension (1×10^7 cells) by centrifugation and remove the supernatant by aspiration. Wash the cells and resuspend the cell pellet in 20 μL PBS-1 and incubate cells on ice for 30 min.

8. Wash and resuspend the cells in 20 μL PBS-1 containing Alexa Fluor 647 conjugated anti-Penta-His antibody (diluted 1:20) and SA-PE (diluted 1:20), followed by incubation on ice for 30 min.

9. Wash the cells and resuspend the cell pellet in 0.5 mL PBS-1 and keep on ice shielded from light until FACS analysis.

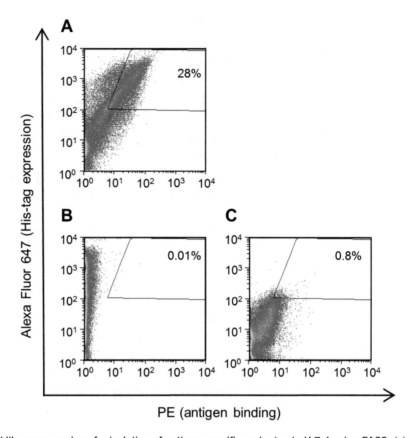

Fig. 3 Initial library screenings for isolation of antigen specific variants at pH 7.4 using FACS. (**a**) *Sort Sample*: Yeast cells expressing library variants are labeled with biotinylated antigen in PBS-1, followed by labeling with SA-PE and Alexa Fluor 647 conjugated anti-Penta-His antibody in PBS-1. (**b**) *Fab Display Control*: Cells are labeled in PBS-1 with Alexa Fluor 647 conjugated anti-Penta-His antibody and SA-PE. (**c**) *Target Binding Control*: Cells are labeled in PBS-1 with biotinylated antigen, followed by labeling with SA-PE

3.4.3 Labeling Strategy for the Isolation of pH-Sensitive Antibody Fragments

A labeling strategy for the enrichment of pH-sensitive variants is described in this section. Cells are initially incubated with unlabeled antigen at pH 7.4 (PBS-1), which allows saturation of antigen-specific antibody variants (Subheading 3.4.1). Afterward, cells are washed and incubated at pH 6.0 (PBS-2), enabling pH-sensitive variants to release unlabeled antigen. As a control, cells are incubated at pH 7.4 (PBS-1) instead of pH 6.0 (PBS-2) to estimate the fraction of pH-sensitive variants. Subsequently, cells are washed and incubated with labeled antigen at pH 7.4 (PBS-1). The rebinding of labeled antigen allows for discrimination of pH-sensitive variants and variants that are still occupied by unlabeled antigen. An exemplary FACS analysis is shown in Fig. 3.

Overview of samples to be prepared:

Sort Sample pH 6.0 (**steps 1–5**): Primary labeling with biotinylated antigen, secondary labeling with SA-PE and Alexa Fluor 647 conjugated anti-Penta-His antibody. Includes incubation step at pH 6.0.

Sample Control pH 7.4 (**steps 6–10**): Primary labeling with biotinylated antigen, secondary labeling with SA-PE and Alexa Fluor 647 conjugated anti-Penta-His antibody. Includes incubation step at pH 7.4 instead of pH 6.0.

FAB Display Control at pH 6.0 (**steps 11–15**): Alexa Fluor 647 conjugated anti-Penta-His antibody and SA-PE. Incubation step at pH 6.0.

Target Binding Control at pH 6.0 (**steps 16–20**): Primary labeling with biotinylated antigen, secondary labeling with SA-PE. Incubation step at pH 6.0.

The following protocol is given for handling 10^7 cells and can be upscaled according to the cell number used (*see* **Note 8**). The number of cells for sorting should exceed the theoretical diversity by at least ten times. By increasing the numbers of cells, volumes and amounts of labeling reagents should be increased proportionally.

1. For preparing the *Sort Sample pH 6.0*, harvest the induced yeast cells by centrifugation (1×10^7 cells) and remove supernatant by aspiration. Wash cells and resuspend the cell pellet in 20 μL PBS-1 containing a saturating concentration of unlabeled antigen and incubate cells on ice for 30 min.

2. Wash the cells and resuspend the pellet in **PBS-2** (0.5 mL/ 1×10^7 cells) and incubate the cells on ice for 30 min (*see* **Note 9**).

3. Centrifuge the cells, and resuspend the pellet in 20 μL PBS-1 containing a saturating concentration of biotinylated antigen (*see* Subheading 3.4.1) and incubate the cells on ice for 30 min.

4. Afterward, wash the cells and resuspend the pellet in 20 μL PBS-1 containing Alexa Fluor 647 conjugated anti-Penta-His antibody (diluted 1:20) and SA-PE (diluted 1:20), followed by incubation on ice for 30 min.

5. Wash the cells and resuspend the cell pellet in 0.5 mL PBS-1 and keep on ice shielded from light until sorting.

6. For preparing the *Sample Control pH 7.4*, harvest the induced yeast cells by centrifugation (1×10^7 cells) and remove supernatant by aspiration. Wash the cells and resuspend the cell pellet in 20 μL PBS-1 containing a saturating concentration of unlabeled antigen and incubate cells on ice for 30 min.

7. Wash the cells and resuspend the pellet in **PBS-1** (0.5 mL/ 1×10^7 cells) and incubate the cells on ice for 30 min.

8. Centrifuge the cells, and resuspend the pellet in 20 μL PBS-1 containing a saturating concentration of biotinylated antigen (*see* Subheading 3.4.1) and incubate the cells on ice for 30 min.

9. Afterward, wash the cells and resuspend the pellet in 20 μL PBS-1 containing Alexa Fluor 647 conjugated anti-Penta-His antibody (diluted 1:20) and SA-PE (diluted 1:20), followed by incubation on ice for 30 min.

10. Wash the cells and resuspend the cell pellet in 0.5 mL PBS-1 and keep on ice shielded from light until FACS analysis.

11. For preparing the *FAB Display Control at pH 6.0*, harvest the induced yeast cells by centrifugation (1×10^7 cells) and remove supernatant by aspiration. Wash the cells and resuspend the cell pellet in 20 μL PBS-1 containing a saturating concentration of unlabeled antigen and incubate cells on ice for 30 min.

12. Wash the cells and resuspend the pellet in **PBS-2** (0.5 mL/1×10^7 cells) and incubate cells on ice for 30 min.

13. Centrifuge the cells, and resuspend the pellet in 20 μL PBS-1 and incubate cells on ice for 30 min.

14. Afterward, wash the cells and resuspend the pellet in 20 μL PBS-1 containing Alexa Fluor 647 conjugated anti-Penta-His antibody (diluted 1:20) and SA-PE (diluted 1:20), followed by incubation on ice for 30 min.

15. Wash the cells and resuspend the cell pellet in 0.5 mL PBS-1 and keep on ice shielded from light until FACS analysis.

16. For preparing the *Target Binding Control at pH 6.0*, harvest the induced yeast cells by centrifugation (1×10^7 cells) and remove supernatant by aspiration. Wash cells and resuspend the cell pellet in 20 μL PBS-1 containing a saturating concentration of unlabeled antigen and incubate cells on ice for 30 min.

17. Wash the cells and resuspend the pellet in **PBS-2** (0.5 mL/1×10^7 cells) and incubate the cells on ice for 30 min.

18. Centrifuge the cells, and resuspend the pellet in 20 μL PBS-1 containing a saturating concentration of biotinylated antigen (*see* Subheading 3.4.1) and incubate the cells on ice for 30 min.

19. Afterward, wash the cells and resuspend the pellet in 20 μL PBS-1 containing SA-PE (diluted 1:20), followed by incubation on ice for 30 min.

20. Wash the cells and resuspend the cell pellet in 0.5 mL PBS-1 and keep on ice shielded from light until FACS analysis.

3.5 Fluorescence-Activated Cell Sorting

Once the labeling procedure is completed, cells can directly be used for sorting. It is important to note that fluorophores with overlapping emission spectra (e.g., phycoerythrin (PE) and fluorescein (FITC)), require proper compensation of the flow cytometer for adequate data interpretation. Furthermore, aggregated yeast cells and debris can be removed by initial gating on forward and side scatter channels. For sorting using a MoFlo cell sorter device, parameters can be set as follows: Side scatter—LOG mode: 650; Forward scatter—LIN mode: 570; FL8—LOG mode (Alexa-Fluor647): 600; FL2—LOG mode (PE): 400; Trigger parameter: side scatter. The sample flow rate is adjusted to an event rate of approximately $10,000–15,000 \text{ s}^{-1}$. The first round of sorting should be done on an enrich mode whereas subsequent rounds can be performed using the purify mode. If other cell sorters are used, settings have to be adjusted.

3.5.1 Sorting of Antigen-Specific Fab Clones

In order to discard variants that do not bind the antigen, cells are incubated with labeled antigen at pH 7.4 (labeling procedure *see* Subheading 3.3, **step 2**), followed by identification and isolation of antigen-specific variants by FACS.

1. Prepare a yeast library (*see* Subheading 3.2) and label the cells including controls according to Subheading 3.4.2.

2. Define a suitable sorting gate that includes the upper right quadrant to isolate all cells that are positive for Fab fragment expression and antigen binding (Fig. 3a). In this step it is important to include great fractions of double positive cells to guarantee full sampling of functional diversity prior to sorting for pH-dependent antigen binding. Gating is performed according to the *Fab Display Control* and *Target Binding Control* samples (Fig. 3b, c). The number of cells analyzed during the sort should exceed at least ten times the theoretical library diversity.

3. Transfer sorted yeast cells to about 10 mL SD -Trp/-Leu medium and expand culture at 30 °C and 225 rpm for 24–72 h. For further sorting, inoculate harvested cells in SG -Trp/-Leu medium followed by 24–48 h of cultivation at 20 °C and 225 rpm (*see* Subheading 3.2). Alternatively, inoculate cells in SD Low -Trp/-Leu medium and apply the procedure for cryoconservation of yeast cells. Consider tenfold oversampling of sorted cells for storage (*see* Subheading 3.3).

3.5.2 Sorting Strategy for the Isolation of pH-Sensitive Fab Clones

This subsection describes the isolation of pH-sensitive variants with high-affinity antigen binding at pH 7.4 and reduced binding at pH 6.0.

1. Label the cell population that have been enriched for target protein binders by consecutive rounds of labeling and FACS as described in Subheadings 3.4.2 and 3.5.1. Consider at least tenfold oversampling of previously sorted cell numbers.

2. Draw an appropriate sorting gate around the double positive cells in the upper right quadrant of the *Sort Sample pH 6.0* (Fig. 4a). The negative gating borders should be aligned according to the *FAB Display Control pH 6.0* (Fig. 4c) and

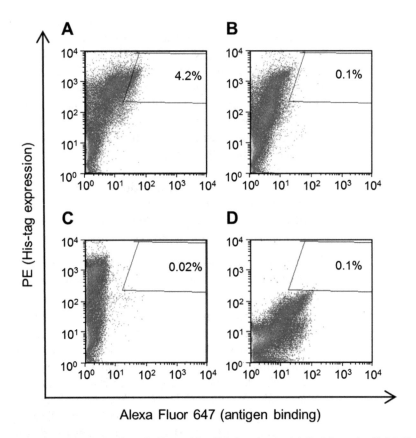

Fig. 4 *Dot plots* illustrating the selection of pH-sensitive Fab fragments. (**a**) *Sort Sample pH 6.0*: Yeast cells are incubated with unlabeled antigen, transferred to PBS-2 (pH 6.0), followed by staining with biotinylated antigen, SA-PE and Alexa Fluor 647 conjugated anti-Penta-His antibody for simultaneous detection of antigen binding (*X*-axis) and surface display (*Y*-axis). (**b**) *Sample Control pH 7.4*: Yeast cells, which are incubated in PBS-1 (pH 7.4) instead of PBS-2 (pH 6.0). (**c**) *FAB Display Control at pH 6.0*: Yeast cells, which are stained with Alexa Fluor 647 conjugated anti-Penta-His antibody. (**d**) *Target Binding Control at pH 6.0*: Yeast cells, which are stained with biotinylated antigen and SA-PE

the *Target Binding Control pH 6.0* (Fig. 4d). In addition, the *Sample Control pH 7.4* (Fig. 4b) is used to precisely quantify cell fractions that refer to variants with pH-sensitive binding (compare Fig. 4a, b). Usually, a less stringent sorting gate can be applied that includes 0.1–5% of cells, thereby reducing the probability of loosing unique clones. The amount of cells that is labeled for sorting should exceed the number of isolated cells from the previous sorting round by a factor of 10 (at least).

3. Regrow selected yeast cells in 10 mL SD -Trp/-Leu medium at 30 °C and 225 rpm for 24–72 h.

4. For further sorting, harvest yeast cells by centrifugation and remove supernatant by aspiration. Herein, cell numbers should exceed the numbers of selected cells from previous sorting round at least tenfold.

5. Resuspend the pellet in SG -Trp/-Leu medium, followed by 24–48 h of cultivation at 20 °C and 225 rpm. Alternatively, inoculate cells in SD Low -Trp/-Leu medium and apply the procedure for cryoconservation of yeast cells (*see* Subheading 3.3). Consider tenfold oversampling of sorted cells for storage.

6. Repeat **steps 1–5** until the enrichment of a population with pH-sensitive binding behavior is observable. This usually requires 3–5 rounds of sorting (*see* **Notes 10** and **11**).

7. It is advisable to plate single clones onto selective SD plates during each sort (usually up to 100 single clones) by flow sorting for subsequent characterization (*see* Subheading 3.4). Colonies are grown on plates by incubation at 30 °C for 48–72 h and grown colonies can be stored for up to 6 weeks at 4 °C. Single clones can also be obtained by plating of serial dilutions from regrown sorted cells onto selective SD plates.

3.6 Characterization of Enriched Clones

Apart from monitoring the phenotype of the yeast population throughout 3–4 rounds of sorting, it is advisable to verify desired properties for enriched library variants by single clone analysis (*see* **Note 12**). Selection of enriched clones can be performed upon sequencing of the Fab coding sequence from single yeast colonies to estimate the diversity of the sorted population. Subsequent characterization of selected single clones by flow cytometric analysis assess specificity as well as pH sensitivity of displayed antibody fragments.

3.6.1 Sequencing of Plasmids from Yeast Cells

1. After flow sorting, single colonies are characterized by sequence analysis. For this, 10–20 randomly picked clones are used to inoculate individually 4 mL of SD -Trp/-Leu medium, followed by 24–48 h cultivation at 30 °C and 225 rpm. Subsequently, heavy and light chain encoding plasmids are isolated from stationary cultures using the RPM plasmid isolation kit (MP Biomedicals) following the manufacturer's protocol.

2. Electrocompetent *E. coli* Top10 cells (Invitrogen) are transformed with heavy and light chain plasmids carrying the ampicillin and kanamycin resistance markers, respectively.

3. Cells are separated by plating on selective agar plates corresponding to the heavy or light chain resistance genes.

4. Furthermore, plasmid DNA is extracted from *E. coli* cells using any commercially available kit and can be used for sequence analysis. Sequence information can be used to identify unique antibody variants for further characterization

3.6.2 Analysis of Single Clones Using Flow Cytometry

This section describes a labeling strategy to validate pH sensitivity of antibody fragments on selected yeast single clones with respect to different off-rates at pH 7.4 and pH 6.0 and also addresses cross-specificity with labeling reagents. In order to avoid the analysis of redundant single clones, identification of unique variants by sequencing should be performed prior to flow cytometric experiments.

Overview of samples to be prepared:

Reference pH 6.0 (**steps 1–4**): Primary labeling with biotinylated antigen, incubation step at pH 6.0, secondary labeling with SA-PE and Alexa Fluor 647 conjugated anti-Penta-His antibody.

Reference pH 7.4 (**steps 5–8**): Primary labeling with biotinylated antigen, incubation step at pH 7.4, secondary labeling with SA-PE and Alexa Fluor 647 conjugated anti-Penta-His antibody.

FAB Display Control at pH 6.0 (**steps 9–12**): Incubation step at pH 6.0, Alexa Fluor 647 conjugated anti-Penta-His antibody and SA-PE.

Target Binding Control at pH 6.0 (**steps 13–16**): Primary labeling with biotinylated antigen, incubation step at pH 6.0, secondary labeling with SA-PE.

For the induction of Fab surface expression, inoculate 4 mL SG -Trp/-Leu medium at an initiate OD_{600} of 1 with a freshly grown single clone in SD -Trp/-Leu culture and cultivate for 24–48 h at 20 °C and 225 rpm.

1. For preparing the *Reference pH 6.0*, harvest the induced yeast cells by centrifugation (1×10^7 cells) and remove supernatant by aspiration. Wash the cells and resuspend the cell pellet in 20 μL PBS-1 containing a saturating concentration of labeled antigen (Subheading 3.4.1) and incubate cells on ice for 30 min.

2. Wash the cells, resuspend the pellet in **PBS-2** (0.5 mL/ 1×10^7 cells), and incubate the cells on ice for 30 min.

3. Centrifuge the cells and resuspend the pellet in 20 μL PBS-1 containing Alexa Fluor 647 conjugated anti-Penta-His antibody (diluted 1:20) and SA-PE (diluted 1:20), followed by incubation on ice for 30 min.

4. Wash the cells and resuspend the cell pellet in 0.5 mL PBS-1 and keep on ice shielded from light until FACS analysis

5. For preparing the *Reference pH 7.4*, harvest the induced yeast cells by centrifugation (1×10^7 cells) and remove supernatant by aspiration. Wash the cells and resuspend the cell pellet in 20 μL PBS-1 containing a saturating concentration of labeled antigen (Subheading 3.4.1) and incubate cells on ice for 30 min.

6. Wash the cells and resuspend the pellet in **PBS-1** (0.5 mL/ 1×10^7 cells) and incubate cells on ice for 30 min.

7. Centrifuge the cells and resuspend the pellet in 20 μL PBS-1 containing Alexa Fluor 647 conjugated anti-Penta-His antibody (diluted 1:20) and SA-PE (diluted 1:20), followed by incubation on ice for 30 min.

8. Wash the cells and resuspend the cell pellet in 0.5 mL PBS-1 and keep on ice shielded from light until FACS analysis.

9. For preparing the *FAB Display Control at pH 6.0*, harvest the induced yeast cells by centrifugation (1×10^7 cells) and remove supernatant by aspiration. Wash the cells and resuspend the cell pellet in 20 μL PBS-1 and incubate cells on ice for 30 min.

10. Wash the cells and resuspend the pellet in **PBS-2** (0.5 mL/ 1×10^7 cells) and incubate the cells on ice for 30 min.

11. Centrifuge the cells and resuspend the pellet in 20 μL PBS-1 containing Alexa Fluor 647 conjugated anti-Penta-His antibody (diluted 1:20) and SA-PE (diluted 1:20), followed by incubation on ice for 30 min.

12. Wash the cells and resuspend the cell pellet in 0.5 mL PBS-1 and keep on ice shielded from light until FACS analysis

13. For preparing the *Target Binding Control at pH 6.0*, harvest the induced yeast cells by centrifugation (1×10^7 cells) and remove the supernatant by aspiration. Wash the cells and resuspend the cell pellet in 20 μL PBS-1 containing a saturating concentration of labeled antigen (Subheading 3.4.1) and incubate the cells on ice for 30 min.

14. Wash the cells and resuspend the pellet in **PBS-2** (0.5 mL/ 1×10^7 cells) and incubate the cells on ice for 30 min.

15. Centrifuge the cells and resuspend the pellet in 20 μL PBS-1 containing SA-PE (diluted 1:20), followed by incubation on ice for 30 min.

16. Wash the cells and resuspend the cell pellet in 0.5 mL PBS-1 and keep on ice shielded from light until FACS analysis

During FACS analysis, gating on double positive cells in the upper right quadrant should be done according to *FAB Display Control at pH 6* (Fig. 5c) and *Target Binding Control at pH 6.0* (Fig. 5d). Double positive cell fractions of *Reference pH 7.4* (Fig. 5a) and *Reference pH 6.0* (Fig. 5b) are analyzed to access the mean fluorescence intensity (MFI) of antigen-binding signals, normalized to display signals. In addition, calculating the MFI ratios of *Reference pH 7.4* and *Reference pH 6.0* allows qualitative estimation of pH sensitivity: The higher the ratio, the faster the off-rate at pH 6.0, compared to pH 7.4 and hence the more pronounced is the pH-sensitive binding (*see* **Note 13**).

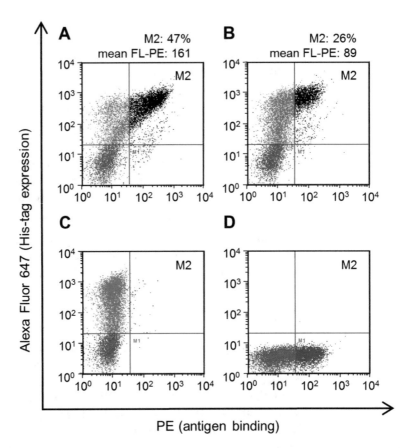

Fig. 5 Single clones analysis for pH-dependent antigen binding. (**a**) *Reference pH 7.4*: Yeast cells are incubated with biotinylated antigen, transferred to PBS-1 (pH 7.4), followed by staining with SA-PE and Alexa Fluor 647 conjugated anti-Penta-His antibody for simultaneous detection of antigen binding (*X*-axis) and surface display (*Y*-axis). (**b**) *Reference pH 6.0*: Cells are incubated with biotinylated antigen, transferred to PBS-2 (pH 6.0), followed by staining with SA-PE and Alexa Fluor 647 conjugated anti-Penta-His antibody. (**c**) *FAB Display Control at pH 6.0*: Cells are incubated in PBS-2 (pH 6.0), followed by staining with SA-PE and Alexa Fluor 647 conjugated anti-Penta-His antibody. (**d**) *Target Binding Control at pH 6.0*: Cells are incubated with biotinylated antigen, transferred to PBS-2 (pH 6.0), followed by staining with SA-PE. Gated cell numbers and mean fluorescence intensities for antigen binding signals are indicated for (**a**) and (**b**)

4   Notes

1. With respect to the antibody library design it needs to be mentioned that the engineering of pH sensitivity into an existing antibody might not always be achieved upon introducing histidine mutations into the complementary determining regions only. Framework residues can be considered as mutational spots as well. However, our suggestion to address interface residues is not limiting for the positioning of histidine residues, since there is a plethora of possible mechanisms involving protonable histidine residues leading to

protein–protein complex destabilization. Hence, mechanisms that are responsible for pH-dependent interactions can differ from case to case.

2. If the library diversity exceeds the maximum capacity of FACS, magnetic activated cell sorting (MACS) should be applied before cells are subjected to flow cytometric sorting. This process allows the rapid isolation of a large number of cells [18] in a relatively short period of time.

3. A variety of labeling reagents are commercially available and different combinations of fluorescent dyes and detection reagents can be used. With respect to fluorophore selection, it should be considered that excitation and emission spectra should not overlap (guidance is given for example on the BD homepage (bdbiosciences.com)). Appropriate labeling reagent concentrations should be assessed in titrating experiments (if no recommendation is given by the manufacturer).

4. It is advisable to test the stability of the antigen at pH 6.0 as well as the intrinsic pH sensitivity of the parental antibody prior to any other experiment.

5. Antigen labeling with 2–3 biotin molecules per protein molecule is sufficient for signal detection using streptavidin–fluorophore conjugates. Higher ratios of biotin/protein enhances the risk for "overbiotinylation," which results in a decreased fraction of functional protein.

6. For FACS analysis and screening, it is helpful to include wild type antibody displaying yeast cells as a control in order to assess binding signals of labeling reagent.

7. Incubation times of about 20–30 min are usually sufficient to gain adequate detection signals, however they can be set up in a way that binding equilibrium between antibody and antigen is achieved. Detailed information is provided by Gera et al. [19].

8. The incubation steps described in this protocol are performed in a volume of $20 \ \mu L/1 \times 10^7$ cells, which allows yeast cells to stay in suspension. At a defined concentration, increased volumes can be used, consequently increasing antigen consumption. In general, variations of volumes (incubation and washing steps) should be adjusted proportionally for control samples.

9. During the labeling procedure for the isolation of pH-sensitive variants, different off-rate stringencies can be applied by varying the incubation time at pH 6.0 (e.g., from 30 to 10 min). When the incubation time at pH 6.0 is decreased, only variants that are able to release the unlabeled antigen in a shortened period of time can rebind the labeled antigen in the following incubation step (increased off-rate stringency, fast off-rate).

10. Different labeling reagents should be alternated in successive screening rounds to avoid the enrichment of antibody fragments that bind to labeling reagents. However, screenings of combinatorial histidine scanning libraries are usually less prone to enrichment of reagent binders, as most of the library variants share high similarity and specificity with the parental antibody.

11. Throughout subsequent sorting rounds, the concentration for antigen labeling can be reduced to enhance the selection stringency which favors selection of high affinity target binding variants. However antigen concentrations should always allow sufficient signal detection during FACS. Details about the optimal antigen concentration for screening are given in [20].

12. Affinity at pH 7.4 and/or pH 6.0 can be easily determined by labeling yeast single clones with varying antigen concentrations. A detailed protocol can be found in [21].

13. Detailed analysis of binding kinetics can be obtained with soluble antibodies by using biolayer interferometry (BLI) or surface plasmon resonance (SPR). For this, antibodies have to be formatted for soluble expression. During binding analysis, binding affinity (K_D), association rate (k_a), and dissociation rate constants (k_d) can be analyzed and dissociation rate constants at pH 6.0 can be measured to assess pH sensitivity. Guidance is provided in [14].

References

1. Chan AC, Carter PJ (2010) Therapeutic antibodies for autoimmunity and inflammation. Nat Rev Immunol 10:301–316

2. Scott AM, Wolchok JD, Old LJ (2012) Antibody therapy of cancer. Nat Rev Cancer 12:278–287

3. Hudson PJ, Souriau C (2003) Engineered antibodies. Nat Med 9:129–134

4. Igawa T, Mimoto F, Hattori K (2014) pH-dependent antigen-binding antibodies as a novel therapeutic modality. Biochim Biophys Acta 1844:1943–1950

5. Igawa T, Ishii S, Tachibana T et al (2010) Antibody recycling by engineered pH-dependent antigen binding improves the duration of antigen neutralization. Nat Biotechnol 28:1203–1207

6. Murtaugh ML, Fanning SW, Sharma TM et al (2011) A combinatorial histidine scanning library approach to engineer highly pH-dependent protein switches. Protein Sci 20:1619–1631

7. Rhiel L, Krah S, Gunther R et al (2014) REAL-Select: full-length antibody display and library screening by surface capture on yeast cells. PLoS One 9:e114887

8. Boder ET, Midelfort KS, Wittrup KD (2000) Directed evolution of antibody fragments with monovalent femtomolar antigen-binding affinity. Proc Natl Acad Sci U S A 97:10701–10705

9. Tillotson BJ, de Larrinoa IF, Skinner CA et al (2013) Antibody affinity maturation using yeast display with detergent-solubilized membrane proteins as antigen sources. Protein Eng Des Sel 26:101–112

10. Doerner A, Rhiel L, Zielonka S et al (2014) Therapeutic antibody engineering by high efficiency cell screening. FEBS Lett 588:278–287

11. Benatuil L, Perez JM, Belk J et al (2010) An improved yeast transformation method for the generation of very large human antibody libraries. Protein Eng Des Sel 23:155–159

12. Weaver-Feldhaus JM, Lou J, Coleman JR et al (2004) Yeast mating for combinatorial Fab library generation and surface display. FEBS Lett 564:24–34

13. Traxlmayr MW, Lobner E, Hasenhindl C et al (2014) Construction of pH-sensitive Her2-

binding IgG1-Fc by directed evolution. Biotechnol J 9:1013–1022

14. Schroeter C, Gunther R, Rhiel L et al (2015) A generic approach to engineer antibody pH-switches using combinatorial histidine scanning libraries and yeast display. MAbs 7:138–151

15. Boder ET, Wittrup KD (1997) Yeast surface display for screening combinatorial polypeptide libraries. Nat Biotechnol 15:553–557

16. Rakestraw JA, Sazinsky SL, Piatesi A et al (2009) Directed evolution of a secretory leader for the improved expression of heterologous proteins and full-length antibodies in *Saccharomyces cerevisiae*. Biotechnol Bioeng 103:1192–1201

17. Arnold FH (1996) Directed evolution: creating biocatalystsfor the future. Chem Eng Sci 51:5091–5102

18. Mattanovich D, Borth N (2006) Applications of cell sorting in biotechnology. Microb Cell Fact 5:12

19. Gera N, Hussain M, Rao BM (2013) Protein selection using yeast surface display. Methods 60:15–26

20. Boder ET, Wittrup KD (1998) Optimal screening of surface-displayed polypeptide libraries. Biotechnol Prog 14:55–62

21. Chao G, Lau WL, Hackel BJ et al (2006) Isolating and engineering human antibodies using yeast surface display. Nat Protoc 1:755–768

Chapter 20

Library Generation and Auxotrophic Selection Assays in *Escherichia coli* and *Thermus thermophilus*

Jörg Claren, Thomas Schwab, and Reinhard Sterner

Abstract

The selection of optimized enzymes from gene libraries is important, both for basic and applied research. Here, we first describe the generation of plasmid-borne libraries using error-prone PCR and highly competent *Escherichia coli* cells. We then provide protocols for the use of these libraries for auxotrophic selection assays with *E. coli* and the extremely thermophilic bacterium *Thermus thermophilus* as hosts.

Key words Auxotrophic selection, Error-prone PCR, Enzyme optimization, Gene libraries, *Thermus thermophilus*

1 Introduction

The optimization of enzymes with respect to stability and activity is of utmost importance for both basic and applied research. Two different strategies are available for the generation of tailored enzymes, rational design and directed evolution. Traditional rational enzyme design uses site-directed mutagenesis to introduce specific amino acid exchanges that are planned on the basis of the visual inspection of high-resolution crystal structures [1]. Modern rational design strategies use sophisticated knowledge-based computational tools, which have allowed to optimize enzymatic activities and even to tweak protein scaffolds such that they can catalyze nonnatural reactions [2]. A directed evolution experiment requires much less information than is necessary for rational design. Here, the gene for the target enzyme is locally or globally randomized, for example by error-prone PCR. The resulting ensemble of mutated genes is then cloned into a plasmid and used to transform the carrier *Escherichia coli*. The plated carrier cells, each containing a plasmid with a specific gene variant, are then collected and propagated, followed by the preparation of the mixture of plasmid DNA molecules. This plasmid-encoded gene library is used to

Uwe T. Bornscheuer and Matthias Höhne (eds.), *Protein Engineering: Methods and Protocols*, Methods in Molecular Biology, vol. 1685, DOI 10.1007/978-1-4939-7366-8_20, © Springer Science+Business Media LLC 2018

transform an appropriate host organism, which is subjected to one or more rounds of high-throughput screening or selection that allow for the isolation of beneficial variants [3]. Screening assays, which are generally based on the spectroscopically monitored conversion of substrate into product, are highly versatile but can be infrastructure-intensive and time-consuming. Moreover, albeit specific screening techniques such as fluorescence-activated cell-sorting allow for the throughput of up to 10^8 variants, the number of variants that can be screened with maintainable effort is generally limited to about 10^4 [4]. Selection assays based on metabolic complementation of auxotrophic cells are not as generally applicable as screening techniques, because they require that beneficial amino acid exchanges can be coupled to the (better) survival of the host organism. However, metabolic selection allows for a very high throughput of up to 10^{10} variants [4]. The most common host organism both for screening and selection is *E. coli*, due to its easy handling and high transformation efficiency. However, other genetically manipulable microorganisms have also been used for metabolic selection. For example, stabilized variants of a metabolic enzyme have been isolated with the help of the extremely thermophilic bacterium *Thermus thermophillus* as host organism [5].

This chapter is devoted to the use of selection assays in directed evolution experiments. We first describe how error-prone PCR can be used to construct plasmid-based gene libraries with *E. coli* as a carrier. We then show how these libraries can be used for the optimization of enzymes by metabolic selection in *E. coli* or *T. thermophilus* as hosts.

2 Materials

2.1 Generation of a Gene Library in E. coli

1. Template DNA encoding the gene of interest.

2. A suitable vector for cloning and expression of the target gene (*see* **Note 1**).

3. Enzymes for conducting PCR, restriction and ligation reactions: Taq polymerase, suitable restriction endonucleases, T4-DNA ligase.

4. Antibiotic stock solution (1000×) (*see* **Note 1**).

5. Plasmid preparation kit (for *E. coli*).

6. Gel extraction kit.

7. SOB medium: 20 g/L tryptone, 5 g/L yeast extract, 0.5 g/L NaCl, 0.186 g/L KCl, 2.4 g/L MgSO$_4$ [6] (*see* **Note 2**).

8. SOC medium: 20 g/L tryptone, 5 g/L yeast extract, 2.4 g/L MgSO$_4$, 0.5 g/L NaCl, 0.186 g/L KCl, 3.603 g/L glucose (*see* **Note 2**).

9. SOC agar plates: SOC medium containing 1.5% (w/v) Bacto agar. Autoclave 3 × 1 L of SOB medium containing 1.5% (w/v) Bacto agar. Add 20 mL of 1 M sterile-filtered glucose (end concentration: 20 mM) and 1 mL of 1000× antibiotic stock solution. 3 L medium is sufficient to spill 40 large (Ø14.5 cm) and 20 small (Ø 8.5 cm) agar dishes.

10. LB medium: 0.5% (w/v) yeast extract, 1.0% (w/v) NaCl, 1.0% (w/v) tryptone.

11. Millipore water, sterile.

12. Thermocycler.

13. Electroporator and electroporation cuvettes with a gap size of 2 mm.

14. Equipment for agarose gel electrophoresis.

2.2 Auxotrophic Selection in E. coli

1. Auxotrophic *E. coli* strain.

2. 1% (w/v) NaCl solution, sterile.

3. SOC medium.

4. M9$^-$ minimal selection medium: Mix 780 mL sterile A. dest., 200 mL of 5× M9 salts, 20 mL of 20% (w/v) glucose, 2 mL of 1 M MgSO$_4$, 0.1 mL of 1 M CaCl$_2$, 1 mL of antibiotic stock solution. Glucose and salt solutions are sterilized by filtration in advance.

   M9 salts: 64 g/L Na$_2$HPO$_4$ · 7H$_2$O, 15 g/L KH$_2$PO$_4$, 2.5 g/L NaCl, 5 g/L NH$_4$Cl. Sterilize by autoclaving.

5. M9$^-$ selection plates: Mix 780 mL sterile A. dest., 200 mL of 5× M9 salts, 15 g Bacto agar and autoclave the suspension. Let cool it down to 50 °C, add 20 mL of 20% (w/v) glucose, 2 mL of 1 M MgSO$_4$, 0.1 mL of 1 M CaCl$_2$, and 1 mL of antibiotic stock solution.

6. LB agar plates: 0.5% (w/v) yeast extract, 1.0% (w/v) NaCl, 1.0% (w/v) tryptone, 1.5% (w/v) Bacto agar. Add 1 mL antibiotic stock solution per L medium before pouring the plates.

7. M9$^-$ positive control medium: M9$^-$ selection medium plus the selection product. Use a final concentration of 20 μg/mL as orientation.

2.3 Auxotrophic Selection in T. thermophilus

1. Auxotrophic *T. thermophilus* strain.

2. Terrific broth (TB) medium: 2.4% (w/v) yeast extract (24 g/L), 1.2% (w/v) peptone/tryptone (12 g/L), 0.8% glycerol (8 g/L). Prepare separately 10× TB-salts: 0.72 M K$_2$HPO$_4$ · 3H$_2$O (164.4 g/L), 0.17 M KH$_2$PO$_4$ (23.2 g/L), both the medium and the salts are sterilized by autoclaving. For use add 100 mL of autoclaved 10× TB-salts to 900 mL of autoclaved TB-medium.

3. TB kanamycin agar plates: TB-medium containing 1.5% (w/v) Bacto agar, 50 mg/L kanamycin (*see* **Note 1**).

4. Synthetic minimal medium (SH⁻) [7]: Mix 900 mL of solution A, 100 mL of solution B, 0.1 mL of solution C, 1 mL of solution D.

Solution A: In 400 mL A. dest., dissolve 20 g of sucrose, 20 g of sodium glutamate, 0.5 g of K_2HPO_4, 0.25 g of KH_2PO_4, 2 g of NaCl, and 0.5 g of $(NH_4)_2SO_4$. Adjust pH to 7.0–7.2 and adjust volume to 500 mL with A. dest. Add 100 mL of 12 g/L $NaMoO_2 \cdot H_2O$, 100 mL of 1 g/L $VOSO_4 \cdot xH_2O$, 100 mL of 5 g/L $MnCl_2 \cdot 4H_2O$ (dissolved in 0.01 M HCl), 60 mg $ZnSO_4 \cdot 7H_2O$ + 15 mg $CuSO_4 \cdot 5H_2O$ (dissolved in 100 mL A. dest.). Final volume: 900 mL. Sterilize by autoclaving.
Solution B: 1.25 g/L $MgCL_2 \cdot 6H_2O$, 0.25 g/L $CaCl_2 \cdot 2H_2O$. Dissolve in A. dest., sterilize by autoclaving.
Solution C: 60 g/L $FeSO_4 \cdot 7H_2O$, 8 g/L $CoCl_2 \cdot 6H_2O$, 0.2 g/L $NiCl_2 \cdot 6H_2O$. Dissolve in 5 mM H_2SO_4, sterilize by filtration.
Solution D: 0.1 mg/mL biotin, 1 mg/mL thiamin, sterilize by filtration.

5. SH⁻ agar plates: SH⁻ medium containing 1.5% (w/v) agar, 50 mg/L kanamycin.

6. Shaking and plate incubators with a temperature range of up to 80 °C. Plates should be wrapped by plastic bags to decelerate the dehydration process.

3 Methods

3.1 Generation of a Gene Library in E. coli

Day 1

1. Prepare the error-prone (ep) PCR to amplify the library fragment (*see* **Note 3**). Make sure to perform sufficient PCRs to get enough of the desired amplification fragment. Three PCR approaches with 50 µL each are best practice.

2. Purify amplification fragment by preparative gel electrophoresis followed by a gel extraction (*see* **Note 4**).

3. Digest the PCR fragment and the desired plasmid with appropriate restriction enzymes (*see* **Note 5**). For later auxotrophic selection in *T. thermophilus*, an *E. coli–T. thermophilus* shuttle plasmid must be used in this step (*see* **Note 1**).

4. Purify digested PCR fragment and vector by preparative gel electrophoresis followed by a gel extraction (*see* **Note 6**).

5. Setup ligation reactions with 1 U of T4-DNA Ligase at 16 °C over night (*see* **Note 7**).

6. Inoculate a suitable *E. coli* strain for preparing the library in a 50 mL over night preculture (*see* **Note 8**).

Day 2

7. Inoculate the 200 mL SOB medium with the *E. coli* preculture from **step 6** to $OD_{600} = 0.1$. Harvest the cells at $OD_{600} = 0.6$ to make fresh electrocompetent cells (*see* **Note 9**).

8. Dialyze the nine ligation preparations on small dialysis membranes against water for 60 min (*see* **Note 10**).

9. Transform *E. coli* cells with ligation preparations in electroporation cuvettes with the electroporator. Use the recommended electric pulse settings, which are for *E. coli* normally 2500 V, 25 μF, and 200 Ω. Use 6 μL of ligation approach per 100 μL of competent *E. coli* cells (*see* **Note 11**). To wash transformed *E. coli* cells out of the electroporation cuvette, use 3 × 500 μL SOC medium and transfer the cells into a 50 mL Falcon tube. In one 50 mL Falcon tube, 10–15 electrotransformation reactions can be pooled. Regenerate the cells for 1 h at 37 °C on a shaking platform (200 rpm). To determine the transformation efficiency of freshly prepared cells, transform 100 μL of cells with 100 ng of undigested empty vector, but handle this transformation strictly separate from library transformations. Regenerate the vector transformation in a 15 mL Falcon tube.

10. After cell regeneration, centrifuge the transformed cells at $3200 \times g$, 10 min, at 25 °C, and resuspend the cells in 8 mL SOC medium.

11. Use large petri dishes (Ø14.5 cm) with SOC agar to plate 200 μL of the library on each dish (40 dishes in total). To determine the size of the library and the transformation efficiency, plate the library and the empty vector control on small SOC agar dishes in the following dilutions, prepared with SOC-medium: $1:10–1:10^7$. Incubate the agar plates at 37 °C, over night.

Day 3

12. Determine transformation efficiency of freshly prepared *E. coli* cells by counting the colonies of the empty vector transformation and determine the size of the library by counting the colonies, both on the small SOC agar dishes. Perform a colony PCR with 20 randomly picked clones from the small dishes with the library to determine the ligation efficiency (*see* **Notes 12** and **13**).

13. Scratch colonies from large agar dishes with a sterile spatula and liquid LB-medium (5–10 mL per plate are sufficient) and collect the cells in 50 mL Falcon tubes.

14. Extract plasmid DNA with a suitable commercial kit along the protocol of supplier.

15. Determine DNA concentration via UV absorption at 260 nm and store the plasmid-encoded library at −20 °C. This library is ready to use for subsequent selection experiments with *E. coli* or *T. thermophilus*.

3.2 Auxotrophic Selection in E. coli

1. Use the desired auxotrophic *E. coli* strain for the complementation assay. In order to obtain high transformation efficiencies, the use of freshly prepared competent cells is recommended (*see* **Note 9** for a convenient protocol). Alternatively, stored cells at −80 °C can be used (*see* **Note 14**).

2. Do the electroporation of the auxotrophic *E. coli* cells with the plasmid-encoded gene library and at least two negative controls (empty plasmid and plasmid with starting construct) and one positive control (plasmid with a gene coding for the desired function of library variants; *see* **Note 15**). Use 100 ng of plasmid DNA for each transformation; if necessary, do several transformations. Wash the transformed *E. coli* cells with 3×500 μL of SOC medium out of the electroporation cuvette and transfer the cells in 15 mL Falcon tubes. Incubate the transformed cells for 1 h at 37 °C on a shaking platform to revitalize the cells.

3. After revitalization of cells, transfer them into 2 mL reaction vessels, centrifuge them at $1500 \times g$ for 1 min at room temperature, and wash the cell pellet with 1 mL of 1% (w/v) NaCl to remove residual rich SOC medium by resuspending the cells with a pipette. Repeat this step for additional two times.

4. Use the prepared large agar dishes (Ø14.5 cm) with M9⁻ medium for plating the cells transformed with the plasmid-encoded gene library and with the controls. Plate on each large agar dish approximately 200 μL of the transformed cells. Divide two M9⁻ selection plates in four equally large sectors, by adding a cross on the backside of the petri dish with a waterproof pen. Label each sector with the desired control, and plate afterward in each sector 50 μL of the two negative controls, the positive control, and the cells transformed with the plasmid-encoded library. Plate the controls in different sectors of a M9⁻ selection plate in two dilutions—undiluted and 1:100 diluted. Additionally, plate 200 μL undiluted cells of the negative controls on a M9⁻ selection plate per control. Plate four M9⁻ selection plates with 200 μL of the undiluted cells transformed with the plasmid-encoded library; optionally, plate several 1:5 or 1:10 dilutions. Start incubation of plates after finishing the plating procedure at 37 °C in an incubator.

To determine the transformation efficiency, plate the controls and the cells transformed with the plasmid-encoded library on plates containing LB agar medium, or on M9$^-$ medium plus the selection product.

5. Optionally, a selection assay can additionally be performed in liquid M9$^-$ medium in a shake flask [8]. To this aim, use 100 mL shake flasks filled with 50 mL M9$^-$ liquid medium supplemented with antibiotic stock solution and inoculate the medium to OD$_{600}$ < 0.1. Use 500 µL of 1:10 dilution of positive and negative control and 500 µL of cells transformed with the plasmid-encoded library undiluted and 1:10 diluted.

6. To follow colony growth, do a visible inspection of each plate and mark grown colonies with date and time. To follow OD increase in case of a liquid complementation assay, take 800 µL samples out of each shaking flask, do an OD$_{600}$ measurement and document the values.

7. To analyze the nucleotide sequence of the relevant gene of grown colonies, pick each grown colony of the library plates and inoculate 5 mL LB medium supplemented with antibiotic stock solution with this colony. Do a DNA preparation and analyze the sequence of the grown variant. To analyze the relevant gene sequence of cells grown in liquid medium, take 100 µL out of the library culture, do different dilutions, plate them on LB agar plates. After overnight incubation, inoculate with at least three colonies from each library culture 5 mL LB liquid medium for a DNA preparation. Afterward, analyze the DNA by sequencing (*see* **Note 16**).

8. To compare the isolated variants directly to each other, a second complementation assay should be performed by transforming the auxotrophic *E. coli* cells with the variants. The procedure is the same for this rescreen of the variants as in the initial screen. Plate the variants in 1:100 and 1:1000 dilutions on M9$^-$ agar plates and in at least one suitable dilution (1:100) on LB agar plates as transformation control. With this second complementation assay a first estimation should be possible, which variants should be in focus.

3.3 Auxotrophic Selection in *T. thermophilus*

1. Use the desired auxotrophic *T. thermophilus* strain for the complementation assay. A glycerol stock culture is applied for inoculation of 5 mL TB medium and cultivated overnight at 70 °C. This preculture is used to seed TB medium, which is cultivated to an OD$_{600}$ = 0.7 at 70 °C. Then 0.5 mL of the cells are transferred to a new cultivation tube and diluted with 0.5 mL of fresh TB medium.

2. After 1 h storage at 70 °C the plasmid-encoded gene library that has been generated in *E. coli* is added to the cultivation

tube, and the mixture is incubated for another 1 h at 70 °C (*see* **Note 17**). Use 2 μg of plasmid DNA for each transformation of 1 mL cells; if necessary, do several transformations with the plasmid-encoded gene library or scale up the respective cultivation volume. Additionally transform the *T. thermophilus* cells with a negative control (plasmid with starting construct). The realization of a positive control (transformation with a plasmid containing a gene coding for the desired function of library variants) is not recommended due to the potential risk for cross-contamination of the selection plates (*see* **Note 18**).

3. After the incubation, centrifuge the tubes at 1500 × *g* for 3 min at room temperature. Wash the cell pellet with 1 mL of 1% (w/v) NaCl to remove residual rich TB medium by resuspending the cells with a pipette. Repeat this step for additional two times.

4. Use the prepared large agar dishes (Ø14.5 cm) with SH⁻ medium for plating the cells transformed with the plasmid-encoded gene library. As a control, SH⁻ medium plates plus the selection product can be applied and tested if available. Plate on each large agar dish approximately 200 μL of the cells transformed with the plasmid-encoded gene library or the negative control; optionally, plate several 1:5 or 1:10 dilutions. Start incubation of plates after finishing the plating procedure at 55–79 °C in an incubator (*see* **Note 19**). The temperature range for selection needs to be assessed based on the thermal stability of the enzyme of interest and the assumed potential of improving its thermal stability by random mutagenesis.

   To determine the transformation efficiency, plate the controls and the cells transformed with the plasmid-encoded gene library on plates containing TB kanamycin agar medium, or on SH⁻ medium plus the selection product.

5. To follow colony growth, do a visible inspection of each plate and mark grown colonies with date and time.

6. To analyze the plasmid sequence of the relevant gene of grown colonies, pick each grown colony of the library plates and inoculate 5 mL TB kanamycin medium with this colony. Do a DNA preparation and analyze the sequence of the grown variant (*see* **Note 20**).

7. To compare the isolated variants directly to each other, a second complementation assay should be performed by transforming the auxotrophic *T. thermophilus* cells with the isolated variants. The procedure is the same for this rescreen of the variants as in the initial screen. Plate the variants in undiluted and 1:10 dilutions on SH⁻ agar plates and in at least one suitable dilution (1:10) on TB kanamycin agar plates

as transformation control. With this second complementation assay a first estimation should be possible, which variants should be in focus.

4   Notes

1. If it is intended to perform selection in *T. thermophilus*, the library should be prepared with an *E. coli*–*T. thermophilus* shuttle vector, such as pMK18 [5, 9]. This shuttle vector contains a kanamycin resistance selection marker. For selection experiments in *E. coli*, any suitable expression plasmid can be employed. Please adopt the required antibiotic throughout this protocol.

2. To increase the competency of *E. coli* cells, use as cultivation volume a maximum of 200 mL SOB medium. Larger volumes result in decreased competency of cells.

   SOB medium is also available commercially (e.g., from Sigma-Aldrich). Add sterile-filtered 1 M glucose (end concentration 20 mM) to autoclaved SOB to create SOC medium.

3. In case of an epPCR, the following protocol with an unbalanced ratio of deoxyribonucleotides and $MnCl_2$ is an approved method [10–12]: 25 ng DNA template, 1 μM of each primer, 0.35 mM dATP, 0.4 mM dCTP, 0.2 mM dGTP, 1.35 mM dTTP (unbalanced dNTP ratio makes sure to get a broader spectrum of transitions and transversions), various concentrations of $MnCl_2$ (in the range of 0.3–0.75 mM; determines the exchange rate—more exchanges with higher concentrations), 1 mM $MgCl_2$, reaction buffer (contains additional 1.5 mM $MgCl_2$) 5 U of Taq DNA-Polymerase, add water to 50 μL total reaction volume.

4. To increase the quality of the amplified PCR fragment after gel extraction, incubate the column with the bound DNA for 1 min at 50 °C to evaporate remained ethanol and elute the DNA afterward with 40 μL of preheated water (50 °C).

5. Use 2 μg of your desired plasmid and digest it in one digestion reaction. Do three separate digestion reactions for the three PCR. Total volume for each approach is 50 μL. Time for digestion is 2–4 h, or overnight, or follow the recommendation of restriction enzyme supplier. Critical step: Use highly efficient restriction enzymes for production of a gene library. Highly suitable is the combination *Sph*I/*Hin*dIII (fast digestion within 2–4 h, low amount of vector religation).

6. Deviating from **Note 3**, use only 30 μL of preheated water for elution of the digested plasmid and for each of the three PCR fragment digestions. After this step, 20–25 μL of digested

plasmid is available (lower volume due to the dead volume of the column after ethanol evaporation) and 50–60 μL of digested PCR fragment.

7. Digested epPCR product is sufficient for ~9 ligation approaches. Best practice is to use 6 μL of insert, 2 μL of digested plasmid, 2 μL of 10× ligation buffer, 1 μL T4-DNA ligase, and 9 μL water to a total volume of 20 μL. Prepare the ligations on ice and incubate the ligation mixtures at 16 °C overnight. After ligation, a heat inactivation step is possible depending on the ligase, which increases transformation efficiency.

8. Highly important is to use an *E. coli* strain, which shows high transformation efficiency. Transformation efficiency of the used strain should be higher than 10^8 cfu/μg with the used plasmid. (Check transformation efficiency of the strain by transforming the undigested vector as control.) A suitable strain is *E. coli* XL1 Blue MRF′, for example. Another important aspect of this step is relevant for subsequent selection experiments for enzyme functions with are present within *E. coli*. To avoid contamination of the primary library with *E. coli* wild-type genes, use in this step directly the auxotrophic *E. coli* deletion strain. For this over night preculture, 50 mL LB are suitable. Use either a single colony from agar plate for this preculture or scratch some cells from a glycerol stock stored at −80 °C.

9. A convenient lab protocol to produce electrocompetent cells is as follows: Inoculate 200 mL SOB medium to an OD_{600} of 0.1 from an overnight preculture and incubate the culture at 37 °C and 220 rpm to a final OD_{600} of 0.6. Cool the culture down on ice for 30 min and transfer the 200 mL in four 50 mL falcon tubes. Centrifuge the tubes at 4 °C, 1500 × g for 10 min. Resuspend the pelleted cells in 50 mL, ice-cold sterile water, store the falcons 15 min on ice. Repeat this step with 25 mL and 10 mL of ice-cold sterile water. Concentrate the freshly produced electrocompetent cells in the last step with the ice-cold sterile water to the required volume of cells for transformation. Aliquot the cells in 1.5 mL reaction tubes with 100 μL per tube.

10. Important for dialysis is to put the ligation approaches for dialysis on the glossy side of the membrane and to use one membrane for each ligation approach. A possibility is to put three dialysis membranes in one petri dish filled with water (room temperature). To avoid contamination, wear gloves and use a tweezer to put the membrane on the water. Put the petri dish cap on the dish for dialysis.

11. During dialysis, approximately 15–20% of ligation volume gets lost (i.e., nine ligation approaches with 20 μL for each

approach—in total 180 µL; dialysis on nine membranes; pooled volume after dialysis approximately 150 µL). Mix 6 µL of ligation approach with 100 µL of *E. coli* cells in one 1.5 mL reaction tube. In case of 150 µL ligation volume, 25 transformation reactions are required to transform all plasmids.

12. After overnight incubation of agar plates at 37 °C, determine the transformation efficiency with the transformant colonies of the empty vector control. Use formula 1 to determine the transformation efficiency. Critical step: Be aware to calculate the dilution factor correctly.

$$T_E = \frac{n \cdot f}{m_{DNA}}$$

Formula 1: Calculation of transformation efficiency

T_E: Transformation efficiency (cfu per microgram DNA); n: number of colonies; f: dilution factor; m_{DNA}: applied DNA [µg]

To determine the size of the library, use the small dilution plates and count the colonies. Determine the amount of transformed cells with formula 2:

$$n_T = n_C \cdot f \cdot V$$

Formula 2: Calculation of amount of transformed cells

n_T: amount of transformed cells; n_C: amount of colonies on dilution plate; f: dilution factor; V: Volume factor (factor to the total plated volume).

To correct the size of the library with the ligation efficiency, do a colony PCR with 20 randomly selected colonies and calculate the ligation efficiency with formula 3:

$$L = \frac{n_f}{n_t} \cdot 100$$

Formula 3: Calculation of ligation efficiency in percent

L: ligation efficiency; n_f: amount of colonies with full length construct; n_t: total amount of analyzed colonies.

The total size of the library G is calculated as a product n_T and L (formula 4):

$$G = n_{\mathrm{T}} \cdot \frac{L}{100}$$

Formula 4: Calculation of the size of a library.

13. With this protocol, library sizes of 10^6–10^7 independent variants are possible. This protocol can relatively easy be adapted to the demand of larger or smaller libraries by up- and downscale the amount of transformations (and consequently the preparation works prior and after transformation). Especially in subsequent screening assays where the throughput is limited by screening capacity, a smaller library size reduces the effort to generate the library.

14. In general, it is recommended to use fresh competent cells for electroporation due to higher transformation efficiencies of these cells. With stored cells at −80 °C, transformation efficiency may become a limiting factor to display the whole size of the library in the complementation test.

15. Obviously, it is highly important to keep the positive control strictly separate from the negative controls and from the library, otherwise, the complementation assay gets contaminated with this positive control.

    Critical step for insoluble enzymes in *E. coli*: If a complementation assay should be done with a library of a gene which produces an insoluble gene product, it could be nearly impossible to identify positive variants, because the hurdle of enzyme solubility has to be overcome first. This can be achieved by doing a fusion construct with the maltose binding protein (MBP) as solubility enhancer. In this case, the gene library is cloned at the C-terminus of MBP [8].
    In general, weak constitutive promoters have been proven to be well suitable for complementation assays like the *E. coli* promoter of the tryptophanase operon [13].

16. The advantage of a complementation assay in liquid medium could be that growth conditions are better for cells compared on agar plates and therefore also weak activities appear. If several cells contain an active gene variant, a mixed culture of different variants could grow, which may produce unclear results. But all gained variants can be reevaluated in a second complementation assay, where the gained variants are plated on M9⁻ medium on large agar dishes to get a better impression for the best performing variants.

17. Unlike *E. coli*, *T. thermophilus* is a naturally competent bacterial system, which is able to take up DNA without specific preparation [14].

18. It has turned out difficult to keep the positive control strictly separate from the negative controls and from the library, for unknown reasons. As a consequence, cross-contamination has been observed. Thus, as an additional safety margin the realization of a positive control (plasmid with a gene coding for the desired function of library variants) is not recommended.

19. The plates applied for auxotrophic selection at 55–79 °C should be packed in plastic foil including some manually added holes enabling circulation. This packaging step enables incubation processes for several days and prevents dehydration of the plates at the elevated temperatures. The incubation at elevated temperatures and the increased evaporation process compared to 37 °C is furthermore the reason to not recommend to apply the selection in liquid medium in a shake flask as for auxotrophic selection in *E. coli*.

20. Isolation of plasmid DNA from *T. thermophilus* is as for *E. coli* cells.

References

1. van den Burg B, Eijsink VG (2002) Selection of mutations for increased protein stability. Curr Opin Biotechnol 13:333–337

2. Hilvert D (2013) Design of protein catalysts. Annu Rev Biochem 82:447–470

3. Bershtein S, Tawfik DS (2008) Advances in laboratory evolution of enzymes. Curr Opin Chem Biol 12:151–158

4. Packer MS, Liu DR (2015) Methods for the directed evolution of proteins. Nat Rev Genet 16:379–394

5. Schwab T, Sterner R (2011) Stabilization of a metabolic enzyme by library selection in *Thermus thermophilus*. ChemBioChem 12:1581–1588

6. Sambrook J, Fritsch EE, Maniatis T (1989) Molecular cloning: a laboratory manual. Cold Spring Harbor Laboratory, Cold Spring Harbor, NY

7. Tanaka T, Kawano N, Oshima T (1981) Cloning of 3-isopropylmalate dehydrogenase gene of an extreme thermophile and partial purification of the gene product. J Biochem 89:677–682

8. Claren J, Malisi C, Höcker B et al (2009) Establishing wild-type levels of catalytic activity on natural and artificial $(\beta\alpha)_8$-barrel protein scaffolds. Proc Natl Acad Sci U S A 106:3704–3709

9. de Grado M, Castán P, Berenguer J (1999) A high-transformation-efficiency cloning vector for *Thermus thermophilus*. Plasmid 42:241–245

10. Drummond DA, Iverson BL, Georgiou G et al (2005) Why high-error-rate random mutagenesis libraries are enriched in functional and improved proteins. J Mol Biol 350:806–816

11. Fromant M, Blanquet S, Plateau P (1995) Direct random mutagenesis of gene-sized DNA fragments using polymerase chain reaction. Anal Biochem 224:347–353

12. Vanhercke T, Ampe C, Tirry L et al (2005) Reducing mutational bias in random protein libraries. Anal Biochem 339:9–14

13. Merz A, Yee MC, Szadkowski H et al (2000) Improving the catalytic activity of a thermophilic enzyme at low temperatures. Biochemistry 39:880–889

14. Koyama Y, Hoshino T, Tomizuka N et al (1986) Genetic transformation of the extreme thermophile *Thermus thermophilus* and of other *Thermus spp.* J Bacteriol 166:338–340

Erratum to: Functional Analysis of Membrane Proteins Produced by Cell-Free Translation

Srujan Kumar Dondapati, Doreen A. Wüstenhagen, and Stefan Kubick

Erratum to:
Chapter 10 in: Uwe T. Bornscheuer and Matthias Höhne (eds.),
Protein Engineering: Methods and Protocols, Methods in Molecular
Biology, vol. 1685, https://doi.org/10.1007/978-1-4939-7366-8_10

The updated online version of this chapter can be found at
https://doi.org/10.1007/978-1-4939-7366-8_10

Uwe T. Bornscheuer and Matthias Höhne (eds.), *Protein Engineering: Methods and Protocols*, Methods in Molecular Biology,
vol. 1685, https://doi.org/10.1007/978-1-4939-7366-8_21, © Springer Science+Business Media LLC 2018

INDEX

Uwe T. Bornscheuer and Matthias Höhne (eds.), *Protein Engineering: Methods and Protocols*, Methods in Molecular Biology,
vol. 1685, DOI 10.1007/978-1-4939-7366-8, © Springer Science+Business Media LLC 2018

Printed in the United States
By Bookmasters